Contributions from Science Education Research

Volume 12

Contributions from Science Education Research is the international, multi-disciplinary book series of the European Science Education Research Association (ESERA). The aim of the series is to synthesize, for the benefit of the scholarly community, the findings of high quality, theoretically-framed research in the domain of science education as well as comprehensive explorations of specific methodological strands in science education research. The series aims to publish books that are innovative in attempting to forge new ways of representing emergent knowledge in the field. The series includes edited collections of chapters, monographs and handbooks that are evaluated on the basis of originality, scientific rigor and significance for science education research. The book series is intended to focus mainly on work carried out in Europe. However, contributions from researchers affiliated with non-European institutions and non-members of the European Science Education Research Association are welcomed. The series is designed to appeal to a wide audience of researchers and post-graduate students in science education. Book proposals for this series may be submitted to the Publishing Editor: Claudia Acuna E-mail: Claudia.Acuna@springer.com

Henriette Tolstrup Holmegaard • Louise Archer
Editors

Science Identities

Theory, method and research

Springer

Editors
Henriette Tolstrup Holmegaard
Department of Science Education
University of Copenhagen
Copenhagen, Denmark

Louise Archer
Institute of Education
University College London
London, UK

ISSN 2213-3623 ISSN 2213-3631 (electronic)
Contributions from Science Education Research
ISBN 978-3-031-17641-8 ISBN 978-3-031-17642-5 (eBook)
https://doi.org/10.1007/978-3-031-17642-5

This Springer imprint is published by the registered company Springer Nature Switzerland AG
The registered company address is: Gewerbestrasse 11, 6330 Cham, Switzerland

Preface

This edited collection brings together current research and thinking around the topic of *science identities*.

The origins of the idea for the collection emerged during the inaugural meeting of the new Science Identities Special Interest Group (SIG) at the 2017 European Science Education Research Association (ESERA) Annual Conference in Dublin, Ireland. As the funders and coordinators of the new SIG, we were not sure what appetite there might be for such a collective. Although we felt there was a clear value and need to bring together and support what we perceived to be a steadily growing field of innovative work using the concept of science identity. On arrival at the designated lecture room, we were a little anxious to see how many conference attendees would actually take up our invitation to attend the meeting and join the new group. We were delighted—and not a little relieved—to find a room full of around 40 excited, lively researchers, representing a range of career stages, drawn from universities across Europe and North America. The discussion and ideas flowed, and by the end of the meeting, we had agreed to put together a proposal for this collection.

It is not uncommon for such stories to end with the popular saying that "the rest is history," but in our case, this history was somewhat prolonged and challenging due to the arrival of the global COVID-19 pandemic in 2020. We are indebted and wish to celebrate the immense efforts of all the chapter authors and reviewers who rose to this challenge and worked so hard in such difficult circumstances for so long to produce the final collection. We wish to thank each and every one of you for your dedication, patience, and generosity in unprecedented times. Thanks also to all the staff at Springer who have supported and helped produce the book.

So why a collection of work on science identities? Why not science identity? Or STEM identities? And what value or purpose do we hope this collection will serve? On the first point, our use of the plural (science identit*ies*) reflects the dynamic, ongoing and processual aspects of identities. But it also illustrate our commitment to a book that values, showcases, and espouses diverse scholarship. Hence, we have tried to encourage submissions from authors across different international contexts, career stages, and who use a range of different interdisciplinary conceptual, analytic, and methodological approaches to understand issues related to science identity across different formal and informal educational settings with a range of actors, including both students and teachers. To a fair extent, many of these aims have been met; however, we are very aware that despite an international spread of contributors, the collection reflects only researchers working within the Global North. This is a limitation and inequality that we hope can be substantially addressed in future works. It is our hope that we with this piece are reaching out to the growing community of researchers who are applying the lenses of science identities in a variety of countries and settings and that it can inspire new applications and ideas of how science identities can be further developed to meet the future challenges of science education.

The book title foregrounds science, rather than STEM, largely because we wanted the book to explore the rich variety of experiences and issues within and between science disciplines and contexts. Hence chapters explore relationships with science not only at a general level but also in relation to specific disciplines, subdisciplines, and areas such as physics, molecular biomedicine, and geology. However, as keen-eyed readers will notice, our plural approach also includes a chapter that focuses on the context of engineering, expanding and testing the boundaries of the framing around "science," adding richness to the collection that we hope will also signal to future work spanning across the fields of STEM, to draw out new synergies and differences.

Our intention is that this collection offers readers a rich and stimulating smorgasbord of ways of thinking about, understanding and researching topics of identity in relation to science and STEM. It aims to provide both a grounding/foundation and a jumping-off point for further work. In particular, we hope that the theme exploring the link between identities, injustices, and science that runs through the core of the book will help sustain and nourish work on these important issues. Since the ideas of the book were proposed at ESERA in Dublin in 2017 and later at the National Association of Research in Science Teaching (NARST) conference in Atlanta in 2018, the science identity community has increased substantially. As a field which is deeply ingrained with intersectional injustices and implicated in the ongoing reproduction of inequity, there is much work still to do in science identity research, scholarship, and activism.

We hope you will enjoy the breadth and depth of the collection and will love reading it as much as we do, and that it will be the platform for discussions, ideas, and dialogue. In particular, we hope that the work showcased within this book will help sustain and move forward the exciting range of research being conducted within the field.

Copenhagen, Denmark — Henriette Tolstrup Holmegaard

London, UK — Louise Archer

Contents

Part I
Introduction

Chapter 1
Understanding and Contextualizing the Field of Science Identity Research

Heidi B. Carlone

About 20 years ago, I had interests in studying the short- and long-term impacts of elementary teachers' pedagogical practices on students' science identities. I consulted with a senior colleague in the field, and he said, "Good ideas. But don't call it identity! That construct is too hard to operationalize." I got the similar warnings and critiques, again and again, from social psychologists, learning scientists, science educators, grant panel reviewers, and grant program officers. My grant proposals related to studying identity were unceremoniously turned down. Truthfully, the critiques were wise and well intentioned. I had not yet operationalized the construct in a way that would produce valid results.

I spent the next couple of years working with Angela Johnson to operationalize the construct, which was eventually published a couple of years later (Carlone & Johnson, 2007). We came up with a grounded theory model of science identity, modeled largely off of Gee's (1999, 2000) work, that emphasized the constructs of science competence, performance, and recognition of self and by others. In 2006, I finally received a grant to study students' science identity work in children ages 10–14. There are other trials related to studying identity in the early days I could mention, but the point here is that the field shifted quickly, from "don't do it" to "everyone's doing it, and maybe you should, too". Identity, about two decades ago, quickly catapulted to one of the field's hot topics.

What accounts for identity's staying power in science education? What's next for identity studies? In this chapter, I offer my perspectives on these questions, and I also address how identity scholars in science education go about operationalizing and studying identity. To help illustrate my points, I draw on the excellent work from the authors in this book, providing readers with immediate resources to

H. B. Carlone (✉)
Vanderbilt, Peabody College, Nashville, TN, USA
e-mail: heidi.carlone@vanderbilt.edu

H. T. Holmegaard, L. Archer (eds.), *Science Identities*, Contributions from Science Education Research 12, https://doi.org/10.1007/978-3-031-17642-5_1

consult as they journey with us on the path to understanding the how's and the why's of studying identity in science education.

1.1 "Seeing" Identity: From the Individual to Lens Toggling

In his book, "Ethnography: A way of seeing", Harry Wolcott (2008) distinguished between looking like an anthropologist and seeing like one. *Looking like an anthropologist* includes using ethnographic methods of long-term observation, in-depth interviewing, and artifact collection. *Seeing like an anthropologist* means focusing on cultural aspects of a learning setting—patterns of activity, group-level meanings that arise from that activity—meanings that shape behavior, values, and logic of individuals in the group. In the next couple of sections, I highlight important, though certainly not exhaustive aspects of *seeing identity* highlighted in the chapters and in my previous work. I note that, although most chapters in this book take up an anthropological, sociological, and/or sociocultural approach to studying identity, clinical and social psychologists would approach the study of identity differently. No single lens can explain all aspects of identity work, so my unapologetic leanings toward sociocultural approaches should not be read as denigrating other perspectives.

1.1.1 The Individual

I begin by laying bare my methodological preferences for seeing identity as more than an individual construct. Identity's ties to micro-, meso-, and macro-educational contexts, as I will argue, is part of its appeal and potential as an explanatory construct. Histories of educational practice and policy typically direct attention towards fixing individuals, e.g., we fix achievement gaps by correcting deficits in knowledge, interest, skills of individuals in the system.

Suppose for a moment one is studying an educational intervention that goes horribly wrong. An individual lens would direct us to ask these kinds of questions: What did the teacher do? What was the teacher's content knowledge? Were the students adequately prepared to learn the content? What was their prior knowledge? What was students' motivation to learn the content? These questions may be relevant to understanding a part of what went awry, but without an account of the teacher's and students' knowledge and practices within the complex organizational, institutional, societal structural, and historical systems, the blame for reform failure is directed toward *individuals* versus the coalescing systems in which individuals function. A theory of learning, knowing, being, and belonging (identity) that focuses solely on individuals is unsatisfying for explaining how and why educational interventions succeed or fail.

None of the chapters in this volume frame identity as a purely individual construct. Johansson and Larsson (Chap. 8) mention that some identity literature in physics education may perpetuate notions of identity as a personal characteristic possessed by individuals versus something negotiated in interaction or practice. For example, their critical review of the physics identity literature surfaced implicit messages that "students do not develop enough of a physics identity", which is a framing of identity as something to be possessed and something that can be quantified. Further, interventions that arise from these perspectives can fall prey to thinking that "fixing individuals" will fix all problems. This kind of thinking leaves disciplinary, institutional, and societal systems that perpetuate injustice intact.

This does not imply that individual perspectives are not part of studying identity. For instance, Dixon, Harris & Ballard (Chap. 12) argue that an individual's narrative identities, or the stories they tell themselves about who they are and want to be as well as the stories others tell about them, offer powerful insights into their ongoing identity work. They do not stop at narrative—the authors pair this approach with an activity-centered lens, which allowed them to see identity in the ways youth "take up, ignore, or transform the practices and labels of science, depending on where, with whom, and for what they are doing science" (p. 23). Their approach is one way of lens-toggling, which I describe below.

1.1.2 Lens-Toggling

Lens-toggling is not a new idea, but it is my descriptive term that highlights a feature of identity research that promotes explicit attention to methodologically connecting local identity work to different levels of systems (e.g., micro-, meso-, and macro-structures). An example of a lens-toggle between the individual and a science learning setting might be a shift from asking "Is this a promising STEM learner?" to "What does it mean to be a promising STEM learner in this setting?" and "Are the criteria for performing and getting recognized as a promising STEM learner accessible, believable and achievable for all learners in the setting?"

Different theoretical perspectives promote lens-toggling. For instance, intersectionality could be viewed as lens toggling by paying attention to the ways race, class, sexuality, *and* gender are intertwined with a Black woman scientist's professional pathway (Crenshaw, 1989). Lens-toggling is evident in work that demonstrates that one's ecological reasoning, which is a psychological lens that examines *how* people think, is connected to a cultural lens that pays attention to *what* people think about the perceived role of humans in nature (Bang et al., 2007).

A lens-toggle between the individual, learning setting, and institutional contexts prompts the following questions: What definitions of promising STEM learner are institutionalized in terms of sorting mechanisms of schooling (e.g., getting chosen for district level STEM enrichment opportunities)? Are promising STEM learners granted opportunities that students without those labels are not? What are the implications of receiving or being denied an institutional label of promising STEM learner on someone's identity work? We can add a more macro-level lens by asking, "How does race, class, or gender affect one's chances of being labeled a promising STEM learner? What are historical demographic patterns of those labeled promising STEM learner? Given histories of participation, is it possible for all learners to be labeled STEM-promising or is that label usually only reserved for some students? What competencies align with promising STEM learner and how does racial, ethnic, linguistic, and socioeconomic privilege intersect with those competencies? How does being a promising STEM learner in this institutional context connect one's trajectory up to a larger network of power (Nespor, 1994)? How do local definitions of promising STEM learner challenge or reproduce status quo meanings of promising STEM learner?

Why lens-toggle? What does it afford? First, it denaturalizes science identity, which means that science identity is not viewed as an innate characteristic of an individual. Instead, lens-toggling draws attention to the ways identity is influenced by multiple contexts and histories of participation that bear down on local sites of becoming. Science competence can be attributed for reasons not at all related to science—e.g., asking and answering questions in a certain way, upholding gender norms, enacting Western science epistemologies, using scientific vocabulary, or behaving with compliance. Similarly, someone with strong science interests and experiences from home might be overlooked for not wanting to or being able to conform to sanctioned ways of being and doing in the classroom. In turn, when science identity is denaturalized, we can begin the important work of theorizing how and why science identity develops and what practices might prompt science identity work.

Lens-toggling also helps researchers theorize how and why some youth's STEM trajectories are smoother than others and how and why some practices promote more inclusive meanings of "science person" than others. With lens-toggling, we can better understand how classroom, institutional, and historical practices and arrangements—i.e., normative ways of doing things—coalesce to make more likely the cultural reproduction of the science person. For example, in a study of fifth-graders engaging in engineering design, the seemingly neutral practice of "using evidence to inform your design" was taken up by youth and the teacher in ways that constrained productive disciplinary practices (Carlone & Lancaster, n.d.). The educators framed the class's data chart as a form of evidence with primary epistemic authority, which inadvertently marginalized evidence from youths' lives and their willingness to test creative solutions or solutions that seemed logical to them. When researchers and educators are aware that seemingly neutral practices of science and science education marginalize and unnecessarily narrow the definition of science competence, there is urgency to envisioning and enacting alternative, boundary-expanding practices.

Finally, lens-toggling allows for consideration of structure and agency. Structures come in many forms—e.g., norms of participation, histories of practice that point to narrow constructions of smartness, normative ways of enacting racial and cultural identities that come into conflict with science or academic identities, ways to "properly talk science", and gender norms implicit in enacting strong, recognizable science identities. Structures also work through exclusionary practices that prevent further study of science, such as A level physics requirements for further study in physics in England or rigid performance expectations on university entrance exams (e.g., gaokao in China; SAT or ACT in the US) (see Wong, Chap. 5). Though identity studies show, again and again, how strong structures bear down on local settings and individuals' trajectories, not all structural arrangements are deterministic. Individuals have a say, though sometimes it is limited, in whether and how to comply with, resist, and transform structural arrangements. Groups of individuals can reconfigure learning arrangements and normative practices to enable the production of new meanings of science, scientific knowledge production, and science person. Agency can vary by scope and time, sometimes resulting in transformations that have staying power and other times resulting in momentary resistances that result in shorter-term changes. These forms of agency are not always conscious push-backs.

For instance, consider a first-grade teacher at an urban school whose administration is pressing for all students to learn to read at a certain reading level by the end of the school year. The curriculum facilitator has determined that teachers should only allot 20 minutes/day to teach science, and they are encouraged to do so through guided reading of expository texts. Further, science time needs to be shared with social studies time. Too, students' reading levels are measured with "benchmark" scores (results of locally issued, interim standardized assessment) once every 9 weeks, which are made public to the grade-level team, school leadership team, principal, and curriculum facilitator. Those with lower-than-expected growth on reading scores are put on an action plan, where administrators will more closely monitor students and the teacher's instruction. This teacher decides that, despite institutionally mandated uses of time, she will regularly integrate science and literacy time slots. Her grade-level teammates notice the energy and enthusiasm her students exhibit during science time, and they ask for her to share her lessons. By the end of the year, the entire first-grade team is integrating science, social studies, and literacy in productive ways, though they have had to answer to school- and district-level administrators because they rearranged institutionally-mandated schedules or strayed from the standardized reading curriculum to fit everything in. Their students perform better-than-expected on standardized reading assessments. These teachers may not interpret their decisions as resistance *per se*; more likely, they will view these decisions as enacting their vision for what teaching and learning should look like. These visions push up against local and historical structures and, in enacting these visions, they produce new cultural forms and local structural arrangements. Their imagination for what could be and motivation to enact their vision represent forms of agency. Following Holland et al. (1998), I view this agency as a part of identity work, navigating social and institutional arrangements to enact practices that align with the kind of teachers they want to be.

Some theoretical perspectives better enable researchers to see and understand structures, while others illuminate improvisation and agency. For example, Archer et al. (Chap. 2) illustrate the ways a Bourdieusian theoretical framework foregrounds structural analyses, but does not capture as well the agentic aspects of identity work that can involve creativity and improvisation. Accounting for the *dialectic* of structure and agency is challenging, but their work points to one tool for doing so. They suggest seeing identity performances as "embodied enactments and dispositions that emanate from the habitus, produced through interactions of habitus, capital and field that are both highly structured (and structuring) while also being, to some extent, agentic and improvised within the moment" (p. 4). Their massaging of the classic Bourdieusian structural lens renders possibility for individuals' agency and, in doing so, they shift the notion of "identity" to "identity performance". This "verbing" of identity is another way to attend to the structure/agency dialectic. When identity is a verb—i.e., identity work, performance, enactment, engagement in practice—actors are not passive recipients of structure, nor are they free of structural constraints in authoring themselves as certain kinds of people. The verbing of identity surfaces identity as ongoing negotiations and achievements.

Another example of lens toggling is the multi-level discourse framework for understanding teacher and student identity development presented by Askew and Wade-Jaimes (Chap. 9). The authors argue that teachers' and students' identities cannot be examined separate from one another and the micro-, meso- (school, classroom), and macro- (society) contexts in which they operate. Their privileging of a lens toggling analysis allowed them to explain that identity work for teachers and students are heavily shaped by social interactions in the form or recognition work, tensions between the accountability culture and teaching science authentically, and historically enduring deficit-based narratives about students and communities in urban schools. Their chapter provides an illustration of Holland et al.'s (1998) points about the structure/agency dialectic inherent in identity work: "The space of authoring, of self-fashioning, remains a social and cultural space, no matter how intimately held it may become. And, it remains, more often than not, a contested space, a space of struggle" (p. 282).

1.2 "Seeing" Identity: From Snapshots to Patterns Over Time

In the previous section, I outlined conceptual lenses one can use to "see" identity. Here, I highlight an ongoing challenge for identity studies—considerations of timescales. Identity has connotations of stability and yet, many studies may only examine identity in snapshots over relatively shorter periods of time. What can one see in "snapshot" studies of identity?

1.2.1 Snapshots

Snapshots are markers or measures of identity in given points in time. Snapshots are used often in social science research, e.g., they allow researchers to compare populations and understand how a given context promotes certain identities and marginalizes others. We can understand these snapshots as examples of "triggered" or situated identity work, a term borrowed from Hidi and Renninger's (2006) four-phase model of interest development. Though they use the term to explain how interest develops, I have adapted the term to understand disciplinary-linked identity development. Triggered identity work is situational, relatively short-term, and often sparked by engagement in novel, affirming, surprising, or emotion-filled experiences. Usually, triggered identity work is prompted externally. It can, but does not necessarily, lead to more enduring, cross-context identity work. These moments are not insignificant. They can tell us a lot about the ways a given setting opens up space for broader types of participation and knowing.

1.2.2 Patterns Over Time

Though commitments to sociocultural theories of identity emphasize its situatedness, identities can become thickened (Wortham, 2006), stabilized (Calabrese Barton et al., 2013), or directional (Johnson et al., 2011) over time. Science educators who take more critical or sociocultural stances when studying science identity over time share commitments to: (1) considering how resources cultivated in one setting are leveraged in other settings; (2) emphasizing structure/agency dialectics by considering how individuals' identity pathways are creatively constructed but also heavily influenced by structural arrangements; (3) toggling between analytic grain sizes such as an *individual's creativity* to author oneself amidst *local structures* that conflate competence with *gender and racial norms* that run counter to one's gendered or racial identities; (4) avoiding artificial linearity in understanding identity work over time (Carlone et al., 2014; Tan & Calabrese Barton, 2020).

Few of the chapters in this book have seriously tackled the question of studying identity over time. Archer et al. (Chap. 2) are the only authors who use longitudinal data, and Dixon et al. (Chap. 12) note longitudinal studies as a needed "next step" in identity research, while Butterfield and Marshall (Chap. 13) also note longer-term studies as a gap in the literature. Such longitudinal studies are, by definition, time-consuming, expensive, and difficult to pull off because of the difficulty of finding and/or following up with past research participants. Butterfield and Marshall's meta-synthesis offers us a starting point, as they introduce a framework for studying a science identity development trajectory, from *novice*, to *intermediate* to *full membership*, along with relevant constructs (e.g., competence, performance, recognition, intergroup inclusion/rejection, and conflicting social identities) at each stage.

1.3 What Do We See When We Use Identity as an Analytic Construct?

1.3.1 Identity Has Helped Us Recognize Mechanisms of Cultural Reproduction/Production

The studies in this book demonstrate the very real consequences of race, class, and gender on people's pursuit, engagement, and persistence in science. On-the-ground, everyday practices do the work of guarding science's boundaries in reproductive, raced, classed, and gendered ways, effectively marginalizing, silencing, and/or making it overtly difficult for those located along various intersections of the matrix of oppression (Collins, 1990).

It is not surprising that the study of identity work in relation to physics came up a number of times in the chapters (e.g., Johansson & Larsson; Rabe & Kressdorf; Wong). Physics has been especially effective in policing its borders; it remains difficult to recruit and retain a diverse physics student and worker population (Wong, Chap. 5). This is true throughout many parts of the world. Yet, its power makes it an excellent context for studying identity work; how do everyday practices, historical legacies of practice, and the organization of physics learning function to reproduce physics as an elite science and physics people as exceptionally brilliant? We can more easily study mechanisms of social and cultural reproduction with a field like physics, which has consistent success in maintaining its status. For example, Wong (Chap. 5) found that high-achieving students studying A-level physics did so, in part, because it held symbolic and exchange value. Physics' status as a subject reserved for the cleverest students, its positioning as a gateway to other high-status fields of interest (i.e., engineering, medicine), and girls' perceptions that their success in physics represents a particularly unique achievement are all factors that play into symbolic exchange. Thus, those most attracted to pursuing A-level physics were most likely to see *added value* to achieving success in physics, which means they have invested in and inadvertently reproduce ongoing cultural meanings of physics as for the elite few.

Askew and Wade-Jaimes (Chap. 9) tackle multi-level analyses from the outset of their chapter: "This chapter argues that teachers and students are both part of a larger context and subject to macro level discourses that constrain both teacher and student science identity" (p. 1). They demonstrate how discourses of science, their students, and accountability shaped the teachers' and students' identity work. For example, the overwhelming deficit-based narratives of students' knowledge, behavior, and experiences limited their enactment of meaningful science in the classroom.

1.4 Everyday Practices Produce Shared Meanings that Influence Identity Work

We can see mechanisms of cultural reproduction clearly through the analysis of the cultural meaning of "talent" in university science and its implications for students in Holmegaard and Johannsen (Chap. 6). To disabuse readers of conceptualizing scientific talent as an inherent, individual trait, the authors highlight practices that render college women's contributions invisible, silence voices, produce competition, and position women below men on the talent barometer. For instance, women who completed meticulous work and were consistently well organized and prepared were deemed helpful and "dutiful" by teachers. Alternatively, the mostly male group of students who were fearless, playful, curious, and perhaps a little less responsible were deemed the ones who could do "real research". Their findings highlight talent as a cultural production, one that students and teachers were complicit in actively constructing in university science practices.

Madsen and Malm (Chap. 7) similarly point to the ways everyday practices, down to what shoes and clothes one wears out in the field for a geology, signal tacit knowledge and access that are markers for belonging in the geology community. Becoming a geologist, they argue, is not merely a cognitive endeavor; it is a process of identity work enacted in sometimes tacit disciplinary practices. They provide three portraits of students whose identities as geology people are tightly connected to field science practices. For instance, as Frida, with supervision, came "to know where to look, how to look, and what to infer from observing" so too did she come to feel a sense of belonging and identity in geology. In another portrait, Madsen and Malm describe the implicit and explicit practices entangled with measuring the dip and strike of observable structures in the field. Doing so accurately requires learning to see like a geologist; the practice is intimately connected with developing an identity and a sense of belonging as a geology person. Fieldwork is filled with so much tacit, interpretive knowledge that students' engagement (i.e., "seeing" what the supervisors see) can be frustrating for them, leaving them feeling incompetent. Alternatively, success in the field fuels their belonging and identities.

1.5 Identity Is a Process of Becoming, Rather Than a Final-Form Achievement

Nearly all chapters emphasize identity as a process of becoming, acknowledge multiple identities, and eschew static considerations of science identity. The authors use different constructs to do so, such as: identity performances (Archer et al., chap. 2; Gonsalves & Rahm, chap. 3), identity negotiations (Rabe & Kressdorf, chap. 4; Ramos-Montañez & Pattison, chap. 15), identity as a process of negotiation (Madsen & Malm, chap. 7; Wong, chap. 5), identity work (Askew & Wade-Jaimes, chap. 9; Seiler & Kwamboka, chap. 10), identities-in-practice (Richmond & Wray, chap. 11), identity configurations (Taconis, Chap. 14), and processes of identification (Dixon et al., chap. 12). The emphasis on identity as a process of self-making versus a reification of self reflects the work necessary to develop a cohesive, stable version of self. In science education, we consider this effort as connected to power, especially in relation to one's social location and relevant communities' histories of practice.

This insight provides alternative ways of understanding learning and development. For example, Richmond and Wray (Chap. 11) use the construct of figured worlds (Holland et al., 1998) to create a model of professional identity development of early career science teachers. This model (p. 9) accounts for the interaction of power and privilege, culture, community, and artifacts on identity work. Their rich account of two beginning science teachers' meanings and experiences of the features of the figured worlds in their schools demonstrated the fluidity of identity as one traverses from one figured world to another.

Building further on identity as a process of becoming, Taconis (Chap. 14) speaks to STEM (science, technology, engineering, mathematics) professional identity configurations which include acquisition of technical knowledge and expertise, and related notions of one's professional self. He argues for the creation of an explicit model and proposes one to account for both of these constructs. Such a tool will give the field a map to study, explore, and test the dynamic relationships of professional STEM and professional STEM educator identity development.

Ramos-Montañez and Pattison (Chap. 15) provide an example of the structural consequences of *in-the-moment* interactions. Drawing on prior work by Goffman (1974), they discuss *activity frames* as one micro-level structure that is consequential for identity work. Activity frames are "the context-specific, emergent understandings or expectations, either implicit or explicit, about the nature and goals of the interaction" and the activity (p. 9). They demonstrate how the framing of an engineering activity as individual/competitive versus collaborative/supportive and failure as desirable or undesirable impacted practice-based identities of participants in an engineering program for girls ages 10–14. Youths' frames can be different from one another and different from the adults' frames, which can cause tension and constrain youths' STEM identity work. Their identity-frame model of situated identity negotiation (p. 12) demonstrates the fluidity of the identity construct quite well.

1.6 Situating the Individual Amidst Structures of Power

Identity studies have produced a lot of evidence to demonstrate the maneuvering individuals have to do to achieve success and belonging in a field that historically has assigned value to narrow, gendered, racialized, and classed constructions of competence and performance. So often, this maneuvering simply is not worth it, is not possible, or is not recognized. Power is a central aspect of defining what and who counts as scientific. My considerations of power are influenced by social practice theory (Holland et al., 1998). I consider power as produced in everyday practices, which are shaped by and shape micro-, meso-, and macro-structures. Structures enable and constrain possibilities for individuals' identity work. For instance, interactional norms at a micro-level shape what is and is not considered a legitimate contribution in a science learning setting. Prized forms of interaction, which often mirror those of the dominant culture, serve as a resource for cultural reproduction. Educational policies about assessment, a practice that concerns meso-level structures shape curricular priorities, which, in turn, inform future educational policies relating to assessment. As policymakers debate about what should be assessed and included in a school science curriculum, they are drawing on ideologies and historical arrangements of schooling and science (macro-level structures). Power is produced in moment-to-moment practices, institutional practices, and enduring legacies of schooling. As individuals participate in these practices, so too are they participants in structure-making and/or disrupting as much as they are being disciplined by those structures.

Many of the chapters in this book demonstrate the power embedded in everyday practices of science, as indicated by the arduous and painful contortions of self they undergo to belong, fit in, get recognized, or to avoid unwanted recognition. We see this in Vanessa's journey (Chap. 2). A young working-class woman of black African (Nigerian) heritage with strong science interests and a family with strong science capital (Archer et al., Chap. 2), Vanessa had a strong foundation for potential success in science. Despite these strengths, the structures that maintain science's elite status, like arbitrary high entry grades for science A levels, sent messages that students like Vanessa are not clever enough to pursue science as a career. As a result, Vanessa opted out of science and reframed her school science experiences as irrelevant to real world concerns and too difficult.

In Chap. 10, Seiler and Kwamboka convey the spiritual and symbolic violence (dehumanization) that accompanies the identity work science teachers from nondominant groups endure to perform themselves in ways that align with Western, Eurocentric models of "science teacher". Too often, their performances of self are rebuffed, denigrated, ignored, and/or deemed inferior to or inauthentic when compared to historically enduring cultural models of science teacher. As nondominant teachers work in colonized and racialized spaces, the resources that they bring to settings are viewed as illegitimate—e.g., intimate knowledge of agricultural practices (Hilda) or knowledge of cultural storytelling that can spark and animate student engagement (Donna). Hilda's and Donna's stories also speak to the ways that White, Eurocentric value judgments of legitimacy become internalized so that nondominant teachers begin to self-police, a practice that denies their full humanity and is traumatizing.

We also see these negotiations between self and cultural models of scientist in Rabe and Kressdorf's study of high school girls' physics identities (Chap. 4). They studied girls' career decision making before, immediately after, 4 months after, and 1 year after a one-week academy designed to introduce high school girls to careers in science and technology. Predictably, various structural meanings bore down on girls' meaning-making of physics and themselves in relation to physics including stereotypes about who does physics, institutional restrictions to physics similar to that described in Vanessa's case above, and the perceived usefulness and relevance of physics for society and for the girls' lives among others. The identity lens allowed the authors to identify structural challenges and to examine how those challenges were differently interpreted and experienced by the girls in the study, highlighting intra-group variation. The authors concluded the study by cautioning against pinning hopes for diversifying the science workforce based on out-of-school, shorter-term interventions for the complexity of identity negotiations warrants longer-term, sustained shifts in the ways high school physics is taught.

These framings situate the individual amidst different structural arrangements that enable and constrain their science identity work. Because of these studies, we more clearly see gatekeeping mechanisms of systems—how science's boundaries are guarded by unnecessary grade requirements to pursue it further (Archer et al., Chap. 2) and arbitrary, masculine definitions of talent (Holmegaard & Johannsen, Chap. 6) are but two examples in this book.

The power of structures is that they sustain a naturalized illusion that science or science learning settings are supposed to be built this way and no other. Scientists are supposed to be smart as evidenced by high grades (Archer et al., Chap. 2), talented as evidenced by doing "more" than the course requirements (Homegaard & Johannsen, Chap. 6) and good science students are supposed to talk and think in techno-rational ways, leaving out emotion and passion, and science is supposed to be objective and devoid of human and political impact. Though there are different mechanisms attempting to disrupt these narrow views (e.g., critical science education research), these dated notions of competence have been confirmed, again and again, in my nearly 25 years of experience conducting ethnographic studies of K-16 science. As a result, individuals who do not or cannot conform to celebrated subject positions (Carlone et al., 2014) are not welcomed into science or actively pushed out of science. When they leave science, these decisions are framed as individual choices rather than evidence of successful machinations of a well-oiled, multi-tiered system. Identity studies have exposed these fallacies, begging questions about how we can reimagine and redesign the system, with its coordinated, multi-tiered structures, to be more just and equitable.

1.7 Changing the Field Versus Changing the Individual: New Forms of (Local) Science

Identity studies do not just illuminate processes of cultural reproduction, but they also show us cracks of possibilities of new ways of organizing learning, positioning learners as knowers and doers of science, and positioning science as a particular way of knowing. A social practice theory perspective points us to new local meanings produced in practice. A few chapters in the book highlight cultural productions, which focus on changing the field of science education versus changing the individual (Archer et al., Chap. 2). For example, Gonsalves and Rahm (Chap. 3) examine teen girls' identity work in relation to science during ConvoClub, an out-of-school program that was part of a community center's programming in Montreal serving large populations of youth who live in poverty. The ConvoClub, during the time of this study, included science programming encouraging teen girls to make connections between their lives and science through digital storytelling and video documentaries. Girls' science-related identity work in the club was facilitated by building solidarity and "bonding capital" around relationships with one another, and they felt comfortable taking on roles as science experts in the community center with youth who were not involved with the club. They effectively reconciled their performances of heteronormative femininity with explorations of science, creating a "local form of science capital" or "non-dominant science capital" that, though valued locally in the ConvoClub, likely would not have exchange value outside of the club. This localness, while creating what the authors call "inside-ness" in the Club, did not quite connect to networks of power in ways that would have been recognizable to powerful others. Nonetheless, the setting produced a recognizably different form of science than what is typically celebrated in prototypical school settings, pointing to the disruptive potential of out-of-school science learning settings.

Johansson and Larsson (Chap. 8) argue for using feminist theory to reconceptualize the physics identity problem. Designing spaces where women and minoritized participants can feel as though they belong, get recognized as being competent in physics is an approach that leaves inequitable physics structures intact, they argue. It is not enough to broaden physics culture simply with the notion of welcoming more people. Using feminist studies of science, physics education could focus on the intersections between knowledge, identity and social justice. The kinds of physics people produced in physics education, they argue, have implications for what physics knowledge is produced and the role of physics in society. When we critically consider intersections of the social environment with physics knowledge, questions and imaginaries like the ones raised by Johansson and Larsson (Chap. 8) bubble to the surface.

1.8 Agency and Spaces of Possibility

A few years ago, Shanahan (2009) argued that identity studies paid too much attention to moments of agency and not enough attention to structural arrangements that bear down on subjects. The chapters in this volume represent a somewhat even distribution, with some chapters featuring agency more prominently and other chapters emphasizing structure.

Dixon et al.'s work (Chap. 12) is one chapter that highlights spaces of possibility. They introduce a new construct, environmental science agency (ESA), to emphasize agentic aspects of identity work in the context of community and citizen science (CCS). They define ESA as "the ability to use experiences in environmental science to make positive changes in one's life, landscape and community" with three practices that prompt its development: (1) fostering participants' understanding of environmental science content, norms, and practices; (2) developing areas of expertise within scientific work, and; (3) using CCS experiences as a foundation for changing perspectives, behavior, and ways of being and seeing beyond the project work (pp. 8–9). This chapter provides a keen focus on the ways learning environments designed for agency and identity for one person (Bryan, in this case) may be constraining for another's identity (Diana). Their work highlights practical applications—i.e., includes concrete design principles that can facilitate multi-dimensional learning. They suggest broadening stakeholders' audiences for youths' work beyond scientists and people youth trust and care about. Further, they recommend engaging youth with socio-ecological issues that encourage consideration of the natural and human-designed worlds. Their recommendations distinguish between "real" and "realistic" work; real work indicates work that feels real and authentic for youth.

1.9 On the Horizon for Identity Studies

I write this while experiencing a heightened sense of unrest in the world, which I think will shape the next generation of identity studies in science education. We are currently in the throes of the COVID-19 pandemic and the rampant, racially motivated police brutality, protests, and civil unrest in the United States. We are feeling the effects of climate change internationally, with increased catastrophic weather events, sea level rise, and paradigms that perpetuate human domination over nature. This feels like a moment of painful, critical unraveling, which presents an opportunity to restitch the fabric. How might identity studies play a role in that re-stitching?

> We will not go back to normal. Normal never was. Our pre-corona existence was never normal other than we normalized greed, inequity, exhaustion, depletion, extraction, disconnection, confusion, rage, hoarding, hate and lack. We should not long to return, my friends. We are being given the opportunity to stitch a new garment. One that fits all of humanity and nature. (Taylor, 2020)

Identity studies would benefit from *additional* criticality. Butterfield and Marshall (Chap. 13), in providing a helpful meta-synthesis of the science identity literature for diverse secondary and post-secondary students, argue that one simply cannot study science identity development of diverse racial and ethnic students without serious consideration of intersectionality— i.e., "how race, ethnicity, and gender intersect with science identity *is obligatory* for students of color" (p. 23, my emphasis). At every point along their science trajectories, minoritized students' competence, for example, gets questioned, and they too "question their ability to succeed" (p. 24). Their meta-synthesis convincingly demonstrates that science is not neutral when it comes to the reproduction of racist systems; stereotypes, systems and policies, microaggressions, and "science ingroups" are all practices that perpetuate the inequitable and unjust status quo in the sciences.

Economic chasms that exist between the Global North and South, grounded in colonialist histories, feed into all aspects of education systems and perpetuate inequities on a global scale. Colonialist structures shape contemporary science and science education (Boisselle, 2016), a point that Seiler and Kwomboka (Chap. 10) argue quite convincingly. They highlight how colonial interests and whiteness play out in most science learning settings around the world. Their chapter illustrates a more critical perspectives for identity studies—taking seriously how histories of enslavement, education's role in assimilating and controlling nondominant people, and economic exploitation of nondominant peoples are part of the fabric of science education all over the world. These histories matter, for they make invisible the resources nondominant youth bring to settings and the ways we and the systems are complicit in and actively reproduce Western hegemony. Historical, progressive, reform-based, and most other globally accepted ways of defining a "good" science are too easily and too often tied up in the world's colonialist past.

These and other chapters in this book remind us that it is not enough to simply uncover and identify racism and colonialism "over there". We must examine our own complicity in rendering invisible and reproducing whiteness in the settings in which we do our work as researchers and teachers. This is difficult work, because of the painful realizations that accompany the work, because many of us have benefited from the norms, beliefs, and values that undergird and actively reproduce systems of oppression, and because one's privilege makes it difficult to see whiteness. The same might be said for classism, ableism, sexism, heterosexism, and scientism. It is important for identity scholars to be continually reflexive about our ontological, epistemological, and axiological assumptions and the ways those assumptions reproduce the structures we have pointed out "over there". They are *here*, too.

Our work as critical identity scholars does not stop with awareness and reflexivity, which are mostly cognitive endeavors. We must move to action. It is no longer enough to point out the mechanisms of cultural reproduction and the violence experienced by nondominant groups at the hands of well-intentioned science education. Science identity studies need to do that, too, but I argue that it is time to move toward designs that combine research, design, and activism. What would it look like to consider disciplinary identity as both analytic construct and design goal? What are models of disciplinary identity that reconfigure models of our colonialist past? Researchers will, I predict, begin to design and study learning settings that disrupt historicized, oppressive narratives about underrepresented groups in science and reimagine and construct new ontologies and epistemologies in science education. This opens space for "future imaginaries" (Johansson & Larsson, Chap. 9) of science education.

A few science educators have been doing this kind of forward-thinking, design-based, activist-oriented research work for years and provide a model for what this work looks like. For example, Angela Calabrese Barton and Edna Tan's research group has been designing and studying youth-centered settings for nearly two decades. Their work teaches us that, rather than bringing youth to science and engineering, we need to bring science and engineering to youths' lives in ways that help them understand and solve problems that are relevant to them. In this work, science education's structures get disrupted, adapted, and harnessed in service of youths' passions, questions, and identities. Irene Rahm's research group has similarly pushed boundaries of normative science education to center youths' identities and epistemologies.

In the decentering of science, this design work almost always points to group-level solutions, signaling the importance of collective agency and solidarity to support youths' science identity work. Most, though not all, of this work also happens in out-of-school spaces, making it no less creative or challenging, but leaving the door open for similar kinds of design-based research to be enacted in future science identity studies in school settings. Louise Archer's group has developed, for example, the science capital teaching approach, which aims to broaden what counts as doing science by: (1) personalizing and localizing science; (2) eliciting, valuing, and linking students' outside of school experiences; and (3) building students' science capital by adapting lesson plans so that students see science in their daily lives and recognize that science can be helpful for any job.

Morales-Doyle (2017) is another researcher who is doing critical, justice-oriented work in schools. Guided by a theoretical framework of justice-centered pedagogy, organized an advanced placement (AP) high school chemistry course around issues of environmental racism (e.g., coal power plant pollution, lead soil contamination; uranium mining on Navajo reservation). The pedagogy supported students' academic achievement and positioned them as "transformative intellectuals" who were able to understand and critique science to expand legitimate ways of knowing and being in an AP chemistry classroom. In other words, they were positioned as "producers of knowledge and culture" (p. 1055). Their focus on locally relevant social and scientific issues was a significant contestation of a typical AP curriculum.

One implication of Morales-Doyle's (2017) work is the importance of epistemic heterogeneity, a construct picked up by Davis and Schaeffer (2019) in their two-year ethnographic study of fourth and fifth grade Black students' (ages 9–11) sensemaking during a long-term study of water justice. While they do not focus specifically on identity as an outcome, they do a deep dive into youths' meanings of this socioscientific issue, which expanded science learning beyond the cognitive to include affective, political, and ethical dimensions of science. Justice-oriented science gave these young learners opportunities to author themselves and get positioned by others as competent, agentic, and able to develop critical scientific capacity.

I imagine that traditional science education will not be sufficient to address problems that are significant for local and international communities (see also Eisenhart et al., 1996, who made this point decades ago). Thus, it stands to reason that our science identity studies, and I include my own work in this plea, cannot continue to have as their only goal to explain why science is so exclusive. Newer research designs provide cracks of opportunity and guidance for work at the intersections of research, action, reflection, and revision, such as design-based research, participatory design research (Bang & Vossoughi, 2016), and social design-based experiments (Gutiérrez & Jurow, 2016). These new methods disrupt historically enduring binaries, like subject/object and action/reflection, while also testing the very limits of a possible science education. This work will likely need to be done in teams, across disciplinary boundaries, and drawing on varied social science traditions.

Identity studies have maintained longevity because of the construct's explanatory power. The studies in this volume may use slightly different terminology, emphasize more agency or more structure, highlight the impact of global or local structures, address youth, teacher, or professional identities, or consider new ways of arranging learning to facilitate science identity. In all cases, however, the authors underscore that identity is an ongoing struggle and a way to organize and make meaning of social and cultural life. They also connect the construct to power, which we would argue is a central aspect of defining what and who counts as scientific, who gets to set the rules for what and who counts, and who is able to and denied opportunities to identify as scientific. We encourage readers to use these chapters as a launching pad to our collective next phase of science identity work, where we leverage our creativity, privilege, and passions to create more just science educations.

References

Bang, M., & Vossoughi, S. (2016). Participatory design research and educational justice: Studying learning and relations within social change making. *Cognition and Instruction, 34*(3), 173–193. https://doi.org/10.1080/07370008.2016.1181879

Bang, M., Medin, D., & Atran, S. (2007). Cultural mosaics and mental models of nature. *Proceedings of the National Academy of Sciences, 104*(35), 13868–13874. https://doi.org/10.1073/pnas.0706627104

Boisselle, L. N. (2016). Decolonizing science and science education in a postcolonial space (Trinidad, a developing Caribbean nation, illustrates). *SAGE Open, 6*(1), 2158244016635257. https://doi.org/10.1177/2158244016635257

Calabrese Barton, A., Kang, H., Tan, E., O'Neill, T. B., Bautista-Guerra, J., & Brecklin, C. (2013). Crafting a future in science: Tracing middle school girls' identity work over time and space. *American Educational Research Journal, 50*(1), 37–75. https://doi.org/10.3102/0002831212458142

Carlone, H. B., & Johnson, A. (2007). Understanding the science experiences of successful women of color: Science identity as an analytic lens. *Journal of Research in Science Teaching, 44*(8), 1187–1218. https://doi.org/10.1002/tea.20237

Carlone, H. B., & Lancaster, M. (n.d.). *"Getting along" and "using evidence" in elementary engineering: Desettling cultural narratives* [Manuscript submitted for publication].

Carlone, H. B., Scott, C. M., & Lowder, C. (2014). Becoming (less) scientific: A longitudinal study of students' identity work from elementary to middle school science. *Journal of Research in Science Teaching, 51*(7), 836–869. https://doi.org/10.1002/tea.21150

Collins, P. H. (1990). Black feminist thought: Knowledge, consciousness and the politics of empowerment. .

Crenshaw, K. (1989). Demarginalizing the intersection of race and sex: A black feminist critique of antidiscrimination doctrine, feminist theory and antiracist politics. *University of Chicago Legal Forum, 1989*(1), 139–167.

Davis, N. R., & Schaeffer, J. (2019). Troubling troubled waters in elementary science education: Politics, ethics and black children's conceptions of water [justice] in the era of Flint. *Cognition and Instruction, 37*(3), 367–389. https://doi.org/10.1080/07370008.2019.1624548

Eisenhart, M., Finkel, E., & Marion, S. F. (1996). Creating the conditions for scientific literacy: A re-examination. *American Educational Research Journal, 33*(2), 261–295. https://doi.org/10.3102/00028312033002261

Gee, J. P. (1999). *An introduction to discourse analysis: Theory and method.* Routledge.

Gee, J. P. (2000). Identity as an analytic lens for research in education. *Review of Research in Education, 25*, 99–125. https://doi.org/10.2307/1167322

Goffman, E. (1974). *Frame analysis: An essay on the organization of experience.* Harvard University Press.

Gutiérrez, K. D., & Jurow, A. S. (2016). Social design experiments: Toward equity by design. *Journal of the Learning Sciences, 25*(4), 565–598. https://doi.org/10.1080/10508406.2016.1204548

Hidi, S., & Renninger, K. A. (2006). The four-phase model of interest development. *Educational Psychologist, 41*(2), 111–127. https://doi.org/10.1207/s15326985ep4102_4

Holland, D., Lachicotte, W., Jr., Skinner, D., & Cain, C. (1998). *Identity and agency in cultural worlds*. Harvard University Press.

Johnson, A., Brown, J., Carlone, H., & Cuevas, A. K. (2011). Authoring identity amidst the treacherous terrain of science: A multiracial feminist examination of the journeys of three women of color in science. *Journal of Research in Science Teaching, 48*(4), 339–366. https://doi.org/10.1002/tea.20411

Morales-Doyle, D. (2017). Justice-centered science pedagogy: A catalyst for academic achievement and social transformation. *Science Education, 101*(6), 1034–1060. https://doi.org/10.1002/sce.21305

Nespor, J. (1994). *Knowledge in motion: Space, time and curriculum in undergraduate physics and management*. The Falmer Press.

Shanahan, M. C. (2009). Identity in science learning: Exploring the attention given to agency and structure in studies of identity. *Studies in Science Education, 45*(1), 43–64. https://doi.org/10.1080/03057260802681847

Tan, E., & Calabrese Barton, A. (2020). Hacking a path into and through STEM: Exploring how youth build connecting pathways between STEM-related landscapes. *Teachers College Record, 122*(2), 1–44. https://doi.org/10.1177/016146812012200211

Taylor, S. R. (2020, April 2). [@sonyareneetaylor] *Normal never was* [Photograph]. Instagram. https://www.instagram.com/p/B-fc3ejAlvd/?hl=en

Wolcott, H. F. (2008). *Ethnography: A way of seeing* (2nd ed.). Altamira Press.

Wortham, S. (2006). *Learning identity: The joint emergence of social identification and academic learning*. Cambridge University Press.

Part II
Student Science Identities Outside and Inside School

Chapter 2
"My Love for It Just Wasn't Enough to Get Me Through": A Longitudinal Case Study of Factors Supporting and Denying Black British Working-Class Young Women's Science Identities and Trajectories

Louise Archer, Spela Godec, and Julie Moote

2.1 Introduction

> I like it [science], don't get me wrong, I loved it, but it was, I just don't think that ... like my love for it just wasn't enough to get me through [...] I wish my grades were a bit better and I found it easier, but [...] I just don't like how I'm feeling. (Vanessa, age 18)

In this chapter we seek to understand and identify the factors that contributed to why Vanessa—a Black working-class young woman in England who from an early age had "loved" science and, with strong family support, aspired to a science career—came to feel by age 18 that her love of science was "not enough" to support her science identity and trajectory, something that she found painful and uncomfortable.

It is well established that students from Black communities have long been under-represented in and excluded from post-compulsory science, technology, engineering and mathematics (STEM) in most western industrialised countries, such as the US and UK (e.g., Campaign for Science and Engineering (CaSE), 2014; National Center for Science and Engineering Statistics (NCES, 2015); National Science Board, 2015). Racialized patterns of participation in STEM are complex in that they differ both between and within different communities, being differentiated by social class, gender, ethnicity and subject area (Mcmaster, 2017). However, it has been noted that, compared with other minoritized communities, Black students experience particular injustices. For instance, Black students are among those least likely to study STEM subjects at post-compulsory levels (Mcmaster, 2017), record higher attrition rates from STEM degrees compared with other ethnic groups (Chen, 2013),

L. Archer (✉) · S. Godec · J. Moote
University College London, London, UK
e-mail: l.archer@ucl.ac.uk; s.godec@ucl.ac.uk; j.moote@ucl.ac.uk

H. T. Holmegaard, L. Archer (eds.), *Science Identities*, Contributions from Science Education Research 12, https://doi.org/10.1007/978-3-031-17642-5_2

are less likely to attain higher degree grades and more likely to attain lower degree outcomes compared with White students, and experience lower rates of STEM career progression compared to students from other ethnic backgrounds (Joice & Tetlow, 2020). Research has also drawn attention to the racial trauma experienced by high attaining Black students who are successful in carving out STEM trajectories (McGee & Stovall, 2015).

The reasons for the exclusion of Black students from post-compulsory science are multiply rooted, with the literature identifying a range of factors, including societal and institutional racisms, the exclusion of minority ethnic groups from scientific knowledge in Western societies (Baker, 1998), racisms within schools, colleges and universities and low teacher expectations of Black students (e.g., Atwater, 2000; Brickhouse & Potter, 2001; Carlone & Johnson, 2007; McGee & Martin, 2011; Rascoe & Atwater, 2005) and exclusionary STEM disciplinary cultures (e.g., Rosa & Mensah, 2016). Dominant norms and associations of science with whiteness and masculinity make it particularly difficult for Black girls to feel included and be recognised as authentic scientists in the classroom. As Atwater (2000) discusses, Black girls experience constant representational violence within science classrooms. Indeed, race, class, and gender intersect in ways that "can make membership in a school science community impossible or undesirable for black girls" (Brickhouse & Potter, 2001, p. 965; see also Brickhouse et al., 2000). These injustices are also found in higher education and STEM workplaces, as exemplified by Carlone and Johnson's (2007) investigation of Black women scientists' experiences, in which "the institutional and historical meanings of being a scientist" equated with "being a white male" (p. 1207). As a result, Black girls and women remain acutely and persistently under-represented in STEM, despite comparatively high achievement and positive dispositions towards science (Young et al., 2017).

Given these immense challenges and injustices, in this chapter we consider: how did one Black British working-class young woman, Vanessa, manage to identify with science over time? And how and why was her trajectory curtailed? We employ Black scholars' extensions of Bourdieusian theory as our conceptual framework for exploring these questions through the case study of Vanessa and her parents, Robbie and Akimi, in the hope of shedding light on the broader factors that enable and constrain Black working-class young women's science identities and trajectories.

2.2 A Bourdieusian Approach to Theorising Vanessa's Science Identity and Trajectory

In this chapter we try to understand the factors, relations and conditions that shaped Vanessa's science identity and trajectory over time. Our focus on her *science identity* involves identifying what supported or mitigated her sense of self in relation to science. Science identity has been shown to be a valuable lens because, as Avraamidou (2019) explains, it is both a personal and political multidimensional concept, which encompasses forms of recognition (e.g., who is/not recognised as being a legitimate 'science person') and how science is represented, configured and produced through intersectional norms, values and systems of oppression. As Avraamidou (2019) argues, "science identity provides not only a valuable lens for exploring science (non)participation but also a lens for better understanding the complexities of becoming a science person which are tied to political, structural, and societal problems" (p. 324). Science identity is not a static, discrete 'thing' but is performed and enacted, a process of becoming that is intertwined with and inseparable from other axes of identity and the conditions and relations of production. Importantly, identity is closely bound up with learning, shaping and mediating what learning is possible and desirable (see also Nasir & Hand, 2008).

Our focus on Vanessa's *science trajectory* considers the extent to which she pursued, or not, science qualifications and a STEM-related career pathway, recognising that trajectories are not solely agentic constructions but are influenced by a range of wider forces, conditions and injustices. As numerous scholars have noted, identity can powerfully mitigate young people's science participation and trajectories (Calabrese Barton et al., 2013; Holmegaard, 2015; Rahm & Moore, 2016). Hence, we seek to understand the factors that shaped Vanessa's changing science trajectory over time, between the ages of 10–18.

We engage with the concept of identity through a Bourdieusian lens (e.g., Bourdieu, 1984, 1990, 1993, 2010), drawing in particular on extensions of Bourdieusian theory by Black scholars, as discussed below. But why use Bourdieu? This question is pertinent given that Bourdieu's work is well known for prioritising social class over considerations of race. However, as various commentators have pointed out, race is not absent in Bourdieu's work (Puwar, 2009) and in recent years Black scholars have been using and extending Bourdieu's conceptual tools to advance theory and develop nuanced understandings of the reproduction of oppressions in the lives of Black young people and families, particularly in relation to the school system, as exemplified by Lofton and Earl Davis's (2015) work on Black habitus among college students and developments of the idea of Black cultural capital among middle-class Black students (Meghji, 2017; Wallace, 2017) and parents (Rollock et al., 2011).

As Wallace (2017) explains: "Bourdieu's conceptual toolkit (habitus, field and capital) offers tools for unearthing the complexities of, and contributions to, social (dis)advantage, including their racialised dimensions" (p. 907). We find Bourdieu's work particularly helpful to exploring science identity because it (i) foregrounds structure and the forces and practices of socially reproduction, helping to explain how and why inequalities in science participation are so enduring and resistant to change, and (ii) highlights the pedagogic work undertaken by schooling – in our case, school science practices – which result in racialized, classed and gendered patterns of participation (e.g., Archer et al., 2020).

At the heart of Bourdieu's theory is his proposal that social life can be understood as produced through the interactions of *habitus, capital* and *field*. *Habitus* bears a resemblance to the notion of identity in that it produces a sense of the self (as both an individual and a member of a collective). Habitus can be thought of as an embodied, socialised framework of dispositions which provides our practical 'feel' for the world and a sense of ourselves in relation to others. Habitus is cultivated over time, through our experiences and interactions and produces the sense of "who I am", in both individual and collective terms (e.g., as a Black, working-class young woman from London). Importantly, habitus contains the imprints of our social positioning and experiences and, in turn, structures our views of and interactions with the world. Hence a young person's habitus both reflects and structures their engagement with the fields in which they have grown up (Bourdieu, 1990). In this way, habitus is shaped by, grounded within, and orientated to engaging with the fields in which an individual has been socialised. Moreover, habitus is not merely a set of dispositions that reside 'in the mind', but is also embodied, such as through a person's taste, bodily gestures, accent, gait, and ways of being. In this way, the formative role of the field and a person's experiences in the world can be detected in their embodied (gendered, classed and racialized) habitus, as discussed by Lofton and Earl Davis (2015) in their analysis of Black college students in the US.

The majority of research that has used the concept of habitus has focused on its reproductive nature, such as its role in producing classed patterns of participation in higher education and the feeling as to whether university (e.g., Archer et al., 2003; Reay et al., 2001) and/or science is "not/for me" (e.g., Archer et al., 2017). However, a smaller body of work has sought to unpick the conditions required for a transformative and reflexive habitus (e.g., Yang, 2014) to explain how and why some people 'go against the grain' of social reproduction, such as White working-class students who access elite universities (Reay et al., 2009), showing how habitus is not produced in a singular, straightforward way, and how variations in young people's experiences between different fields can produce distinctive forms of habitus that are not simply aligned with social reproduction and, in turn, how students' encounters with a new, unfamiliar field, such as that of elite higher education, can lead to aspects of their habitus being 'restructured' and 'transformed' (Bourdieu, 2005; Travers, 2017).

Habitus does not exist in isolation, but is formed and operates in conjunction with *capital* and *field*. *Capital* refers to the economic, cultural, social and symbolic resources that a person (or collective) possesses, the value of which is determined by the field. Over the years, considerable attention has been given to delineating different forms of capital and the role that such capital plays in producing unjust patterns of educational participation and outcomes (e.g., Lareau & Horvat, 1999; Noble & Davies, 2009). Here, we focus in particular on work that has sought to explicate the idea of Black cultural capital, which has been proposed and explored as an important counterpoint to the tendency for Black students and communities to be defined in deficit terms as 'lacking' cultural capital. Wallace (2017) defines Black cultural capital as "dominant tastes and expressions adopted and adjusted by the Black middle classes to signal class status and racial identity simultaneously" (p. 907). His conceptualisation includes both use value and exchange value forms of cultural capital (or dominant and non-dominant, in Wallace's (2017) terminology), including specific forms of symbolic Black cultural capital (such as knowledge of Black composers) and knowledge of how to navigate racialized relations (e.g., in educational spaces). Wallace (2017) suggests that "Black cultural capital reflects the simultaneous negotiations of 'race' and class in the acquisition of resource advantages that more often than not are defined by white middle class interests". The concept of Black cultural capital has predominantly been explored in relation to Black middle-class students and parents (for instance, showing how it can be deployed to challenge White hegemonic practices and knowledge within schools, Wallace, 2017), but in this chapter we also consider its role in relation to Vanessa and her parents, as a Black working-class family and, specifically, the potential of Black science capital in supporting her science identity and trajectory.

Capital does not exist or hold value in isolation – the nature, volume and types of capital that someone has will be shaped by the *field* (socio-spatial contexts of power relations that set the 'rules of the game'), in that the *field* determines the value of *capital* and the potential for its translation and accrual. In this way, the structure of *capital* represents the "immanent structure of the social world" (Bourdieu, 1986, p. 242). For instance, our prior work noted that changes in the field can open up or close down the value and potential of particular science-related forms of *habitus* and *capital* in shaping the likelihood of a young person seeing themselves as a science person and/or continuing with science after the age of 16 (e.g., Archer et al., 2015). Likewise, field shapes the ways in which social identities and inequalities are articulated and understood, so that, for instance, race functions differently across different socio-historical and spatial fields (Bourdieu & Wacquant, 1992).

We thus understand expressions and enactments of science/identity as socio-spatially and temporally produced structured agency (i.e., they can be spontaneous and improvised – but also always constrained in possibility and desirability by embodiment and structure produced through the interaction of habitus, capital and field). That is, the form and possibilities of a young person's identity performances will be shaped by their racialized, classed and gendered (etc.) socialised dispositions, experiences, social positioning and embodiment in interaction with the power relations and socio-spatial organisation of the *field* in question.

In this chapter, we additionally draw on Bourdieu's ideas about how societal inequalities are cemented and perpetuated through *symbolic violence*, that is, "violence which is exercised upon a social agent with his or her complicity" (Bourdieu & Wacquant, 1992, p. 167). Symbolic violence refers to how people are inculcated to accept the values and interests of the dominant as 'natural', justified and 'just the way things are'. In this way, the many different advantages enjoyed by dominant (e.g., White, male, middle-class) groups are accepted as a reflection of their superior intellect and talents, and the oppressed and disadvantaged positions of Others are represented as due to a lack of, for instance, intelligence, motivation, character and so on. In other words, symbolic violence is a "process whereby individuals, through their experience of the social world and of the various institutions and structures that compose it, come progressively to develop taken-for-granted ways of thinking and behaving that reflect this lived experience" (Connolly & Healy, 2004, p. 16). For instance, we have previously found that the pedagogic work that is undertaken by school physics in England to 'naturalise' the association of physics with cleverness and masculinity makes it difficult for many young women to identify with and progress in the subject (Archer et al., 2020). In this chapter, we extend these ideas to consider how dominant practices and representations of science may exclude Vanessa from a science identity and trajectory through a process of symbolic violence that hides the racialized, gendered and classed nature of the exclusion so that Vanessa and her family come instead to attribute the causes to a personal 'lack' of love and attainment.

2.3 Data Sources

The chapter discusses data and analyses from the ASPIRES study, an ongoing, 13 year, mixed methods, longitudinal study of young people's science and career aspirations, conducted with students in England between 2009–2023. To date, the study has collected survey data from a total of 40,000 young people students through five waves (Wave 1 (age 10–11) = 9319; Wave 2 (age 12–13) = 5634; Wave 3 (age13–14) = 4600; Wave 4 (age 15–16) = 13,421; Wave 5 (age 18–19) = 7013), and has conducted over 700 longitudinal interviews with a subsample of students and their parents (with 60 students tracked over the 10–18 period). Surveys and interviews were conducted at the same five time periods, when students were age 10–11, 12–13, 13–14, 15–16 and 18–19; survey measures related to aspirations in science and the science self-concept (see Moote et al., 2019).

We start by providing some brief contextualising analyses from the survey data and then focus in depth on one of the interview students, Vanessa, and her parents, Robbie and Akimi, who are described in more detail below. We selected Vanessa's case study in order to explore in-depth and over time the challenges faced by young Black working-class women in identifying and participating in science. Vanessa's case was particularly apt (and potentially able to shed light on the quantitative findings) because Vanessa reported a strong interest in and linking of science through the early years of the study, yet over time came to decide that science was 'not for her'. In discussing Vanessa's case study, we draw on four interviews conducted with her between the ages of 10–18 (in the last year of primary school, age 10; twice during secondary school, age 13 and 16; and once during sixth form (post 16) college, age 18) and four interviews with her parents (Akimi and Robbie) at each of these time points.

2.4 Survey Findings: Black Students Do Not Lack Science Interest, Aspiration or 'Identity'

We recognise that it is difficult, futile even, for quantitative surveys to attempt to capture the richness and complexity of science identity. However, survey data can be useful for helping to identify broad patterns and trends in how people construct and represent their identities and to help situate more in-depth qualitative work. Accordingly, before focusing on Vanessa's case, we want to first consider some broad-brush patterns that we noted in the ASPIRES project survey data, which help to show how Vanessa's case, while necessarily unique, also relates to and exemplifies issues facing Black students in England more broadly.

The point that we would like to highlight from the survey data is that analysis of each of the waves shows that between the ages of 10 and 18, Black students reported comparatively higher levels of science aspiration and science self-concept than White students. Specifically, odds ratio analyses show that Black students were approximately one and a half times more likely than White students to aspire to become a scientist and/or work in science. Across all survey waves/age points, Black students also reported significantly higher self-concepts and more confident learner identities in science (e.g., feeling that they do well in science, learn things quickly in the subject (M = 17.65, SD = 4.21), compared to White students (M = 16.70, SD = 4.33; $t(31{,}809) = 9.615$, $p < .001$), with Cohen's d.22 indicating a small effect size.

In other words, our analyses of the survey data raise some intriguing questions and challenges to mainstream policy assumptions and approaches, which tend to assume that a key factor underlying the under-representation of Black students in post-compulsory science is due to a lack of aspiration and self-confidence (e.g., Cabinet Office, 2011; Department for Education (DfE, 2010), see also numerous interventions aimed at 'raising aspirations' and mentoring Black students). Instead, our data indicate indeed that Black students are more likely than their White peers to express science aspirations – yet these aspirations do not seem to translate into science participation and retention (e.g., Mcmaster, 2017). We now turn to Vanessa's case study to further explore the potential reasons for this pattern.

2.5 Introducing Vanessa

Vanessa is a young woman of Black African (Nigerian) heritage, who comes from a science-keen, working-class family. She lives with her mother, Akimi, in a council flat in a tower block in inner London. Her father, Robbie, who lives locally, has a degree in pharmacy from Nigeria. For reasons that we would interpret as due to societal and institutional racisms, he had not been able to realise the full potential exchange value of his qualifications in Britain and worked in a relatively low paid job as a science technician in a "very challenging" local London secondary state boys' school. Vanessa's mother completed secondary schooling in Nigeria, studied for 2 years in a teacher training college and now works as a care assistant in a residential care home.

2.5.1 Vanessa Age 10

When we first met her, at age 10, Vanessa aspired to become a doctor ("I like helping people ... when they're hurt and I like looking at blood"), which she saw as a potential route towards "a better life". At home, Vanessa's family expressed a strong interest in and valuing of science. Robbie held a degree in pharmaceutical science from Nigeria, although as he reflected the first time we met him, "I still cry for not climbing on education-wise". Akimi strongly motivated Vanessa to do well in order to have a well-paying job to become more socially mobile. Both parents valued education and qualifications as the way to get on in society, but especially so for members of the African diaspora (e.g., "You see, for us coming from Africa, we had no option. For us, education is the only way", Robbie).

From an early age, Vanessa described how her family had influenced her own aspirations towards a science profession, explaining:

> Because my dad's a scientist he would like me to get a bit interested in science … and I do find science a bit interesting […] He seems to buy quite a lot of science things for me as well. (Vanessa, age 10)

Vanessa felt that her family was not unique in their valuing of science, but reflected a wider common valuing of science among families of African heritage:

> But actually, some African people like science a bit more, because science in Africa seems to be what's getting more money. At the moment because science people like to look at germs and sicknesses, because in Africa … they get sick a bit quickly, so they like doctors and scientists who can find out how it's happening and it's going around. (Vanessa, age 10)

Robbie also talked about an African cultural valuing of science (e.g., "because back home in Africa, we know that science, it's something very good"). The family also invested what little spare money they had in paying for extra tutoring to support Vanessa's academic attainment, a practice that has been found to be common among minority ethnic families in the UK (Shah et al., 2010).

Vanessa attended a local inner London primary school. From age 10, she loved school science and named science as her favourite subject ("I like it a lot"). She maintained this love of science despite finding the school science curriculum of variable interest ("sometimes … a bit boring, sometimes it's interesting") and being relatively alone among her peers in this respect ("I think that most people don't like science because the noise we get when we hear we've got science is just so horrible", age 10). Indeed, Vanessa told us "When they [students] upset the teacher, I feel upset too … I don't like the way they treat our teachers". Vanessa also described herself and her close friends as different to the other girls in her class, who she categorised as "just a bit too girly".

She felt that the national Standard Assessment Tests (SATs; taken by children at the start of their final term in primary school) dominated teaching content and timetabled lessons and she was disappointed by the increased time that was given to literacy and numeracy (the two areas that are nationally assessed), in preparation for the tests. She also expressed her frustration that even once the tests were over, science was still not taught ("And now we've finished our SATs we don't do science anymore").

2.5.2 Vanessa Age 13

At age 13, Vanessa had shifted a little in her aspirations and now wanted to become a forensic scientist ("I want to be a forensic scientist [...] I don't really want to be a doctor anymore. Because I found the forensic thing more interesting"). At home, her family science interest and support remained strong, and Vanessa described how Robbie was able to help support her science learning due to his science knowledge and expertise:

> [In science class], if I don't get it, it's like better because my dad can then just come and help so it's easier for me to learn. (Vanessa, age 13)

The family still also paid for Vanessa to receive additional tutoring support to help her attainment.

At this age, Vanessa also recounted how her science interest and aspirations were further fostered and shaped at home through her consumption of forensics-related television programmes, notably the US popular drama series, *CSI: Miami.* She explained that she had really enjoyed a forensics-themed school session in Year 7 (age 11/12), and when they had heard of her interest, her aunts had recommended that she should watch the programme. In explaining her enjoyment of *CSI: Miami,* Vanessa commented specifically on the diversity of characters she had seen in science roles on the programme (e.g., a lead character in the series is a female African American forensic scientist).

By the time of her age 13 interview, Vanessa attended a local London secondary school, where science remained Vanessa's favourite subject:

> There's just something about it like I look forward, like 'Oh, I've got Science, yeah' it's the lesson you kind of look forward to going to ... I work harder in science than anything else. (Vanessa, age 13)

While she still enjoyed the subject, Vanessa suggested that most of her peers disliked school science due to a lack of practical science ("Because they find science really, um, boring because these days we don't do that much experiments anymore."). At this age, she started to express some common gendered views of science:

> If you're really girly you just don't see yourself working in Mechanics, do you, and you don't see yourself doing Physics. If you're girly, you probably see yourself more with Biology and Chemistry. (Vanessa, age 13)

However, Vanessa still continued to draw distinctions between herself and other girls, in that she wore "my trainers and jogging bottoms" and more "relaxed" clothes whereas her girly friends and peers wore "the skirts and the dresses and the nail varnish", adding "I'm not really the skirt and ... heels kind of person".

Like the majority of other young people in our study, Vanessa suggested that "I think it's the clever people" that do science and she thought of science as a "hard" subject – a point that was also consistently repeated across her later interviews. She identified being "really focussed and determined" and "wanting to do better" as legitimate ways to do, and be recognised in, science:

> I don't think you have to be clever; you just have to be really focussed and determined. Cos I wasn't like amazing … I'm not amazing at science, but I do have the mind of wanting to be better in it. (Vanessa, age 13)

Whereas Vanessa felt that 'hard work' would enable her to continue with science, Robbie expressed concerns:

> One thing about science is you need to be very, very good, yeah? It's almost as good as nothing having just an additional science at GCSE. You need to be able to do the triple sciences if you are want to, you know, carry on big time in science […] I still feel I probably am not doing enough. (Robbie)

In particular, Robbie worried that not only is it necessary to study the prestigious triple science (equivalent to three GCSE qualifications) route but a student needs to be not just good but "very, very good" in the subject in order to succeed. He described how he was conducting additional after school study sessions with Vanessa and paying for her to have a tutor, to try to raise her attainment. But he still worried that he might be still "not doing enough".

Robbie expressed his frustration with what he perceived to be noisy and disruptive classes and a lack of science teacher specialism at Vanessa's school, conditions which he described as being "worse than in most countries in the third world". He found the state of science teaching and learning in English schools particularly perplexing given his experience of growing up in Nigeria, which had originally been colonised by Britain, when discipline and respect had been key discourses within colonial propaganda ("We were colonised by these people! They came to our country and told us 'now listen, your teacher is your second God, you have to respect them'. Well now coming here [to England] … it's not like that. And sometimes we ask, what is happening? You know?").

2.5.3 Vanessa Age 16

At age 16, Vanessa still aspired to be a forensic scientist:

> I'd like to be a forensic scientist… I'd like to work with the police actually and kind of do all that forensic stuff – finding out how the crime happened, what happened, the time, weapons used and all of that stuff. (Vanessa, age 16)

Vanessa's parents continued to be interested in and support her science trajectory, a point that Vanessa recognised as helping maintain her interest and attainment in the subject:

> Science was like the one thing I had a parent that could really help me with. So, it's kind of like Science was the one thing I was a bit okay with, cos I had someone that could really help me in science. And then I think … it started from that to actually liking it, and then … [My dad] knew all the sciences, but he's strong on his Chemistry. (Vanessa, age 16)

However, Vanessa explained that Akimi encouraged her to consider a medical career rather than forensic science ("My mum, she still wants me to do Medicine, she kind of wants me to work like in a pharmacy. My dad, he just kind of wants me to do what I like, but something good that would obviously pay the bills in the end"). At home, Vanessa continued to love watching *CSI: Miami*:

> I've been probably just watching a lot of CSI. That's probably my one big thing. Like I can't stop! CSI: Miami, that's just the only one I watch. (Vanessa, age 16)

The programme's content and characters also shaped her view of tAAhe profession as an ethnically diverse and gender-balanced field (e.g., "I think it's quite a … I think it's a diverse … it's quite unisexual. Like from what I've seen there's the same amount of women in Forensic as men, from what I've watched").

At school, Science was still Vanessa's favourite subject and she studied for the prestigious Triple Science route ("I liked it, it was interesting") which could potentially enable her to apply for Advanced level science courses. Vanessa obtained "mostly B grades" in her science General Certificate of Secondary Education (GCSE) exams at the age of 16, which Robbie described afterwards as "average". Vanessa applied to take Advanced level (A Level, the 2 year, national post-compulsory academic qualification) courses in Biology, Chemistry, English Language and Psychology, however, while B grades are widely accepted for entry to courses like English and Psychology, as Robbie explained, the sixth form college that she attended would not allow her to study Chemistry A level with a B grade (a situation that is common across schools in England). As a result, she applied instead for Sociology, and was accepted on to this course. Both Vanessa and her father were disappointed and "not exactly pleased" with this outcome, with Robbie lamenting that she was "not strong enough" to continue to A level in Chemistry.

Usually, schools in England are expected to provide 16 year-old students with careers education, advice and guidance. However, Vanessa recounted that she received very little relevant or useful careers education support and in particular, had not experienced any support relating to science careers ("they bring people into school but they don't have many people that are like scientific or anything").

2.5.4 Vanessa, Age 18

By age 18, Vanessa's ambitions had changed, and she now planned to study for a degree in Criminology ("I've chosen Criminology to study at uni"), an aspiration that she still related to her interest in solving crimes but which, much to her parents' regret, had moved away from a science-based approach to social science. Vanessa felt that a Criminology degree would enable her to pursue her interest in solving crimes without requiring formal science qualifications or content. At the time of her last interview, she had received a number of offers from UK universities.

Her interest and identification with science had also changed substantially compared with former interviews, as she emphasised "I don't really want to do anything science based". When asked to explain further, she replied:

> I think I don't know I just find it difficult. I like it, don't get me wrong, I loved it, but it was, I just don't think that … like my love for it just wasn't enough to get me through […] I wish my grades were a bit better and I found it easier, but […] I just don't like how I'm feeling. (Vanessa, age 18)

In a poignant echo to Vanessa's quote, (although the two interviews were conducted entirely independently, at different times and locations), Robbie similarly concluded in his final interview: "She loved the sciences, but just loving a subject is not enough".

Vanessa also described how she no longer focused on science-related content within TV programmes in the same way that she had previously:

> Like when I'm watching like CSI and like criminal cases and stuff, like and documentaries on people that have gone to jail and finding out stuff like that, there is like, you do see the Science, bits of it, like the Science aspects coming through, but that's not what I focus on anymore. (Vanessa, age 18)

When asked what she felt she had gained from her experiences of school science, Vanessa reflected: "What have I gained? Um, not much, nothing that I remember now, so yeah". She continued to explain that she did not feel it had given her much 'useful' knowledge that she could apply in her current life ("science doesn't help you a lot in life, but it does help a tiny bit"). Vanessa explained that she had started studying for Biology A level, but a year later she dropped the subject because:

> I kind of looked at my grades and I thought what would match, like what subjects would be best? […] Because that was my lowest, I got the lowest grade in that, so I dropped that […] It was harder than I thought. … Because I got, I mean I got Bs in everything really, but I just thought like English and my writing skills are better, so I thought I'd be better off with something like Sociology and Psychology, which are essay-based". (Vanessa, age 18)

When we asked Vanessa about her subject choices and the decision to drop Biology, and whether she had spoken to anyone else before making the decision, she replied "No, not really, I kind of just said what I was going to do and just did it".

In both their interviews at this time, Vanessa and Robbie expressed a strong regret that Vanessa had not been able to continue with science and both said that they wished they had known and been told about alternative science qualification routes, such as the more vocationally-orientated Business and Technology Education Council (BTEC) qualification that might have enabled her to continue with science (e.g., "I wish I'd took BTEC", Vanessa; "If we had the knowledge we have now, we probably would have taken the BTEC route, you know, through applied science or so. But I mean it was too late", Robbie). Robbie was particularly disappointed that, by not taking any post-16 science qualifications, the door to Vanessa pursuing a science degree was now closed – a situation that he strongly wished were otherwise ("I had wanted her to do a degree in pure science, but that's no longer possible now"). Vanessa also explained that she had not received any individualised careers advice or guidance sessions in the previous 2 years and had only been offered the occasional job fair or generic speaker, which was, she noted "not actually personalised for me".

2.6 Discussion

We now consider (i) how and why Vanessa was able to develop and sustain over time an identification with and aspirations for science and (ii) how and why she ultimately came to feel that her "love" of science was "not enough" to sustain a science trajectory. We argue that it was her home and family habitus and capital (notably her Black science capital) that enabled the former, while the school system produced the latter, in ways that we interpreted as symbolic violence.

2.6.1 Supporting Vanessa's Trajectory: Black Habitus, Black Cultural Capital and Black Science Capital

In various prior studies, family support has been identified as a key factor supporting the STEM trajectories of Black (Rosa & Mensah, 2016; Russell & Atwater, 2005) and Hispanic (Brown, 2002) students. Similarly, in Vanessa's case, her science identity and trajectory were substantially fostered, maintained and possibilised by her family, particularly her parents. We interpret Vanessa's parents as not only having engaged and motivated their daughter towards science (through cultivation of a science rich habitus), they also generated and deployed capital in key ways that supported and enabled Vanessa's science identity and trajectory.

First, the family diasporic habitus shaped Vanessa's own habitus towards a strong valuing of science. For instance, Robbie and Akimi's hopes and dreams for intergenerational social mobility and a "better life" for their daughter (generated through their experiences of migration, racism and poverty in Britain) had strongly shaped Vanessa's early aspirations. Robbie's belief that "education is the only way" to achieve social mobility echoes a discourse and strategy that have been noted among various diasporic communities in Britain (e.g., Archer & Francis, 2007). Second, despite their working-class social positioning in the UK, following Wallace (2017) we interpreted the family as mobilising forms of Black cultural capital and, more specifically, Black African cultural capital, and in particular, Black (African) science capital, to support Vanessa's science identity and trajectory. For instance, despite limited finances, the family deployed economic and cultural capital to support Vanessa's engagement with science and to foster her interest in and connection with science – as illustrated by Robbie's purchasing of numerous science kits for Vanessa throughout her childhood and his constant motivation, expectations and sharing of his own science interests and expertise with his daughter over the years. His academic support, tutoring and monitoring to support Vanessa's science attainment can also be understood as deployments of capital to try to secure educational advantage and 'success' – albeit from a position of trying to enable social mobility rather than social reproduction (as in the case of the middle-classes). Robbie and Vanessa's narratives around the valuing of science within African culture might also be interpreted as examples of Black cultural capital that draw on rich cultural values and expertise with the aim of challenging or overcoming class and racial disadvantage. In other words, we understand Vanessa's diasporic, Black African family habitus and capital as not just valuable but indeed essential to sparking, fostering and sustaining Vanessa's science identity and trajectory – hence in contrast to many popular policy assumptions, Vanessa's cultural context and resources were assets, not deficits, that supported her engagement with science.

In addition to the family habitus and capital, we suggest that Vanessa's consumption of *CSI: Miami* might also be interpreted as a form of Black cultural capital in that the programme was specifically recommended by her aunts (as an example of Black family social capital) and the identity of the Black female scientist lead character in the series appeared to be an important cultural resource that helped foster and possibilise Vanessa's own forensic science identity and aspirations in relation to both race and gender. In line with our Bourdieusian framework, we understand the series as not constituting a fixed form of capital per se (in which the 'value' is intrinsic to the TV programme would be universally accrued by any viewer), but rather that the value of the programme as a potential source of capital is determined by the field. Hence in this case, Vanessa is able to leverage capital from her consumption of the series – and this leveraging is enabled through specific interactions of habitus and capital. We interpret the value of Vanessa's consumption and love of the TV series as underlining the importance and value of inclusive media representations of science and scientists – while also noting that this was the only popular representation of a Black, female scientist that any participant in the study identified, suggesting that they remain rare and elusive. Moreover, we suggest that the presence of diverse representations (and role models) alone is insufficient – we interpret Vanessa's case as illustrating the range of additional 'work' that needs to be done through interactions of habitus, capital and field, to leverage and realise the potential value of these representations within the science identities and trajectories of young people.

While Vanessa's engagement with *CSI: Miami* and the science kits that her parents bought her constitute examples of informal, everyday science learning and engagement, we also noted that Vanessa did not additionally benefit from experiences of informal science learning through designed or community settings (e.g., she did not attend any science clubs or visit science centres and so on). While we cannot judge the impact of this on Vanessa's identity and trajectory, we do note that a range of studies of 'successful' Black science students (particularly those who went on to study for a science degree) draw attention to the value and role of culturally rooted, informal science learning contexts for supporting Black students' identity and trajectories in science (e.g., Calabrese Barton & Tan, 2010; Rahm, 2010;

Rosa & Mensah, 2016) and mathematics (e.g., Nasir, 2002; Nasir & Hand, 2008; Walker, 2006, 2009, 2011) For instance, Ortiz et al. (2019) examined identity formation among Black students, highlighting the importance of both formal and informal STEM educational experiences over time, emphasising how experiences as early as kindergarten can be important.

2.6.2 Closing Down Vanessa's Science Identity and Trajectory: The Role of Schooling

Whereas we identified family and home support and experiences as being instrumental in developing and supporting Vanessa's science identity and trajectory, we found that school science and the education system were closely implicated in closing down and denying her science identity and trajectory, a theme that is also present in the narratives of the six women detailed in Rosa and Mensah's (2016) study. For instance, throughout Vanessa's time at school, national assessments seemed to play a key role in shaping the possibilities for her to access and engage with science, restricting the amount of time that her primary school gave to science teaching and lessons and directly preventing her entry and participation in science qualification routes in secondary school through the requirement for higher grades in science than is required for many other subjects. While Vanessa's family support, habitus and Black (African) cultural, social and science capital significantly supported and sustained her science identity and trajectory over a number of years, in the end these could not mitigate the powerful educational systems, structures and practices that maintain the elite status of science through restricted entry routes to science and through their action (through symbolic violence, as discussed below) on the habitus and their mitigation of Black capital that lead Vanessa to the painful conclusion that her 'love' for the subject is not enough.

From a Bourdieusian perspective, we understand the education system as fundamentally designed to support and enable the reproduction of privilege and relations of oppression and domination (Bourdieu & Passeron, 1977). That is, the education system is based on and designed to reproduce the interests, values and social positioning of the privileged, which means the system is designed to reproduce White, male, middle-class privilege and supremacy. We can see this underlying impetus play out in various ways in Vanessa's case.

First, we interpret Vanessa's experiences of urban schooling as an example of how the education system in England is differentially resourced and distributed, a process that has been termed the 'rationing' of education along class and ethnic lines (Gillborn & Youdell, 1999). As a working-class, Black student, Vanessa is not unusual in experiencing under-resourced, disruptive classrooms with high science teacher turnover and an under-supply of specialist science teachers (Manning, 2017) and unequal provision of careers education support (Moote & Archer, 2018) – a point that Robbie highlights as exemplifying the "myth of colonialism" (in which the educational vision that has been "sold" is restricted to the privileged and not shared with the colonised).

Second, we note that one of the conditions of possibility for many young women to identify with and pursue science trajectories is a negotiation of femininity and the performance of 'different', but specifically 'non-girly' femininity (see also Francis et al., 2017; Gonsalves, 2014), which we interpret as highlighting the continued alignment of science with masculinity and the reproduction of gender privilege.

Third, we suggest that while Vanessa's Black habitus and Black capital were valuable and transformative in that they challenged dominant power relations to produce and sustain her science identity and trajectory, the normative and socially reproductive dominant power of the field (to value and leverage dominant forms of capital) still remained, as exemplified by the ways in which Vanessa was disadvantaged at various points in her trajectory by not having specific forms of dominant cultural knowledge and resources to support her decision-making (as exemplified by her and Robbie's "regret" and poignant reflection "I wish I'd known").

Fourth, we note the powerful ways in which educational gatekeeping (e.g., requirements to access to high status science qualification routes and advanced level science courses) is designed to 'keep out' particular students in order to reproduce the elite status of the subject (Archer et al., 2017). As we have discussed elsewhere, we interpret the popular and pervasive construction of science as 'hard' and 'only for the clever' as reproductive of racialized, gendered and classed science participation, through constructions of 'cleverness' as aligned with White, male, middle-class identity, in which the educational attainment of Others tends to get explained away (e.g., as due to hard work, rather than intelligence or aptitude, see Archer, 2008; Archer & Francis, 2007; Bourdieu & Passeron, 1977). Hence, we see how dominant associations of science with cleverness mitigated Vanessa's relationship with science, as her good grades in science were deemed insufficient by a system that restricts entry to exceptional/best grades only. We see such practices as part of how the field undertakes social reproduction – reproducing the elite status of science and maintaining unjust patterns of participation – albeit at the cost of the aspiring individuals who are excluded and to the detriment of the field itself, which misses out on the talents of young Black women, such as Vanessa.

Finally, using our Bourdieusian lens, we identify the ways in which these practices of distinction operate to a considerable extent through symbolic violence, as exemplified by Robbie and Vanessa 'blaming' themselves (rather than systemic and structural injustices) for Vanessa's 'failure' to maintain a science identity and trajectory. We see this, for instance, in Robbie's concern that he is "not doing enough" to support his daughter and Vanessa's statement that her love for science was "not enough" and her "personal" decision to withdraw from Biology A level on the basis that it is "too hard" (see also Robbie's own regrets at "not climbing on education-wise"). As Bourdieu explains, symbolic violence is a powerful technology of oppression that works through intersectional injustices. While Vanessa's decision to drop Biology is arguably a logical and strategic response to injustices within the *field* (to maximise her chances of academic success within a subject that disadvantages her), we suggest that it highlights how practices of distinction in science (whereby science subjects achieve status through their demands for higher levels of attainment than other subjects) are both arbitrary and play a role in reproducing unjust patterns of participation. Finally, we note the pain that these experiences inflict on Vanessa and the identity work that she must now, as exemplified by her account of how she now watches *CSI: Miami* in a different way, not focusing on the science aspects and in her description of how she does not like how she now feels about science.

We also note that Vanessa's account notably does not mention the presence or role of caring and kind teachers, and particularly caring teachers of colour, factors that have been noted as important in various studies of successful Black STEM students (e.g., Brown, 2002; Mensah, 2019; Mensah & Jackson, 2018; Nasir et al., 2019; Russell & Atwater, 2005). We also note the absence of any specific educational support programmes within Vanessa's trajectory – which, again, have been identified as important facilitators of Black STEM students' trajectories, such as advanced school preparation programmes (Rosa & Mensah, 2016) and college preparation and support programmes (Brown, 2002; Maton et al., 2000; Russell & Atwater, 2005), particularly those aimed at supporting minoritized students on to STEM degree programmes. Notably, a number of these studies also identify the crucial role of financial assistance for students to undertake specific programmes – which again, Vanessa did not benefit from. While we cannot speculate as to whether the presence of these factors would have resulted in Vanessa completed advanced level science qualifications and entering her desired forensic science degree, we suggest that their absence arguably increased her vulnerability and that had such support been in place, it would have likely offered valuable additional resilience and support to her science identity and trajectory.

2.6.3 The Science Debt?

Our analysis of Vanessa's case has highlighted intersectional injuries and injustices within the lives of Black, working-class young women and the ways in which these injustices are 'hidden' by symbolic violence, such that the 'blame' and cause of injustices in STEM participation become located within Black individuals (e.g., as a lack of attainment) rather than being attributed to unjust educational practices and systems. Importantly, Vanessa's case shows how Black families' habitus and capital are important and valuable resources that can support a young person's science identity and trajectory – providing an important counterbalance to popular policy assumptions (e.g., Commission on Race and Ethnic Disparities, 2021).

We propose that the disjuncture identified by our quantitative and qualitative analyses, in which Black school students tend to express high aspirations for science and high levels of science self-concept yet record comparatively lower rates of science participation and high attrition, can be conceptualised as an example of the 'science debt'. The notion of science debt borrows directly from Gloria Ladson Billing's (2006) foundational notion of the 'education debt', in which she powerfully reconceptualised educational 'gaps' in attainment (between Black and White students) as a debt that is accumulated over time, through relations of injustice and experiences of both overt racisms and repeated micro-aggressions, and that is consequently 'owed' by the system to racialized communities. We suggest that Vanessa's case powerfully exemplifies the 'science debt', underlining how, despite possessing considerable family science capital and a habitus orientated towards science, social and educational inequalities worked to exclude, demoralise and persuade her away from science, denying her 'love' of the subject and rendering her science identity and aspirations 'impossible'.

2.6.4 Thinking Otherwise: Changing the Field of School Science

So how might the field of school science better support young people like Vanessa's science identity and participation? Bourdieu proposes that field determines the value accorded to students' capital, setting the 'rules of the game' regarding what and who gets valued within a given space. From this perspective, we understand that supporting the science identities and trajectories of students requires changing power relations and inequities within the field of science education (see Archer et al., 2017; Godec et al., 2017), rather than seeking to change the student. This approach contrasts considerably with many existing interventions in the UK that are aimed at diversifying participation in STEM, which tend to adopt a deficit approach that are premised upon addressing a 'lack' of interest, motivation, knowledge and/or skills within the young people in question.

Instead, our call to change the field focuses on transforming unjust power relations and practices of privilege/oppression and distinction, changing educational systems and practices, for instance, drawing on justice-orientated, culturally relevant (Ladson-Billings, 1995; Mensah, 2011) and assets-based pedagogies within teaching, learning and outreach work that challenge dominant binaries (e.g., Jammula & Mensah, 2020) and develop relations of trust and care between teachers and young people (Nasir et al., 2019). Changing the field also needs to happen not just in the classroom but through the entire education system and particularly at the level of national education policy, as education policy produces the conditions that require schools to enact unjust practices and is a key way in which the continued 'science debt' is produced and sustained and which results in the symbolic violence and exclusion experienced by young women like Vanessa.

Finally, returning to Vanessa's opening quote, we suggest that love also constitutes a valuable and important resource in the reimaging of science education policy and practice. Love is a long-standing principle and practice within the civil rights movements and many contemporary movements, enabling dialogue and connection and supporting educators to recognise and understand the hurt and oppression experienced by Black students (Hooks, 1994). As Freire (1970) argued, a politics and practice of love can enable the breaking down of unequal power relations between teachers and as hooks explains, "emphasizes the importance of service to others" (Hooks, 1994). Hence, we suggest that in addition to recognising, valuing and supporting a students' love for science, educators can also centre love in their pedagogy, "softening the power differential between students and teachers in support of greater relational equity in the space" (DiGiacomo & Gutiérrez, 2016) and using love as a means for supporting students' science identities and trajectories (e.g., Morrell, 2015).

2.7 Conclusion

In this chapter we have argued that 'problem' of Black students' under-representation in science is created by the operation of societal and systemic injustices that are enacted through everyday science teaching, learning and education policy. Contrary to deficit assumptions that are evident in many popular policy texts and outreach initiatives, which assume that Black families and students 'lack' information, motivation and/or aspiration towards science, our survey analysis found significantly higher science aspirations among Black students compared to White students and our qualitative analysis identified how Vanessa's home and family habitus and capital, particularly her family's deployment of Black African science capital, were important and valuable to fostering and sustaining her science identity and trajectory. Yet, education system was closely implicated in closing down and denying Vanessa's science identity and trajectory, ultimately leading her to decide that science was not for her. This exclusion not only negatively impacts Vanessa, but speaks to the wider exclusion of Black students from science – an injustice that is to the detriment of science and society and requires urgent change. To conclude, we argue that productive ways forward should focus on changing the field, rather than young people, to challenge injustices in STEM education policy and practice and to support and embed culturally responsive pedagogies of love within school science teaching.

References

Archer, L. (2008). The impossibility of minority ethnic educational 'success'? An examination of the discourses of teachers and pupils in British secondary schools. *European Educational Research Journal, 7*(1), 89–107. https://doi.org/10.2304/eerj.2008.7.1.89

Archer, L., & Francis, B. (2007). *Understanding minority ethnic achievement: Race, gender, class and 'success'*. Routledge.

Archer, L., Hutchings, M., & Ross, A. (2003). *Higher education and social class*. Routledge.

Archer, L., Dawson, E., DeWitt, J., Seakins, A., & Wong, B. (2015). "Science capital": A conceptual, methodological, and empirical argument for extending Bourdieusian notions of capital beyond the arts. *Journal of Research in Science Teaching, 52*(7), 922–948. https://doi.org/10.1002/tea.21227

Archer, L., Moote, J., Francis, B., DeWitt, J., & Yeomans, L. (2017). Stratifying science: A Bourdieusian analysis of student views and experiences of school selective practices in relation to 'Triple Science' at Key Stage 4 in England. *Research Papers in Education, 32*(3), 296–315. https://doi.org/10.1080/02671522.2016.1219382

Archer, L., Moote, J., & Mac Leod, E. (2020). Learning that physics is 'not for me': Pedagogic work and the cultivation of habitus among advanced level physics students. *Journal of the Learning Sciences, 29*(3), 347–384. https://doi.org/10.1080/10508406.2019.1707679

Atwater, M. M. (2000). Females in science education: White is the norm and class, language, lifestyle and religion are nonissues. *Journal of Research in Science Teaching, 37*(4), 386–387. https://doi.org/10.1002/(SICI)1098-2736(200004)37:4<386::AID-TEA6>3.0.CO;2-M

Avraamidou, L. (2019). Science identity as a landscape of becoming: Rethinking recognition and emotions through an intersectionality lens. *Cultural Studies of Science Education, 15*, 323–345. https://doi.org/10.1007/s11422-019-09954-7

Baker, D. (1998). Equity issues in science education. In B. J. Fraser & K. G. Tobin (Eds.), *International handbook of science education* (pp. 869–896). Kluwer.

Bourdieu, P. (1984). *Distinction*. Harvard University Press.

Bourdieu, P. (1986). The forms of capital. In J. Richardson (Ed.), *Handbook of theory and research for the sociology of education* (pp. 241–258). Greenwood.

Bourdieu, P. (1990). *The logic of practice*. Polity Press.

Bourdieu, P. (1993). *The field of cultural production*. Columbia University Press.

Bourdieu, P. (2005). Habitus. In J. Hillier & E. Rooksby (Eds.), *Habitus: A sense of place*. Ashgate.

Bourdieu, P. (2010). The myth of 'globalization' and the European welfare state. In G. Sapiro (Ed.), *Sociology is a martial art: Political writings by Pierre Bourdieu*. New Press.

Bourdieu, P., & Passeron, J. C. (1977). *Reproduction in education*. Sage.

Bourdieu, P., & Wacquant, L. J. D. (1992). *An invitation to reflexive sociology*. Polity Press.

Brickhouse, N. W., & Potter, J. T. (2001). Young women's scientific identity formation in an urban context. *Journal of Research in Science Teaching, 38*(8), 965–980. https://doi.org/10.1002/tea.1041

Brickhouse, N. W., Lowery, P., & Schultz, K. (2000). What kind of a girl does science? The construction of school science identities. *Journal of Research in Science Teaching, 37*(5), 441–458. https://doi.org/10.1002/(SICI)1098-2736(200005)37:5<441::AID-TEA4>3.0.CO;2-3

Brown, S. W. (2002). Hispanic students majoring in science or engineering: What happened in their educational journeys? *Journal of Women and Minorities in Science and Engineering, 8*(2), 123–148. https://doi.org/10.1615/JWomenMinorScienEng.v8.i2.20

Cabinet Office. (2011). *Opening doors, breaking barriers: A strategy for social mobility*. HM Government. https://www.gov.uk/government/publications/opening-doors-breaking-barriers-a-strategy-for-social-mobility

Calabrese Barton, A., & Tan, E. (2010). We be burnin'! Agency, identity, and science learning. *The Journal of the Learning Sciences, 19*(2), 187–229. https://doi.org/10.1080/10508400903530044

Calabrese Barton, A., Kang, H., Tan, E., O'Neill, T. B., Bautista-Guerra, J., & Brecklin, C. (2013). Crafting a future in science: Tracing middle school girls' identity work over time and space. *American Educational Research Journal, 50*(1), 37–75. https://doi.org/10.3102/0002831212458142

Campaign for Science and Engineering (CaSE). (2014). *Improving diversity in STEM*. https://www.sciencecampaign.org.uk/resource/ImprovingDiversityinSTEM2014.html

Carlone, H., & Johnson, A. (2007). Understanding the science experiences of successful women of color: Science identity as an analytic lens. *Journal of Research in Science Teaching, 44*(8), 1187–1218. https://doi.org/10.1002/tea.20237

Chen, X. (2013). *STEM attrition: College students' paths into and out of STEM fields (NCES 2014-001)*. National Center for Education Statistics, Institute of Education Sciences, US Department of Education. https://nces.ed.gov/pubs2014/2014001rev.pdf

Commission on Race and Ethnic Disparities. (2021). *Commission on race and ethnic disparities: The report*. HM Government. https://assets.publishing.service.gov.uk/government/uploads/system/uploads/attachment_data/file/974507/20210331_-_CRED_Report_-_FINAL_-_Web_Accessible.pdf

Connolly, P., & Healy, J. (2004). Symbolic violence, locality and social class: The educational and career aspirations of 10-11-year old boys in Belfast. *Pedagogy, Culture and Society, 12*(1), 15–32. https://doi.org/10.1080/14681360400200187

Department for Education. (2010). *White paper 91: The importance of teaching*. https://assets.publishing.service.gov.uk/government/uploads/system/uploads/attachment_data/file/175429/CM-7980.pdf

DiGiacomo, D. K., & Gutiérrez, K. D. (2016). Relational equity as a design tool within making and tinkering activities. *Mind, Culture, and Activity, 23*(2), 141–153. https://doi.org/10.1080/10749039.2015.1058398

Francis, B., Archer, L., Moote, J., de Witt, J., & Yeomans, L. (2017). Femininity, science, and the denigration of the girly girl. *British Journal of Sociology of Education, 38*(8), 1097–1110. https://doi.org/10.1080/01425692.2016.1253455

Freire, P. (1970). *Pedagogy of the oppressed*. Routledge.

Gillborn, D., & Youdell, D. (1999). *Rationing education: Policy, practice, reform, and equity*. McGraw-Hill Education.

Godec, S., King, H., & Archer, L. (2017). *The science capital teaching approach: Engaging students with science, promoting social justice*. UCL Institute of Education. https://discovery.ucl.ac.uk/id/eprint/10080166/

Gonsalves, A. J. (2014). "Physics and the girly girl – There is a contradiction somewhere": Doctoral students' positioning around discourses of gender and competence in physics. *Cultural Studies of Science Education, 9*(2), 503–521. https://doi.org/10.1007/s11422-012-9447-6

Holmegaard, H. T. (2015). Performing a choice-narrative: A qualitative study of the patterns in STEM students' higher education choices. *International Journal of Science Education, 37*(9), 1454–1477. https://doi.org/10.1080/09500693.2015.1042940

Hooks, B. (1994). *Teaching to transgress: Education as the practice of freedom*. Routledge.

Jammula, D. C., & Mensah, F. M. (2020). Urban college students negotiate their identities to dis/connect with notions of physics. In A. Gonsalves & A. Danielsson (Eds.), *Physics education and gender: Identity as an analytic lens for research* (pp. 81–96). Springer.

Joice, W., & Tetlow, A. (2020). *Baselines for improving STEM participation: Ethnicity STEM data for students and academic staff in higher education 2007/08 to 2018/19*. Royal Society. https://royalsociety.org/-/media/policy/Publications/2021/trends-ethnic-minorities-stem/Ethnicity-STEM-data-for-students-and-academic-staff-in-higher-education.pdf

Ladson-Billings, G. (1995). Toward a theory of culturally relevant pedagogy. *American Educational Research Journal, 32*(3), 465–491. https://doi.org/10.3102/00028312032003465

Ladson-Billings, G. (2006). From the achievement gap to the education debt: Understanding achievement in US schools. *Educational Researcher, 35*(7), 3–13. https://doi.org/10.3102/0013189X035007003

Lareau, A., & Horvat, E. M. (1999). Moments of social inclusion and exclusion race, class, and cultural capital in family-school relationships. *Sociology of Education, 72*(1), 37–53. https://doi.org/10.2307/2673185

Lofton, R., & Earl Davis, J. (2015). Toward a Black habitus: African Americans navigating systemic inequalities within home, school, and community. *Journal of Negro Education, 84*(3), 214–230. https://doi.org/10.7709/jnegroeducation.84.3.0214

Manning, A. (2017) *Urban science teachers: Exploring how their views and experiences can influence decisions to remain in post or not*. Doctoral dissertation, King's College London. https://kclpure.kcl.ac.uk/portal/files/83046301/2017_Manning_Alex_0029050_ethesis.pdf

Maton, K. I., Hrabowski, F. A., III, & Schmitt, C. L. (2000). African American college students excelling in the sciences: College and postcollege outcomes in the Meyerhoff Scholars Program. *Journal of Research in Science Teaching, 37*(7), 629–654. https://doi.org/10.1002/1098-2736(200009)37:7%3C629::AID-TEA2%3E3.0.CO;2-8

McGee, E. O., & Martin, D. B. (2011). "You would not believe what I have to go through to prove my intellectual value!" Stereotype management among academically successful Black mathematics and engineering students. *American Educational Research Journal, 48*(6), 1347–1389. https://doi.org/10.3102/0002831211423972

McGee, E. O., & Stovall, D. (2015). Reimagining critical race theory in education: Mental health, healing, and the pathway to liberatory praxis. *Educational Theory, 65*(5), 491–511. https://doi.org/10.1111/edth.12129

Mcmaster, N. C. (2017). Who studies STEM subjects at A level and degree in England? An investigation into the intersections between students' family background, gender and ethnicity in determining choice. *British Educational Research Journal, 43*(3), 528–553. https://doi.org/10.1002/berj.3270

Meghji, A. (2017). Positionings of the black middle-classes: Understanding identity construction beyond strategic assimilation. *Ethnic and Racial Studies, 40*(6), 1007–1025. https://doi.org/10.1080/01419870.2016.1201585

Mensah, F. M. (2011). A case for culturally relevant teaching in science education and lessons learned for teacher education. *The Journal of Negro Education, 80*(3), 296–309.

Mensah, F. M. (2019). Finding voice and passion: Critical race theory methodology in science teacher education. *American Educational Research Journal, 56*(4), 1412–1456. https://doi.org/10.3102/0002831218818093

Mensah, F. M., & Jackson, I. (2018). Whiteness as property in science teacher education. *Teachers College Record, 120*(1), 1–38. https://doi.org/10.1177/016146811812000108

Moote, J., & Archer, L. (2018). Failing to deliver? Exploring the current status of career education provision in England. *Research Papers in Education, 33*(2), 187–215. https://doi.org/10.1080/02671522.2016.1271005

Moote, J., Archer, L., DeWitt, J., & Mac Leod, E. (2019). Comparing students' engineering and science aspirations from age 10 to 16: Investigating the role of gender, ethnicity, cultural capital, and attitudinal factors. *Journal of Engineering Education, 109*(1), 34–51. https://doi.org/10.1002/jee.20302

Morrell, E. (2015). Toward a critical pedagogy of race: Ethnic studies and literacies of power in high school classrooms. *Race and Social Problems, 7*(1), 84–96. https://doi.org/10.1007/s12552-014-9142-1

Nasir, N. S. (2002). Identity, goals, and learning: Mathematics in cultural practice. *Mathematical Thinking and Learning, 4*(2–3, 213), –248. https://doi.org/10.1207/S15327833MTL04023_6

Nasir, N. S., & Hand, V. (2008). From the court to the classroom: Opportunities for engagement, learning, and identity in basketball and classroom mathematics. *The Journal of the Learning Sciences, 17*(2), 143–179. https://doi.org/10.1080/10508400801986108

Nasir, N. S., Chatmon, C. P., Givens, J. R., & (Eds.). (2019). *"We dare say love": Supporting achievement in the educational life of Black boys*. Teachers College Press.

National Center for Science and Engineering Statistics. (2015). *Women, minorities, and persons with disabilities in science and engineering*. National Science Foundation. https://ncses.nsf.gov/pubs/nsf21321/report

National Science Board. (2015). *Revisiting the STEM workforce: A companion to science and engineering indicators 2014*. National Science Foundation (NSB-2015-10). https://www.nsf.gov/pubs/2015/nsb201510/nsb201510.pdf

Noble, J., & Davies, P. (2009). Cultural capital as an explanation of variation in participation in higher education. *British Journal of Sociology of Education, 30*(5), 591–605. https://doi.org/10.1080/01425690903101098

Ortiz, N. A., Morton, T. R., Miles, M. L., & Roby, R. S. (2019). What about us? Exploring the challenges and sources of support influencing black students' STEM identity development in postsecondary education. *The Journal of Negro Education, 88*(3), 311–326.

Puwar, N. (2009). Sensing a post-colonial Bourdieu: An introduction. *The Sociological Review, 57*(3), 371–384. https://doi.org/10.1111/j.1467-954X.2009.01856.x

Rahm, J. (2010). *Science in the making at the margin: A multisited ethnography of learning and becoming in an afterschool program, a garden, and a math and science upward bound program*. Sense Publishers.

Rahm, J., & Moore, J. C. (2016). A case study of long-term engagement and identity-in-practice: Insights into the STEM pathways of four underrepresented youths. *Journal of Research in Science Teaching, 53*(5), 768–801. https://doi.org/10.1002/tea.21268

Rascoe, B., & Atwater, M. M. (2005). Black males' self-perceptions of academic ability and gifted potential in advanced science classes. *Journal of Research in Science Teaching, 42*(8), 888–911. https://doi.org/10.1002/tea.20077

Reay, D., Davies, J., David, M., & Ball, S. J. (2001). Choices of degree or degrees of choice? Class, 'race' and the higher education choice process. *Sociology, 35*(4), 855–874. https://doi.org/10.1177/0038038501035004004

Reay, D., Crozier, G., & Clayton, J. (2009). 'Strangers in paradise'? Working-class students in elite universities. *Sociology, 43*(6), 1103–1112. https://doi.org/10.1177/0038038509345700

Rollock, N., Gillborn, D., Vincent, C., & Ball, S. (2011). The public identities of the black middle classes: Managing race in public spaces. *Sociology, 45*(6), 1078–1093. https://doi.org/10.1177/0038038511416167

Rosa, K., & Mensah, F. M. (2016). Educational pathways of Black women physicists: Stories of experiencing and overcoming obstacles in life. *Physical Review Physics Education Research, 12*(2). https://doi.org/10.1103/PhysRevPhysEducRes.12.020113

Russell, M. L., & Atwater, M. M. (2005). Traveling the road to success: A discourse on persistence throughout the science pipeline with African American students at a predominantly white institution. *Journal of Research in Science Teaching, 42*(6), 691–715. https://doi.org/10.1002/tea.20068

Shah, B., Dwyer, C., & Modood, T. (2010). Explaining educational achievement and career aspirations among young British Pakistanis: Mobilizing 'ethnic capital'? *Sociology, 44*(6), 1109–1127. https://doi.org/10.1177/0038038510381606

Travers, M. C. (2017). *White working-class boys: Teachers matter*. Trentham Books.

Walker, E. N. (2006). Urban high school students' academic communities and their effects on mathematics success. *American Educational Research Journal, 43*(1), 43–73. https://doi.org/10.3102/00028312043001043

Walker, E. N. (2009). "A border state": A historical exploration of the formative, educational, and professional experiences of Black mathematicians in the United States. *International Journal of History in Mathematics Education, 4*(2), 53–78.

Walker, E. N. (2011). Supporting giftedness: Historical and contemporary contexts for mentoring within Black mathematicians' academic communities. *Canadian Journal for Science, Mathematics, and Technology Education, 11*(1), 19–28. https://doi.org/10.1080/14926156.2011.548899

Wallace, D. (2017). Reading 'race' in Bourdieu? Examining black cultural capital among black Caribbean youth in South London. *Sociology, 51*(5), 907–923. https://doi.org/10.1177/0038038516643478

Yang, Y. (2014). Bourdieu, practice and change: Beyond the criticism of determinism. *Educational Philosophy and Theory, 46*(14), 1522–1540. https://doi.org/10.1080/00131857.2013.839375

Young, J. L., Feille, K., & Young, J. R. (2017). Black girls as learners and doers of science: A single-group summary of elementary science achievement. *Electronic Journal of Science Education, 21*(2), 1–20.

Chapter 3
"It Was Always About Relationships and It Was Awesome": Girls Performing Gender and Identity in an Out-Of-School-Time Science Conversation Club

Allison J. Gonsalves and Jrene Rahm

Girls' engagement with science in structured afterschool programs has received much attention, particularly in relation to opportunities that these programs offer to engage girls in science learning that is meaningful to them and helps them to claim identities in science. (e.g., Calabrese Barton et al., 2013; Carlone et al., 2015; Gonsalves et al., 2013; Thompson, 2014). This chapter documents an attempt to bring science into an afterschool space, and to determine what contributions, if any, it made towards girls' science literacy development for life. Recent research on girls' participation in science suggest that reasons for girls dis-identification with science could be related to gendered representations of scientists that limit opportunities for girls to see themselves as scientists (Archer & DeWitt, 2017; Francis et al., 2017). From this perspective, out-of-school-time (OST) science programs for girls have the potential to disrupt representations of science as disconnected from girls' lives, and can create possibilities for girls to begin to make claims of identities as "insiders to science", at least temporarily (Rahm, 2010). Following these concerns about girls' identification with science this study explored the tensions between girls' performances of femininity and their engagement in conversations about science in a girls group activity, *ConvoClub*, held in a youth community program. In this chapter, we offer a critical analysis of engagement with science in the ConvoClub, paying particular attention to the manner the girls negotiated that space, its activities and science. We document opportunities that the girls had to position themselves as insiders to science in ways that demonstrated how local forms of

---

A. J. Gonsalves (✉)
McGill University, Montreal, Canada
e-mail: allison.gonsalves@mcgill.ca

J. Rahm
University of Montreal, Montreal, Canada
e-mail: jrene.rahm@umontreal.ca

H. T. Holmegaard, L. Archer (eds.), *Science Identities*, Contributions from Science Education Research 12, https://doi.org/10.1007/978-3-031-17642-5_3

capital (Bourdieu, 1986; Yosso, 2005) are valued in the OST space, and the exchange value these may have in other educational spaces.

ConvoClub is a program at the Cartier Community Centre (CCC) embedded in a Montreal community with a long history of poverty. This research site was part of a larger multi-sited ethnography in which we explored science literacy development and identity work among diverse youth as they engaged with science outside of school. The CCC has been in existence since 1956, with deep roots in the Montreal Anglophone community. While the neighbourhood is gentrifying rapidly and the population it services is changing, it is still predominantly an Anglophone centre, in an increasingly Francophone community. The neighbourhood is historically a working-class Irish-Quebec community. As the community changes, it has welcomed a growing immigrant population, due to its (previously) affordable housing. Thus, the demographic of youth attending the community centre has changed, and has become increasingly diverse, both ethnically and linguistically. The CCC offers bilingual (French and English) programing to youth ages 0–17, and invites alumni of their programs to act as volunteers or paid counsellors for camps and youth groups. Often youth enter in the toddler program and stay until they age out of the programming. Darlene, the program coordinator described the program in the following way:

> We see different people come and go, they graduate, and some older people pass and it's very [focused] on the community, and that means family, and a lot of the youth and a lot of kids and some of the older ones, they don't have family and CCC becomes their family. So, it makes our job more difficult, but much more, I guess, rewarding. (INT 1)

One of the goals for the program coordinators at CCC is to create a positive space for youth where they can build self-esteem, positive images of themselves, and develop community or "family". But Darlene suggests that these goals are secondary to ensuring the youths' safety:

> The biggest need is for them] to eat. They're all poor. They- poverty, yeah, they are poor. To eat. To eat and learn. 'Cuz right now it's survival. You know, it's not even about self-esteem, it's survival. I'm talking about having more than one meal. 'Cuz a lot of them are siblings of seven. And the oldest one usually don't eat. So, the biggest need I find is for them to have a place to feel safe, to have somebody to believe in them and to eat. It's a family, that's the biggest need. If they ca- if their own family can't offer that. At CCC, during these four hours, if we have muffins or spaghettis and we cook and that's their meal and then we check their homework to make sure it's done, ask about their day, that's it. It's survival They're struggling. (INT 1)

At the time of the study, CCC was male-dominated. Seeing a need to support girls in the club Darlene proposed the creation of ConvoClub to support girls as they navigate the pressures of adolescence and "sit in a circle and just talk about life and ask questions from a hat, anything like that" (Darlene INT-1). We approached Darlene and the ConvoClub with the intent to offer youth driven science programming and to provide human and material resources to engage in personal and science explorations in the form of digital storytelling and video documentary. The girls were receptive to these opportunities and we spent 12 weeks documenting their engagement with stories about self and self in science (see Gonsalves et al.,

2013 for details of activities). Previously, CCC had offered science activities occasionally, by relying on instructors brought in from local community organizations, or volunteers who assisted youth with science fair projects. Infrequently, science activities were offered during day camps or on weekends. Those science activities, like the ones we engaged the girls in, were often offered in the contexts of other on-going activities and were driven by larger educational goals than science literacy.

The structure of the program sought to flip the traditional view of adolescents as consumers of science, and instead attempted to capture emergent forms of engagement with science that demonstrated instances of youth voice. In this chapter, we explore to what extent the program became a place supportive of "youth-powered science" where science begins from questions youth bring to the program (Seiler & Gonsalves, 2010). Youth-powered forms of science engagement can provide critical opportunities for exercising "youth voice" by allowing youth choice and ownership of science (Basu, 2008) which can broaden youths' concepts of what counts as science and what identities in science are available to them. These opportunities can make insider identity work possible for youth, often for the first time (Gonsalves et al., 2013; Rahm, 2010).

3.1 Theoretical Framework

This study draws on three related theoretical framings. First, to understand how girls *do* identity work and position themselves in relation to science, we draw on Holland et al.'s (1998) theory of identity-in-practice. This framing views learning and identity work in STEM as an ongoing and dynamic process and thus we are interested in documenting "persons taking form in the flow of historically, socially, culturally, and materially shaped lives" (Holland et al., 1998, p. 5). Thinking of girls as constantly doing "identities-in-practice" (Holland & Lave, 2001) gives us the possibility of exploring the multitude of ways girls are positioned but also position themselves in relation to science. This theoretical focus to our analysis gave us the possibility to identify moments when certain enactments (e.g., performances of femininities) permitted the girls to position themselves as insiders to the ConvoClub, but also as "insiders to science" (Rahm, 2010). Considering simultaneously how girls navigate discourses of gender and its intersections with ethnicity and social class can offer still further insights into who can be in or outside of science (Carlone et al., 2015). For example, Godec (2018) argues that science identities are considered appropriate for girls in some contexts, but inappropriate in others. In her study exploring working-class girls' identity work in secondary science, "scientist" is configured as an appropriate subject position for South Asian girls in certain cultural and social contexts (e.g., Baker et al., 2008), but is regarded as culturally inappropriate for White working-class girls. Thus, to understand the identity work of girls in the ConvoClub we need to account for the ways that identity performances as insiders-to-science were recognized both in the context of the club, and in light of their ethnicity and social class. To understand performances of femininity, we draw on post-structural feminism and the work of Judith Butler (1999).

Butler (1999) envisions gender as something that is *done* through behaviour, dress, forms of speech, their preferences, interests and other enactments. Following Butler (1999), we see gender as constructed through a "stylized repetition of acts" like behaviours, dress, speech, interests and other enactments (p. 140). Thus, in this orientation towards gender, we understand performances as situated in local contexts (Paechter, 2007). As such, masculinities and femininities are often understood as ideal, typical forms that are associated with how to 'do' boy/man or girl/woman (Paechter, 2007). We caution that this orientation to gender does not suggest that performances of masculinity or femininity are signifiers of male-ness or female-ness. Furthermore, enactments of masculinities and femininities might be acceptable or intelligible in some contexts, but not in others (Butler, 1990). For instance, femininity has been constructed as incompatible with science in its female embodied form (e.g., Francis et al., 2017; Gonsalves, 2014). Francis et al. (2017) have demonstrated that youth and parents alike identify constructions of femininity as "superficial" and associated with an overall denigration of girly/super-feminine girls. In a recent study, Godec (2020) describes forms of hyper-femininity that involve an investment in personal appearance, flirtatiousness and popularity. Godec (2020) argues that this form of femininity is often reprimanded and at odds with science, and therefore in contexts where performances of hyper-femininity are rewarded, girls will reject science. Dawson et al. (2019) have also described a form of hetero-femininity which interacts with science identity performances. Hetero-femininity in this case is a form of popular femininity invested in being heterosexually attractive through appeals to fashion, beauty and sociability (e.g., Dawson et al., 2019). Taking this into account, we are concerned with understanding how science identity performances are intelligible in an OST science learning context like the ConvoClub where performances of hetero-femininity are valued (Read et al., 2011). We agree with Archer et al. (2019) that intelligibility is a helpful tool for understanding femininity in relation to youths' science identity performances, as it may reveal the normative or dominant values that are recognized or valued in a given context. Applied to this study, this led us explore how the young women *do* gender in relation to local, valued forms of popular femininity (Read et al., 2011), and how the intersections of these with science produces locally intelligible performances in the context of ConvoClub (e.g., Godec, 2018). Performances are considered to be 'intelligible' (Butler, 1990) when they align with the expected appropriate behaviours for girls or boys in various contexts.

Third, we are interested in exploring what forms of capital (Bourdieu, 1986) are valued, produced, and exchanged in the ConvoClub and within the context of youth programming at CCC generally. We sought to parse out the various ways that girls locally accrue various forms of capital *in the context of the* ConvoClub, and what exchange value these have for recognition outside of the club (Gonsalves, 2014; Gonsalves et al., 2013). We are curious whether dimensions of *science capital* (Archer et al., 2015) are produced and valued in the ConvoClub, and whether they are leveraged in and beyond the club. We are also interested in what forms of capital render youths' identity performances intelligible (e.g., Butler, 1990) in this context. Science capital is understood as science-related forms of social and cultural capital which students may accrue both in and out of school. To examine the forms of science capital valued in the ConvoClub, how they intersect with the girls' performances of femininity as intelligible identity positions, we consider the possibility that these would have currency in mainstream science education spaces. Thinking more broadly about forms of capital in the club, we also draw on the work of Nasir and de Royston (2013) who discuss the value of *bonding capital* among groups of African American students, and how it served as a source of status and facilitated identity work in relation to mathematics in the program they studied. Bonding refers to the value assigned to the development of social networks among groups. Nasir and de Royston (2013) have suggested that this happens within groups that have some homogeneity in social identity, but we also suggest that bonding may happen over common interests held by group members. Bonding capital emerged as salient in this study as we considered how identity performances (involving performances of femininity) were rendered intelligible within the group in ways that were valued among group members. Thus, we mobilize the concept of capital in the study to explore instances when the girls engage in identity performances that accrue locally valued forms of capital among club members. We then consider the possibility that the forms of capital valued among club members can be exchanged for intelligibility and recognition as insiders within the club, and the possibilities for exchange-value outside of the club.

Taken together, these three dimensions of our theoretical framework enable us to explore girls' engagement in identity work through science conversations in the ConvoClub. We use the term identity work to signal that we understand identities as produced in practice, by being positioned and by positioning themselves relative to science and to others in the club or the world (e.g., Holland et al., 1998). In the context of ConvoClub, science conversations may have enabled girls to position themselves as certain kinds of people (Gee, 2011; e.g., science people, as girls, as friends, as insiders to the club, and a variety of other subject positions that are valued within the ConvoClub context). Identity performances in this context are constituted in relation to locally recognizable and valued forms of femininity (Paechter, 2007), which also work to signal *inside-ness* to the club. Inside-ness in the club may also be constructed through the accumulation of locally-valuable forms of capital – taking the form of bonding capital or science capital, with both having exchange value for recognition within the bounded context of the ConvoClub. This framework led to two research questions which then guided our analysis:

1. What identity performances are valued among girls in an OST science conversation club?
2. What do these performances reveal about the potential for the development of durable science identities in these contexts?

3.2 Research Context and Methods

The study took place over a semester, and entailed co-planning events with the ConvoClub coordinator, and at times with the girls. As we were interested in doing more than just describing and interpreting the events of the club, we endeavoured to work with participants to arrive at science-related experiences that might be transformative for the youth involved. Thus, we sought to identify aspects of science learning that were alienating to the girls, and aimed to reframe those aspects through methods that would encourage the expression of the girls' voices (e.g., mini-documentary making; Furman & Calabrese Barton, 2006), and thereby make the science relevant their lives (e.g., Seiler, 2001). This chapter draws on ethnographic data (video data of meeting sessions, fieldnotes, group discussions, individual interviews) collected from all the activities we engaged the girls in throughout the semester-long program (weekly meetings of 2 h over 16 weeks). Seven youth participated in this study, with six youth completing all the activities over time. Three of the participating youth were attending ConvoClub for the first time, and four had participated in the group previously. Participating girls all came from working class backgrounds, one had just moved out from the foster care system. The oldest of the group, Shanice, was of Black Caribbean descent, and worked as a volunteer in the youth program. She had recently moved to a different, slightly more affluent neighborhood, but she continued to travel to the youth program at CCC because she had become interested in a career as a social worker and wished to gain experience working with youth. Sharon, Kelly, and Caileigh are all between the ages of 17- and 18-years old, from Irish-Canadian backgrounds. They each participate in the teen program, but also volunteer their time in the 5–12 program. Sarah is the youngest of the group at 13, and is also of Irish Canadian heritage. Karen is biracial (Irish-Canadian and Jamaican-Canadian background), 14-years old at the time of the study, and like Sarah, participates in the Teen program only. Sarah and Karen had been involved in youth programming at CCC since they were 2-years old, and were best friends. Both of their mothers were also graduates of the youth program at CCC. Pseudonyms are used for the youth, the program facilitator and volunteer, the program and the centre to ensure confidentiality.

The activities at ConvoClub took many different forms, and ranged from sharing personal narratives through digital storytelling (Robin, 2006), to mini-documentary making (e.g., O'Neill, 2005). The following table (Table 3.1) details the activities offered on a weekly basis over the course of a 12 week semester. In the table, we describe the activities that took place each week that we visited the ConvoClub, and the manner the data source, once analysed, touched upon one of the three themes summarized in results in this chapter. Data collected in weeks 1–12 include video data and artifact data, whereas in weeks 16–18 we collected interview data. The final column indicates the themes (1, 2, or 3) in which the data figured most prominently.

Table 3.1 Timeline of activities associated themes

Week	Activity	Description	Data source
1	Postcard storytelling	Youth drafted a story about their hair to fit on a postcard. We chose to write stories about hair because they were stories of concern to the group members, and can touch on issues of culture/ethnicity, sexuality, gender identity and self-esteem. This was a 'practice-run' for drafting a personal story that was short and succinct	
2	Storytelling circle	Youth shared personal stories with others in the group, the group spent time asking questions and giving suggestions to refine and develop the stories for the digital storytelling activity	Themes 1, 2
3	Polishing stories	Youth worked with each other and the research team to polish stories and collect images/music for digital story	Themes 1, 2
4	Story recording	Stories were audio-recorded / finalized image selection/ populated Windows Movie Maker with images	Themes 1, 2
5	Workshop on video editing	Lesson on cross fades, volume control, zoom and pan functions, title overlay, credits. Hands-on time to edit sound files, and images	
6	Final cut and film screening	Last minute changes to stories, then we ordered pizza and screened the stories in the computer room at the centre	Theme 1, 2
	Break	Research team arrived at CCC[a] but youth were overwhelmed with school work and needed a break	
7	Collage activity	Where do we see science in our everyday lives? Research team brought in magazines, and youth created collages to discuss the everyday ways that science is important in their lives	Themes 1, 2
8	Presentation and concept mapping	Mapped out the ways that science is present in our everyday lives	Themes 1, 2
9	Brainstorm for mini-documentary	Youth decided they wanted to continue working with digital technology, and make a mini-documentary. We discussed the goal of the project and its form. Decision was taken to interview other youth at the centre, development of interview questions	Themes 1, 2
10	Streeters	Finalized interview questions for the video project. Quick lesson on filming and then youth conducted interviews with volunteers and other youth at CCC. As a group, we viewed and discussed interview footage	Themes 2, 3
11	Story boarding and editing	Youth created the storyboard for the film, and then worked in groups to edit footage for three different 'scenes' of the film	Themes 2, 3
12	Final cut and screening	Merged the three scenes and screened the final production!	Themes 2, 3
16–18	Interviews	Individual interviews were conducted with each of the participating youth, and the group coordinator	Themes 1, 3

[a] Cartier Community Centre (CCC) = XX

3.3 Data Sources

As we were concerned with the identity work the girls engaged in, over the course of the ConvoClub, we designed our data collection to capture various aspects of identities-in-practice. For example, video data was collected in the manner of Baker et al. (2008) describe, as we intended to explore the participants' "discourse-in-use" to then infer the girls' identity work by exploring how they perform gender and identity in science, yet also how they are positioned and in return position themselves in light of discourses around gender and science. Each session of ConvoClub (12 weeks) was video-recorded, leaving us with approximately 23 hours of video data. Written fieldnotes accompanied the video data (Hammersley & Atkinson, 2007), which helped us to plan events from week to week. Field notes were also mobilized analytically, to identify episodes that we interpreted as meaningful to the girls' identity work (e.g., moments that illustrated identity work around science; positioning and being positioned around locally-valued forms of femininity). Episodes of interest were then transcribed verbatim. Semi-structured videotaped interviews (Kvale, 1996) were conducted with each of the girls and the program coordinator at the end of club year. They were transcribed verbatim. Artifacts were also collected such as collages, storyboards, informal notes and the video documentary. These artifacts became important sources for our analysis of identity work as they allowed us to investigate identity expressions that went beyond the discursive. For example, the girls' collage activity yielded science-related themes that were also connected to their persistent interests as adolescent girls. Analysis of the collage activity data entailed extracting themes from the collages (often related to forms of 'hyper-femininity' and 'hetero-femininity', e.g., Dawson et al., 2019; Francis et al., 2017), and integrating these with the analysis of the conversation around the collages.

Analysis entailed a bricolage of the multiple data sources (Kincheloe & Berry, 2004). First, data was coded in NVivo9 and an initial content analysis of fieldnotes and video data was conducted to identify performances across the various activities. We began with open-coding (Saldaña, 2015) using primarily in vivo codes that we organized into three broad categories: (1) goals of ConvoClub; (2) descriptions of "doing science", and; (3) positioning hetero-femininity in opposition to science. Within each of these three broad categories, we then drew on our theoretical framework to examine which of these emphasized relationships (Nasir & de Royston, 2013), revealed aspects of science capital (i.e., experiences with science and opportunities to do science out of school; Archer et al., 2015), and episodes when the youth used science conversations to perform the forms of femininity they were invested in (Butler, 1990). This yielded the three themes, situated around various activities the girls engaged in during our time together. We then performed a more theoretically focused round of analysis and paid attention to broader meanings of gender, identity and science as constructed through talk and performed in action (Gee, 2011). We identified performances of femininity that seemed to be valued in the group, and what affordances these performances gave for positioning oneself as an "insider to science" (Rahm, 2010) and gaining recognition as a science kind of person (Carlone & Johnson, 2007). Finally, we queried whether these performances yielded affordances tied to locally-valued forms of capital.

3.4 Results

Our analysis revealed three different themes tied to the girls' identity performances in the group and their orientations towards science:

1. Building solidarity around 'relationships'
2. Co-opting science to advance the goals of ConvoClub
3. Positioning selves as science experts in the club

Across each of these themes, we saw the girls performing identities that seemed to have significant value within the context of the ConvoClub. Data analysis demonstrates that performances of 'pretty popular femininity' (Read et al., 2011) influence how the girls position themselves in relation to science, and how they author themselves as members of the ConvoClub group.

3.4.1 Building Solidarity and Bonding Capital Around Relationships

We saw the girls' investment in relationships as driven by their performances of 'popular femininity' (Read et al., 2011) which emphasized heterosexual relations, and a strong focus on appearance. In an interview with Caleigh, we asked her what she enjoys most about being in ConvoClub. She responded:

> Probably... Probably relationships. Like everyone was always talking about how "oh, I'm (now into) a relat- relationship with this guy, and they changed my life". It was always about relationships and it was just like- it was awesome. It was the most fun to hear everybody talk about something that bothered them, or something that pissed them off, made them upset, 'cuz- like I have my own problems and to hear everyone else's problems in my mind was just- I just get really upset, but then it was like- It was also funny to hear them, 'cuz they were funny (INT- C1).

This theme was most apparent in the digital storytelling activity, an introductory project which was meant to provide a way for the girls to connect to each other personally, and to "dig deep" as the teen group coordinator put it. Table 3.2 offers a list of the titles and subjects the digital stories addressed. At the same time, as researchers we introduced the activity to familiarize youth with digital media production and story-writing. In the end, the emphasis placed on relationships in the digital stories seemed to set the tone for the rest of the activities conducted in ConvoClub, as the girls kept coming back to this topic, wishing to explore relationship-themed activities further. We suggest that this focus on relationships

Table 3.2 Digital storytelling title and topic

Name	Title	Subject
Shanice	When I turned 16…	Shanice's experience with an abusive ex-boyfriend
Karen	"Siblings share childhood memories and grown-up dreams"	Karen's relationship with her younger brother who was born prematurely
Caleigh	[Brother's name]	Caleigh's brother's involvement with drugs and subsequent incarceration, and her evolving relationship with him
Sarah	The Beatles is our connection	Sarah's relationship with her mother and step-family
Sharon	My story	Sharon's relationship to her own body, self-esteem and a relationship with a boy from summer camp
Kelly	It all started like this	Childhood medical trauma and its impact on Kelly's life

provided the girls with opportunities to co-construct "bonding capital" (Nasir & de Royston, 2013), which created solidarity among the group members. Bonding capital in this case emerged as a desire and willingness to "dig deep" and engage in storytelling about relationships given that kind of talk had currency within this group. We found many instances of bonding occurring alongside science conversations during the ConvoClub activities.

Video data from the storytelling circle (during Week 2) contains several examples of instances where group members cried in solidarity with the storyteller, got up to hug them, or whisper "We love you" in supportive ways (FN-022411).

Darlene, the group leader, described the digital storytelling activity in the following way:

> To have them open up and they finally did, it was a beautiful thing…it took a lot of strength from a lot of them, a lot of courage to speak on certain things and to open themselves up. You know, so was- 'cuz to talk about themselves is very, very hard, and they went through it, they did it, and finally, and it was emotional, and I think it brought the group together. (INT 1)

At no time did the youth discuss science during these activities, nor did the researchers raise science topics. However, the bonding capital that developed through digital storytelling eventually shaped the science conversations that emerged in the subsequent meetings. The theme of "relationships" was so salient to the group identity that it framed the "science in my life" collage activity, and subsequently influenced the themes the youth wished to explore when planning the documentary. We see the bonding capital emerging in these activities as forming the foundations for a "thick place" (Duff, 2010) that facilitated place-making work. Duff (2010) argues that thick places are made through the *affective force* of practices that are given meaning while presenting opportunities for personal enrichment. Thick places enhance one's sense of belonging thereby "forging a series of affective and experiential connections in place" (Duff, 2010, p. 882). In this sense, place-making happens through affect and practice. The bonding practices we observed in ConvoClub generated a sense of meaning and belonging, by providing opportunities for the girls to share affective experiences that drive their connections within and to the space of the ConvoClub. Duff (2010) argues that thick places can structure young people's experiences of self and belonging through an "intensification of the affective pull of place" (p. 882), and thick places can provide resources for young people to facilitate personal enrichment, nurture ambitions and help to realize their own capacities. We argue that the bonding capital emerging from the digital storytelling activities facilitated the formation of a thick place in which the girls could push and pull on each other to "go deep" and share aspects of their lives that they struggled deeply with. Membership in this group required the girls to be emotionally vulnerable with each other, but also supportive. Many of these girls expressed feelings that they were "not good" in science, and strong feelings of alienation from the discursive and material practices of classroom science (e.g., Gonsalves et al., 2013). When we approached them initially with the possibility to engage in a science club, the girls showed very little interest in participating. However, as the group evolved into a place with affective pull, they showed increasing interest in having conversations about science, if they could relate those science conversations back to meaningful topics in their lives. In this way, they established a temporally and spatially located membership in a science group.

3.4.2 Co-Opting Science to Advance the Goals of ConvoClub

The foundations for co-opting science to advance the goals of ConvoClub were formed in the digital storytelling activity and persisted through the subsequent activities. The affective pull of the private space of ConvoClub was established through "going deep" in the digital storytelling activity, which seemed to influence the goals the youth had for the science activities. We discuss the following activities as co-opting science to advance these goals, but we do not mean to frame co-opting in a negative sense. Rather, we suggest that talking about science in a thick place provided possibilities for the girls to momentarily construct inside-ness in relation to STEM. Talking science into their lives (e.g., via "relationships") became a pivotal point around which the girls explored the affective dimensions of this thick place and could signal their belonging to ConvoClub. Thus, rather than a science learning community forming the basis of girls' engagement with each other, the girls' place-making and bonding were in focus, and these touched on science in meaningful ways, supporting the girls' identity work.

This orientation to ConvoClub was galvanized in the collaging activity and the mini-documentary planning meetings. These activities provided many opportunities for science conversations that were usually co-opted by the girls' preferences for talking about issues deeply connected with performances of 'popular' or 'girly-girl' forms of femininity, with an emphasis on boys, sex, beauty, and relationships (Read et al., 2011). Femininity in this group was closely connected to compulsory heterosexuality (Rich, 1980), although conversations about same-sex relationships did occasionally occur, with assurances from the girls that "you guys aren't homophobes, right?" (FN 021011). Since relationships were generally assumed to be heterosexual, much of the talk in the group focussed on boys and sex. Data from this activity demonstrates instances when girls attempted to connect with science through performances of hetero-femininity. The following conversation, recorded during one of the planning meetings for the mini-documentary, details conversational evidence of youths' attempts in reconciling their investment in hetero-femininity with science.

Karen: Is love science?
Sharon: Yes.
[Other girls say on the back: "Yes!" "Yes!"]
Darlene: Lord have mercy. That's ok, Karen. That's ok, sweetie.
Karen: That could be science.
Darlene: That's ok. That's the male body.
Volunteer: How do you… How do you know you love someone?
Darlene: How you feel inside.
Kelly: Yeah.
Volunteer: Do you feel different physically at all?
Darlene: The chemicals in your body.
Kelly: Yes.
Darlene: The hormones. (TRANS4040711)

To connect the "male body" with love and feelings, caused by "chemicals in your body" and "hormones" is an example of the form co-opting of science took in the club. Evidence of such work can also be found in the collage activity. The girls created poster boards themed around "What science means to me". Each of the girls' collages was unique in that it was also themed around issues that were of interest to them, as shown in Table 3.3.

Table 3.3 Collage topics and descriptions

Name	Theme	Content
Shanice	Science of beauty and relationships	Numerous images of bodies, cosmetics, and various images of heterosexual "love" inspired by a course she took on sexuality and relationships
Karen	Justin Bieber and science (music)	Numerous images of Justin Bieber and references to music, along with "scientific placeholders" (Gonsalves et al., 2013) like the formula for Boyle's law
Caleigh	Science of relationships	Images that represent friendships because you have to "click" with someone so you can become friends. The click requires chemistry, which explained images depicting chemistry
Sarah	Glee and communication technology	Images of the TV show Glee because "music is science". Also included images of cellphones because "cellphones are also science"
Sharon	Science is making mistakes	People make mistakes; used images of words themed around love and change
Kelly	Toxic relationship	Poem about a relationship that ended; uses words like "chemicals", "exploded" and "poison" to connect with science

The collage activity was intended to explore how science is situated in the girls' everyday lives, but the girls co-opted this activity to instead collage themes related to popular culture, beauty, and relationships. These performances on paper (the collages) can be read as the kind of 'cool girl' performances described by Dawson et al. (2019): they describe the manner the working-class girls' performances of hetero-femininity in museums drew on resources from youth culture rather than the resources available in the science museum. Like Dawson et al. (2019) we found that the girls in this setting also combined these performances of hetero-femininity with classed performances of being loud, funny and dismissing school and especially school-based science. The collaging activity made evident the girls' desires to draw on their own resources (e.g., cellphones, friends, music, cosmetics) to build bonding capital in the group, rather than drawing on the material and human resources available in ConvoClub (e.g., science-related magazines and texts, activities and conversations). As described previously (Gonsalves et al., 2013), these girls had all previously indicated disengagement from school science, and little interest in science beyond the occasional fun science activity at summer camps or other occasional programs at the Centre. Many of the girls expressed negative experiences with science in school, and feelings that science was not relevant to their lives or topics that they were invested in. However, talk in the ConvoClub that repositioned some of their interests as science-related resulted in the refiguring of some of their own ambitions for participating in the group. Framing their expressions of femininity as related to science, caused them to demonstrate an interest in pursuing these conversations.

Following Dawson et al. (2019) we argue that these performances of "cool girl" hetero-femininity provide opportunities for the girls to perform insider identities in the ConvoClub, and suggest that the collage activity in particular illustrates the challenges the girls face in assuming an identity as insiders to science. It can be read as an agentic act of pushing back on canonical science and in that sense, Convoclub offered them an opportunity to play with a form of science that lend itself to be co-opted with their lives and current challenges tied to relationships and love. The co-opting looked different across the girls in that Sarah and Karen for instance, make attempts to illustrate scientific interests or concepts in their collages (e.g., direct references to science-related content), while others like Shanice, Caleigh and especially Kelly make only cursory references to scientific placeholders (a molecule). Instead, they discuss how science words emerge in their persistent interests. All references to 'science' are couched in language that the girls encounter in school like "wires and resistors" (for a discussion, see Gonsalves et al., 2013), which they consider as the kind of science that has capital, yet not in their own lives as girls on the margins of Western science, and specifically, school science.

3.4.3 Positioning Selves as Science Experts in the Club

The mini-documentary making was an activity designed by the girls in ConvoClub to expose the other youth in the community centre to science topics that had relevance to their everyday lives. Through this activity, the girls in ConvoClub took on the roles of science experts in the CC Centre. They conducted "streeters" (video-recorded on-the-spot interviews) with the other youth, volunteers and coordinators at the centre, and chose the clips to include in the final product. To choose which clips to include in the documentary, we thematized the interviews. As the girls embarked on their documentary project, particularly their engagement in streeters (a journalistic practice wherein they stopped other youth in the centre and quizzed them about their knowledge of science and whether they knew they "did science every day?"), we noticed that they began to position themselves as insiders to science in relation to the boys in the club. Our observations of the girls interviewing the boys in the community centre raised questions for us about whether they were accumulating a locally valuable form of science capital (e.g., Archer et al., 2015). Their focus on science as tied to their lives suggests the girls noticed that it had local power. The girls seemed to invest significant emotional energy in this activity (e.g., Gonsalves, 2014) as it provided opportunities to engage in scientific knowledge in new and meaningful ways, but also to position themselves in relation to boys that were powerful for them in the context of the club. Girls were enabled to enact identities as insiders-to-science through this documentary activity in ways that afforded them capital in the space of the club. In this way, the club provided them opportunities to test out identities as science-savvy people (in relation to the boys), which is something they seemed to play with and enjoy. For example, the following discussion happened during the editing process when we tried to identify a common theme among the streeters':

Allison: Most people [interviewed by the girls] think science is boring, most people… don't know that they do science in their everyday lives. So, did you wanna keep all those little clips, where people say "I don't know", "I really have no idea"…
Sharon: I think so.
Shanice: I think it's a bit true.
Sharon: Yeah.
Sarah: The truth.
Shanice: I don't think people actually think about using science every day. Like you go to school 'cuz they're supposed to teach you that there's science all the time, and blah blah blah.
Shanice: But nobody remembers science at school.
Sharon: Unless you have to sit and think about it. 'Cuz even when you asked us, we were like "Oh…".
Allison: Yeah. I know. It is kind of a difficult question to just bring on people.
Sharon: That's why [Bear] said he felt stupid. (TRANS042811).

Despite the outputs from the collaging activity that were largely disconnected from science, the mini-documentary making afforded the girls opportunities to position themselves as insiders to science, particularly as knowledgeable about how we engage in science in our everyday lives. We argue that this facilitated a chance for girls to accrue a locally valuable form of science capital. An example of how the girls positioned themselves in relation to others in the club can be seen in video data collected at the time of the documentary editing:

Allison: What do all of these interviews tell you about what people think about science?
Shanice: They think that science is boring, and they don’t know very much about science.
Allison: OK.
Shanice: That’s what I got, anyway.
Caleigh: I know, eh?
Sharon: There’s not one definition of science.
Kelly: People don’t know very much.
Sharon: What science means to one person is not necessarily what it means for the other. (TRANS042811).

We found that within the space of ConvoClub youth valued not just knowledge about science, attitudes or exposure to science—all key dimensions of *science capital*—but specifically the ability to talk about science with others in the club, and the ability to connect everyday engagement with science and the persistent interests of the youth in the club. Thus, we suggest that some dimensions of science capital (related to talking about science and connecting science to their lives) can be valued highly and exchanged locally for recognition as insiders-to-science (Carlone & Johnson, 2007; Gee, 2011) in the context of the club. However, we caution that the local accrual of science capital in this form, does not seem to have exchange value outside of the club, or at least, the youth do not see its exchange value. Interviews with the girls in interviews after the activities were completed revealed their persistent positioning as outsiders-to-science. Many of the girls articulated that the kinds of science conversations we had in ConvoClub did not correspond with aspects of school science which they interpreted as “real science” (Kelly, INT 1). Additionally, despite positioning themselves as insiders-to-science in the context of the ConvoClub, follow-up interview data saw the girls persistently positioning femininity in opposition to school science. To illustrate, the following exchange provides an answer frequently given by the girls in interviews in response to questions about who does science.

Allison: Do you think that most girls are interested in science?
Caleigh: Probably not.
Allison: Why not, do you think?
Caleigh: ‘Cuz they’re more interested in like their hair, and like their makeup, and like being popular, ‘cuz like science doesn’t necessarily go with being popular, being like in the cool clique, so girls like being in the cool clique.

Allison: Do you think that's more normal for boys to do science than girls?
Caleigh: Ah. Probably. I think maybe yeah. 'Cuz girls are more like I think we're more like: we don't wanna like get our nails ruined, or we don't wanna like get something on our clothes, or like we don't want this chemical doing something with my hair, or like if it's humid and you don't iron them, they'll be puffy, like guys are just "I don't care", do whatever you do. (XX)

In this excerpt, Caleigh reifies the notion that femininity is incompatible with science (Francis et al., 2017; Gonsalves, 2014). Her statement that girls are interested in hair and makeup leads directly to the assumption that they would thus not be interested in science. Caleigh herself later on suggests that she is still not interested in school science, despite her interest and engagement in the science conversations at ConvoClub. This suggests that although the performances of gender in ConvoClub aligned with forms of science capital valued in the context of the OST space, the girls themselves did not see the exchange value for these outside the club. We suggest that the forms of capital generated in the club, were only locally valued and did not have any significant exchange value outside of the club. They could be exchanged for insider status among the girls, and generated insider-ness at the club, but they did not seem to create any notable shifts in the way the girls talked about school science or science outside of the club context, or their future aspirations in science. We did not spend a significant amount of time explicitly discussing the gendering of science with the girls in ConvoClub. In retrospect, more explicit conversations about the under-representation of women in science (e.g., Hazari et al., 2013), and the gendering of scientific knowledge may have contributed greatly to the girls' reflections on their own positionality in relation to science.

3.5 Discussion

In this study, we saw the girls' engagement in the ConvoClub activities evolving over time, with increasing identification towards science, as evident in attempts to reconcile their engagement in science conversations with their performances of popular femininity (Read et al., 2011). Rather than seeing this as a constraint, we regard this an innovative way to accrue a local form of science capital (Archer et al., 2015) in the space of the ConvoClub, where an investment in performances of popular femininity have currency. This suggests that the concept of science capital can have local value, as a form of currency that girls exchange for momentary recognition as insiders-to-science (Carlone & Johnson, 2007). Hence, activities and conversations about science may contribute to youths' "holdall" of science capital (DeWitt et al., 2016) in ways that are more congruent with their investment in popular forms of femininity, and as such may offer a 'way in' to engage girls over time and possibly across practices, in science learning. In this sense, we do not see girls' engagement in science talk in the ConvoClub as contributing to their accumulation of science capital that will have significant exchange value outside of the club; nor do we imagine that girls gain science capital in the forms it has been previously conceptualized (e.g., Archer et al., 2015; Archer & DeWitt, 2017). The forms of science participation that emerge in ConvoClub may be better understood as examples of non-dominant science capital (e.g., Carter, 2003). Others have argued that cultural capital has been conflated with dominant cultural practices, and biases towards White, middle-class cultural productions that ignores the cultural capital of non-dominant groups (Carter, 2003; Ehret & Hollett, 2016; Yosso, 2005). Carter (2003) suggests that we may be served by understanding various forms of dominant and non-dominant cultural capital, where non-dominant forms of cultural capital are typically undervalued in educational spaces.

We find it important to stress that the girls in the ConvoClub demonstrated aspects of science capital that were specific to their community and thus valued locally (Yosso, 2005). However, the girls themselves were well aware that their locally valued productions of science capital would not yield any exchange value outside of the CCC space. We suggest that this was not the goal of ConvoClub, and that the girls' investments in their own bonding and place making via the ConvoClub was of greater value to them than the possibilities to yield science capital with exchange-value. As we reflect back on the girls' agency in the group and their actions of co-opting science to advance the goals of ConvoClub, we recognize that bonding and support was the primary activity of the club, and that this led to brief encounters with science, rather than the other way around. As researchers, we were first concerned to provide high quality science experiences in the context of the ConvoClub, but instead what emerged was the possibility for the emergence of a "thick space" (Duff, 2010) for youth engagement. The thick space that comprised the ConvoClub was grounded in the girls' performances of popular femininity (e.g., Read et al., 2011), which in some ways they still regarded as incompatible with science. However, these performances were meaningful to them, and the local productions of science capital that emerged in tandem with these gender performances created fleeting opportunities for the girls to see themselves as insiders to science. Perhaps more importantly in this context, though, was the affective placemaking (Ehret and Hollett 2016) the girls engaged in. Placemaking has been associated with young people's positive development, sense of agency and position and purpose in community life (Duff, 2010). Placemaking has also been described as an affective practice that creates a sense of belonging, but also an "action-potential" wherein agency can be enacted. Reflecting back on ConvoClub, we realize that while we thought the girls' investments in "relationships" was initially a distraction from science, it actually created the possibility for a "thick place" and set the groundwork for the possibility of learning. While we did not observe any long-lasting identity shifts for the participating girls in relation to science, we argue that further opportunities to engage with the girls in ways that emphasized bonding and group membership might allow for these. The study was limited by the time in which we could engage with the girls. Had we more time to develop science conversations about femininity and the under-representation of women and girls in STEM disciplines, we may have further developed possibilities to contribute to their identity-work in relation to science. We argue here that the placemaking work we engaged in was not insignificant for the girls' experiences with science. By approaching our project with a clear goal to prioritize relationship-building over science learning, we created the possibilities to develop an "affective pull" even when engaging in conversations about science.

These findings have consequences for how we think about girls' identity performances in relation to science. We found that through the affective work of bonding and through their multi-modal expressions of self in relation to science, the girls created a thick space for themselves to engage in identity work around science on their own terms. While they managed to position themselves as insiders to science in this context, they may encounter struggles to be recognized as science insiders outside of the ConvoClub. In schools and in their everyday lives, youth, especially girls, are likely to encounter meaningful others (teachers, instructors, family members, friends) who don't recognize girls' resources as valid contributions to science (e.g., Wade-Jaimes & Schwartz, 2019), or their identity performances as congruent with science (e.g., Francis et al., 2017). Therefore, broadening what we consider to be identity work in relation to science, and attending to the affective dimensions of this might also help to support not only youth's science learning, but also their continued identification with science.

References

Archer, L., & DeWitt, J. (2017). *Understanding young people's science aspirations*. Routledge.

Archer, L., Dawson, E., DeWitt, J., Seakins, A., & Wong, B. (2015). "Science capital": A conceptual, methodological, and empirical argument for extending Bourdieusian notions of capital beyond the arts. *Journal of Research in Science Teaching, 52*(7), 922–948. https://doi.org/10.1002/tea.21227

Archer, L., Nomikou, E., Mau, A., King, H., Godec, S., DeWitt, J., & Dawson, E. (2019). Can the subaltern 'speak' science? An intersectional analysis of performances of 'talking science through muscular intellect' by "subaltern' students in UK urban secondary science classrooms. *Cultural Studies of Science Education, 14*(3), 723–751.

Baker, W. D., Green, J. L., & Skukauskaite, A. (2008). Video-enabled ethnographic research: A microethnographic perspective. *How to Do Educational Ethnography*, 76–114.

Basu, S. J. (2008). How students design and enact physics lessons: Five immigrant Caribbean youth and the cultivation of student voice. *Journal of Research in Science Teaching: The Official Journal of the National Association for Research in Science Teaching, 45*(8), 881–899. https://doi.org/10.1002/tea.20257

Bourdieu, P. (1986). The forms of capital. In J. Richardson (Ed.), *Handbook of theory and research for the sociology of education* (pp. 241–258). Greenwood.

Butler, J. (1990). *Gender Trouble*. Routledge.

Butler, J. (1999). *Gender trouble: Feminism and the subversion of identity*. Routledge.

Calabrese Barton, A., Kang, H., Tan, E., O'Neill, T. B., Bautista-Guerra, J., & Brecklin, C. (2013). Crafting a future in science: Tracing middle school girls' identity work over time and space. *American Educational Research Journal, 50*(1), 37–75. https://doi.org/10.3102/0002831212458142

Carlone, H. B., & Johnson, A. (2007). Understanding the science experiences of successful women of color: Science identity as an analytic lens. *Journal of Research in Science Teaching, 44*(8), 1187–1218. https://doi.org/10.1002/tea.20237

Carlone, H. B., Johnson, A., & Scott, C. M. (2015). Agency amidst formidable structures: How girls perform gender in science class. *Journal of Research in Science Education, 52*(4), 474–488. https://doi.org/10.1002/tea.21224

Carter, P. L. (2003). "Black" cultural capital, status positioning, and schooling conflicts for low-income African American youth. *Social Problems, 50*(1), 136–155. https://doi.org/10.1525/sp.2003.50.1.136

Dawson, E., Archer, L., Seakins, A., Godec, S., DeWitt, J., King, H., Mau, A., & Nomikou, E. (2019). Selfies at the science museum: Exploring girls' identity performances in a science learning space. *Gender and Education, 32*(5), 1–18. https://doi.org/10.1080/09540253.2018.1557322

DeWitt, J., Archer, L., & Mau, A. (2016). Dimensions of science capital: Exploring its potential for understanding students' science participation. *International Journal of Science Education, 38*(16), 2431–2449. https://doi.org/10.1080/09500693.2016.1248520

Duff, C. (2010). On the role of affect and practice in the production of place. *Environment and Planning D: Society and Space, 28*(5), 881–895. https://doi.org/10.1068/d16209

Ehret, C., & Hollett, T. (2016). Affective dimensions of participatory design research in informal learning environments: Placemaking, belonging, and correspondence. *Cognition and Instruction, 34*(3), 250–258. https://doi.org/10.1080/07370008.2016.1169815

Francis, B., Archer, L., Moote, J., de Witt, J., & Yeomans, L. (2017). Femininity, science, and the denigration of the girly girl. *British Journal of Sociology of Education, 38*(8), 1097–1110. https://doi.org/10.1080/01425692.2016.1253455

Furman & Calabrese Barton. (2006).

Gee, J. P. (2011). *An introduction to discourse analysis: Theory and method* (3rd ed.). Routledge.

Godec, S. (2018). Sciencey girls: Discourses supporting working-class girls to identify with science. *Education Sciences, 8*(1). https://doi.org/10.3390/educsci8010019

Godec, S. (2020). Home, school and the museum: Shifting gender performances and engagement with science. *British Journal of Sociology of Education, 41*(2), 147–159.

Gonsalves, A. J. (2014). "Science isn't just what we learn in school": Interaction rituals that value youth voice in out-of- school-time science. *Canadian Journal of Education, 37*(1), 185–208.

Gonsalves, A. J., Rahm, J., & Carvalho, A. (2013). "We could think of things that could be science": Girls' re-figuring of science in an out-of-school-time club. *Journal of Research in Science Teaching, 50*(9), 1068–1097. https://doi.org/10.1002/tea.21105

Hammersley, M., & Atkinson, P. (2007). *Ethnography: Principles in practice*. Routledge.

Hazari, Z., Potvin, G., Lock, R. M., Lung, F., Sonnert, G., & Sadler, P. M. (2013). Factors that affect the physical science career interest of female students: Testing five common hypotheses. *Physical Review Special Topics-Physics Education Research, 9*(2), 020115. https://doi.org/10.1103/PhysRevSTPER.9.020115

Holland, D., & Lave, J. (2001). *History in person: Enduring struggles, contentious practice, intimate identities*. SAR Press.

Holland, D., Lachicotte, W., Jr., Skinner, D., & Cain, C. (1998). *Identity and agency in cultural worlds*. Harvard University Press.

Kincheloe, J. L., & Berry, K. S. (2004). *Rigour and complexity in educational research: Conceptualizing the bricolage* (E-book). http://site.ebrary.com/id/10510839.

Kvale, S. (1996). *Interviews: An introduction to qualitative research methodology*. Sage.

Nasir, N., & de Royston, M. M. (2013). Power, identity, and mathematical practices outside and inside school. *Journal for Research in Mathematics Education, 44*(1), 264–287. https://doi.org/10.5951/jresematheduc.44.1.0264

O'Neill, T. (2005). Uncovering student ownership in science learning: The making of a student created mini-documentary. *School Science and Mathematics, 105*(6), 292–301. https://doi.org/10.1111/j.1949-8594.2005.tb18130.x

Paechter, C. (2007). *Being boys; being girls: Learning masculinities and femininities*. McGraw-Hill Education.

Rahm, J. (2010). Science in the making at the margin: A multisited ethnography of learning and becoming in an afterschool program, a garden and a math and science upward bound program. *Sense*. Publications.
Read, B., Francis, B., & Skelton, C. (2011). Gender, popularity and notions of in/authenticity amongst 12-year-old to 13-year-old school girls. *British Journal of Sociology of Education, 32*(2), 169–183. https://doi.org/10.1080/01425692.2011.547304
Rich, A. (1980). Compulsory heterosexuality and lesbian existence. *Signs, 5*(4), 631–660. https://doi.org/10.1086/493756
Robin, B. (2006, March). The educational uses of digital storytelling. In *Society for information technology & teacher education international conference* (pp. 709–716). Association for the Advancement of Computing in Education (AACE).
Saldaña, J. (2015). *The Coding Manual for Qualitative Researchers*. Sage.
Seiler, G. (2001). Reversing the "standard" direction: Science emerging from the lives of African American students. *Journal of Research in Science Teaching, 38*(9), 1000–1014.
Seiler, G., & Gonsalves, A. (2010). Student-powered science: Science education for and by African American students. *Equity & Excellence in Education, 43*(1), 88–104. https://doi.org/10.1080/10665680903489361
Thompson, J. (2014). Engaging girls' sociohistorical identities in science. *Journal of the Learning Sciences, 23*(3), 392–446. https://doi.org/10.1080/10508406.2014.888351
Wade-Jaimes, K., & Schwartz, R. (2019). "I don't think it's science:" African American girls and the figured world of school science. *Journal of Research in Science Teaching, 56*(6), 679–706.
Yosso, T. J. (2005). Whose culture has capital? A critical race theory discussion of community cultural wealth. *Race Ethnicity and Education, 8*(1), 69–91. https://doi.org/10.1080/1361332052000341006

Chapter 4
Young Women's Identity Work in Relation to Physics at the Transition from School to Further Educational Pathways

Thorid Rabe and Freja Kressdorf

At the end of high school young people are expected to think about their further education and have to come to terms with their ideas about their professional future. The school-leaving certificate marks the end of this stage of education. It opens up certain educational pathways while blocking others. On completion of high school young people enter a transitional phase that offers socially established and institutionalised opportunities for studying and making career choices.

The importance of study and career choices of young people is twofold. They are central to an individual's life as the choices can influence how satisfied a person is with their life and what position he or she can take in society (e.g., welfare in terms of health, livelihood and socio-economic status). In this respect, educational decisions are closely linked to individual development opportunities. At the same time educational choices are of high relevance for the social and economic needs of a society. In Germany, there have been complaints for years about a shortage of labour, especially in technical and scientific occupations (Pohlmann & Möller, 2010). Furthermore, there is a need for a diversity of professionals, and questions of empowerment of under-represented groups can be raised that are closely related to issues of justice (Bøe et al., 2011).

In Germany – similar to other countries – there is the (perceived) societal need to attract more young people and especially women to study physics or to choose professions linked to physics. However, the number of young women participating in physics is only moderately increasing, drop-out numbers at universities in Germany are large (Heublein & Schmelzer, 2018) and women continue to be under-represented (for instance, twice as many male as female first-year students enrolled in physics in Germany in 2018; Statistisches Bundesamt, 2019).

T. Rabe (✉) · F. Kressdorf
Martin-Luther-Universität Halle-Wittenberg, Halle, Germany
e-mail: thorid.rabe@physik.uni-halle.de

H. T. Holmegaard, L. Archer (eds.), *Science Identities*, Contributions from Science Education Research 12, https://doi.org/10.1007/978-3-031-17642-5_4

Thus, decisions about educational pathways in science, technology, engineering and mathematics (STEM), and specifically in physics, are a recurring issue both in research and in the public sphere because of a perceived shortage of STEM professionals in general and women in particular (Anger et al., 2011). Currently there is a major funding initiative by the Federal Ministry of Education and Research in Germany to recruit more young people, especially women, to STEM-area subjects (cf. https://www.komm-mach-mint.de).

In the following section, research on educational pathways and transitions in the field of science education is shown not to be new, but limited in its definition of the object of investigation, its scope and methodology.

In the following Chapter we will first clarify the concepts of educational choices and (physics) identity that are central to the present study and draw consequences for the methodological approach. On this basis, we will specify the research interest and then present the study design, data collection procedure and the procedure for data analysis. In the results section, we present four case studies in which young women negotiate their identity in relation to physics at the end of high school. From the case analyses and their comparison, dimensions of identity work in the context of educational pathway decisions can be derived, which are finally discussed in terms of their implications for research and intervention.

4.1 Research Approaches to Decisions on Educational Pathways

Research on educational choices in the German-speaking countries so far has predominantly followed the quantitative paradigm, in which decisions are examined in large cohorts and a few psychologically valid measurable constructs are collected in order to make predictions. Decisions are usually examined retrospectively rather than while the process of decision-taking is still going on. These 'mechanistic approaches', to put it more pointedly, tend to see decisions as rational and predictable (Rabe & Krey, 2018). Although these approaches provide interesting and relevant descriptive results, there are also blind spots. Typically, transition phases are not examined in their temporal extension, because decisions are often only examined retrospectively. This means that the procedural nature of decision-making is not taken into account. In addition, the specific group of scientifically interested girls is not examined. Research in Germany has methodologically been one-sided and in particular there has to date been little research done in the field of science education that takes the perspective of identity as its focus.

Against this background it is relevant to supplement current findings relating to decisions about educational pathways at the transition from high school to further education. Our primary interest is on decisions and negotiations with regard to physics as a future field of study and profession. The study at hand therefore aims to undertake an in-depth analysis of educational choices of young women at the transition from high school to further education, to gain deeper insights into the complexity of educational choices.

We argue that a methodological shift based on the construction of identity would help to better understand the procedural and negotiated character of decisions on educational pathways in the field of STEM. Quantitative research should therefore be supplemented by further methodological approaches. Quantitative studies with a basis in psychology are able to describe the overall situation of STEM choices by identifying distinctive factors and their correlations that influence the educational choices of young women on a larger scale. However, they fail to map the process-oriented and complex nature of educational choices at the level of the individual. As long as these characteristics of educational choices are not considered or are portrayed only to a limited extent, a deeper understanding of young women's decision-making processes with regard to physics is not possible. One way to overcome these shortcomings is through research that follows an identity approach that is now widely established in the field of science education.

In the following, two theoretical starting points – educational choices and the construction of identity – will be further explained.

4.2 Educational Choices: Theoretical Foundation

We assume that educational choices – leading for example to enrolment in a study programme but also to dropping out – occur in a complex negotiation process that does not end once a decision becomes manifest. Choices are therefore to be seen as long-term processes, starting in early childhood and transforming into career choices in adulthood. They are not isolated events that can be assigned to a fixed point in time (Holmegaard et al., 2014). On the one hand educational choices follow an internal logic and on the other they result from a complex set of interactions with other people and the environment. As a result, from an external perspective this decision process often does not appear to follow a linear, rational logic but might include continuous phases as well as abrupt and seemingly unpredictable moments. Finally, from our perspective educational choices are part of or are embedded in the so-called identity work of a person: "They are what they choose" (Loeken, 2015, p. 291). Numerous examples in the relevant literature show how decisions on educational pathways can be linked to identity negotiations (Archer et al., 2010; Carlone & Johnson, 2007; Carlone et al., 2014; Cleaves, 2005; Hsu et al., 2009; Stokking, 2000). In our case educational choices are therefore a matter of matching one's own identity or facets of this identity with perceptions of physics or of people related to physics.

Physics has a cultural image as being male, difficult, heteronomous and rigid, while people who pursue educational or career paths relating to physics are seen as particularly intelligent, gifted, but also less social and more "nerdy" (Allegrini, 2015; Archer et al., 2010; Carlone, 2004; DeWitt et al., 2013; Kessels et al., 2006; Lyons & Quinn, 2010; Osborne et al., 2003; Whitehead, 1996). In view of these ideas, there often seems to be an incompatibility between self-image and the image of physics. As Andersen et al. (2014) point out, "many young people hold rather stereotypical images of scientists, making it hard for them to see themselves as future scientists" (p. 439). Such stereotypical images or prototypes, which are seen as typical representatives of a certain group of people – such as physicists or students interested in physics – are particularly relevant in so-called self-to-prototype matching, in which the self-image is compared with the prototype in terms of similarities and differences (Hannover & Kessels, 2004; Lykkegaard & Ulriksen, 2016).

The transition from high school to further education that we focus on in our study is a central biographical phase of course-setting for educational choices as well as for identity (re)construction. Not only will issues of interest in physics be negotiated, but ideas about future life in general (family, children, partnership, work-life balance, income, security, etc.) will be of interest. The individual positions him or herself in relation to these issues and balances the tension between individual agency on the one hand and the influence of social structures (requirements, norms, ethics) on the other in a unique manner.

Factors that influence educational choices are other people like parents, teachers and peers who serve as "significant others" (Sjaastad, 2012). Institutional factors like school types (Bennett et al., 2013) or the curriculum (Stokking, 2000) with particular options and access conditions (Lyons & Quinn, 2010) can also restrict or promote educational choices in specific ways.

4.3 Conception of (Physics) Identity and Methodological Issues

Identity is repeatedly characterised as a "messy construct". In a first approximation, identity can be seen as the subjective feeling of stability, coherence and individuality and the notion of being distinct from others as an individual (Lee, 2012; Morf & Koole, 2014). In other words, identity is the way we think about ourselves, which involves the need to perceive oneself as a unified and consistent person. At the same time we see identity not as a characteristic inherent in the individual and coherent in itself, but as the result of constructions that are shaped by contexts (Archer & DeWitt, 2015; Archer et al., 2010; Regan & DeWitt, 2015; Smith & Sparkes, 2008). We thus follow a theoretical approach that is informed by both sociology and psychology. Due to our research interest in educational decisions *on physics*, we focus our attention especially on the processual production and negotiation of *physics* identity. We define physics identity as derived from external factors and from self-perception as well as manifested in the (linguistic) behaviour of people with regard to physics-related contents, topics and activities.

So-called "identity work" – the negotiation of one or more identities – takes place in an interplay of individual agency and the influence of social structures (Carlone, 2012). Identity has a preliminary character because it results from ongoing construction processes through introspection, self-reflection, social comparisons and social interactions (Archer & DeWitt, 2015; Sjaastad, 2012) and identity work can therefore be considered a lifelong task.

Identity should be established in the tension between individuality and group membership (Smith & Sparkes, 2008) and to be balanced between stability versus change or variability (Shanahan, 2009).

Put simply, identity work is based on a reciprocal relationship between the individual and society. As identity is constructed and reconstructed in social practices we assume that plural identities arise within each individual (cf. Gee, 2000, p. 99) which are undergoing constant change (Shanahan, 2009, p. 58). Each identity reflects different life contexts of the individual with different properties – for example, someone can have a gender identity as well as a physics identity that are independent of each other in certain areas, but also overlap in other contexts and become hybrid: "one of the characteristics of science identity is that it is relational to other multiple identities, for example, gender identity, religious identity, and ethnic identity" (Avraamidou, 2020, p. 328).

As mentioned above, in the context of educational choices we are specifically interested in physics identity that we conceive to be a specific facet or component of one's identity. In this sense physics identity has the same basic characteristics as identity in general, but focuses on those aspects of identity that are negotiated in relation to physics – or what the person associates with physics. In line with this definition, physics identity reflects what kind of person we think we are in relation to physics, how we see ourselves in relation to physics and to people working in the field of physics (self-perception), and how we are seen by others in relation to physics (external perception). Following Carlone and Johnson's (2007, p. 1190) definition of science identity, (Hazari et al., 2010 have adapted and extended this definition), we see physics identity at the interface between performance, competence and recognition. To us, however, an issue still worth discussing theoretically is whether only a positive self-image in relation to physics should be understood as a physics identity or whether a demarcation and distanced attitude towards physics also reflects a specific – negative – form of physics identity. The difference we try to hint at is between 'identifying with something' and 'identifying who I am in relation to' Gender identity is central both for the individual and for society, and is constantly recreated in socially and culturally shaped performative practices (Allegrini, 2015; Archer et al., 2013). At the same time the perception of physics or typical physics identities in terms of prototypes or stereotypical images often conflicts with common and/or desired gender identities (Bøe et al., 2011), and this can have a lasting impact on educational decisions.

There is ample research evidence for the relevance of gender identities in the negotiation of physics identities. However, other aspects of identity, such as those resulting from the educational background of the family or from peer group membership, can also be significant. Therefore, in the context of our study focusing on physics identity and gender identity, we need to continue to be sensitive to those other facets of identity that might interact with physics identity.

4.4 Specification of the Research Interest

Based on the research background and the theoretical considerations above, we can formulate our research question:

How do young women negotiate their identity with regard to physics ("physics identity") at the transition from high school to further education?

The following questions explicate aspects of the main question and serve to structure the planning and evaluation of the study:

1. How do the young women present themselves as individuals, and which self-images do they display, including their ideas about their future?
2. How do they position themselves towards their idea of physics and to "physics people"?
3. In what ways does gender and their own gender identity play a role in relation to physics?
4. Who are significant others and what roles do they play?
5. What other aspects are visible as relevant to identity negotiations in the context of educational choices?

As mentioned above, the construct of physics identity is an appropriate theoretical approach to gaining a deeper understanding of educational choices (in contrast to studies using more quantitative approaches), because it allows us to take seriously the complexity of the decision-making processes behind these choices and enables us to understand educational choices from the perspective of the individual.

4.5 Methodological Considerations

People make sense of their lives by telling stories or by telling *the* story of their life.

Accordingly we assume that identity work amongst other settings takes place and becomes accessible in (autobiographical) narratives (cf. Gee, 2010, p. 161), so that a specific manifestation of identity – the narrative identity – can be reconstructed from these narratives.

In autobiographical narration, talking about ourselves and events from our lives, we interpret our experiences, feelings and thoughts from the perspective of our present identity. In narratives we represent and produce situationally relevant aspects of our own identity (Lucius-Hoene & Deppermann, 2002). We reflect on the past, ascribe meaning to it and make sense of ourselves and others (Kane, 2012). In addition, the narrator positions and compares him or herself to and with other people, such as a particular physicist.

As a consequence, in our empirical study we have collected interview data that allows us to reconstruct the narrative identities of young women at the transition from high school to further education.

4.5.1 Investigation Setting and Method

Organisationally the study is affiliated to three academies of the project 'helpING!' (cf. https://www.helping.academy) promoted by the German Federal Ministry of Education and Research as part of a broader funding programme aiming to attract more women to STEM subjects. Three so-called 'orientation academies' have been set up that address young women at the end of their high school time or before their entry into higher education. The academies are geared towards providing orientation about possible career paths into science and technology and promoting these. In the best case the academies (viewed from the perspective of the funders) can even convince some girls that it is worth considering these professions for themselves.

The academies took place in different places in Germany over the span of 3 years, each academy lasting for about 1 week and all of them having a strong focus on social aspects in the pursuit of science. The young women have to actively apply for participation (while no application has been rejected so far), so they usually bring a certain interest in the natural sciences with them already. Besides, all participating girls are heading for the German *Abitur*, the qualification for university entrance. Thus, the participants represent a specific selection of adolescents and there is a chance that some of them are considering studying physics or related subjects. There are no costs of participation in the academies, so that the influence of anyone's socio-economic background might at least be mitigated.

Despite the fact that the academies might influence the educational choices of the participating girls, our intention with the present study is not to measure the impact of the academies; it is explicitly not an intervention study. The academies above all offer a good opportunity to pursue our research interests.

The longitudinal interview study presented here follows a qualitative approach supplemented by questionnaires and group discussions which are more related to the academies. These data are not included in the evaluation of the interviews but will be published after completion of the project. Only demographic information and data relating to interest in science from the questionnaires (conducted at similar times to the interviews) are considered.

The qualitative interview study focuses on single cases drawn from a longitudinal design. With four to five participants per academy, narrative interviews are conducted at four points in time. A first in-depth interview takes place at the beginning of the academy. Three more interviews are conducted at the end of the academy, 4 months and 1 year after the academy respectively. Participation in the study is at all times voluntary. We refer here to the first ($n = 4$) and fourth interviews ($n = 3$) of the first academy, which took place in spring 2018, with one participant dropping out after the second interview.

The selection of interviewees is based on willingness to participate. No further theoretical sampling strategies or selection criteria (Przyborski & Wohlrab-Sahr, 2014) were applied beyond the precondition of participation in the academy. The organisational framework of the academy and data protection considerations did not allow for a more targeted selection.

In several steps, an interview guideline was developed based on an established procedure (Helfferich, 2011; Kruse, 2015), refined with colleagues from a physics education department, revised and piloted with students. All interviews follow the scheme of starting as openly as possible and then passing into a guided interview ensuring that the topics of interest are adequately addressed. The first interview begins with open-ended prompts about the interviewee's biography, interests and people who are important to her. Subsequently, further prompts are given on the topic of physics: how was physics a topic in one's own biography, which people were important in this, how is one's own relationship to physics seen, what role did physics lessons play in this? Another interview section deals with the image of physics and physicists. Finally, questions are asked about future perspectives and especially about professional ideas. The aim throughout is to draw the interviewee into a mode of storytelling and to allow them to set their own parameters of relevance (Lucius-Hoene & Deppermann, 2002).

The second interview focuses on experiences during the academy. The girls' experiences in school and private life since the academy are at the centre of the third interview and it also serves to maintain contact with them throughout the year. The fourth interview provides both a review of the past year and an outlook for the future and focuses on the young women's ideas and reflections on their plans and wishes for the future. At the time of this last interview the young women are about to finish high school or a voluntary service and enter a phase of biographical changes. We have to assume, and we can also see from the first round of surveying, that the process of decision-making is not complete within this time and with the first choice made. In this paper we limit the presentation of the interpretation to the first and last interview due to the scope (also of the analysis), but this allows us to look at the individual cases over a long and biographically relevant period of time (Fig. 4.1).

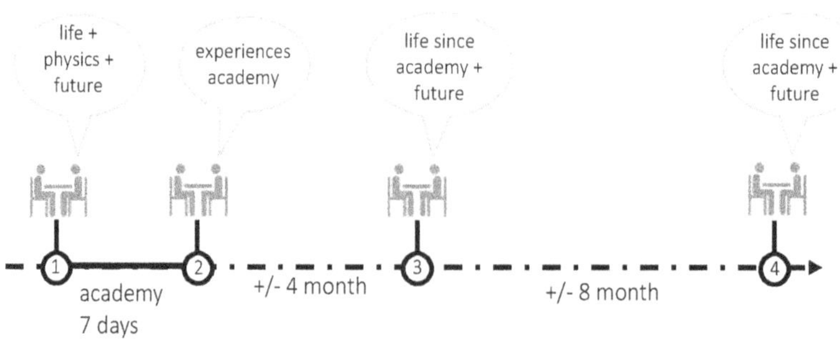

Fig. 4.1 Timeline and main topics of the interviews

4.6 Analysis Procedure

The interviews were transcribed and then analysed according to the work of Lucius-Hoene and Deppermann (2002), an approach that seeks to reconstruct narrative identity through sequential text interpretation. The approach is closely related to discourse analysis as described by Gee (2010, 2011), for example.

In analysing both the content of what is said and the specific ways in which it is said, the guiding questions are: What is presented? How or in which way is it represented? Why is it represented in this way here and now? The researchers also analysed how the narrator positions herself in relation to (relevant) others and which values and norms she refers to. Self-positioning and perceived external positioning by others are considered. The constellation of the narrating self (at the time of the interview) towards the 'narrated self' (at the time of the story told) has to be taken into account, because the interviewees naturally talk about themselves in the past from the perspective of the present. Consequently, the interpretation should take into account whether the interviewees are returning to the perspective of the past self or whether they are talking about the past retrospectively from today's perspective. In line with our research interest, we pay particular attention to how the girls talk about themselves and their lives, how they present themselves, how they relate to physics and physics teaching, how they relate to other people involved in physics, ideas about physics and gender, and what ideas about their professional future are addressed. Special features on the linguistic level are considered in the interpretation. These include the register in which the language is spoken (for example, language indicating membership of a peer group), the specific choice of words (technical terms like "theoretical versus experimental physics" indicating knowledge about the structure of physics institutes in Germany; metaphors "to hang in", exclamatory words "uh"), the use of past or present tense or formulations in the first person ("I" as opposed to an impersonal "one" or "you"). The language used gives clues to the contextualisation, to the background knowledge, implicit valuations and positioning of the narrator and also to the narrator's self-image. Since narratives are influenced by the context in which they are constructed (Kane, 2012, p. 30), the interpretation should be attentive to the interview situation (especially as the interviewer is part of the mutual construction of identity). Therefore, the interviewer's statements and the interviewees' reactions to these are also included in the interpretation.

The interviews were first interpreted sequentially, partly supported by the analysis software MAXQDA (https://www.maxqda.com): sequence by sequence, analysing what is said in terms of content and how it is formulated. Subsequently, the sections relevant to the research question are clustered thematically. The following focal points of analysis can be derived from the research question and the resulting interview guide: autobiographical self-representation, approach to physics and to physics lessons, perspectives on "physics people", significant others, views on women and physics, ideas about their (professional) future. Finally, the interviews are compared with each other in order to work out similarities and contrasts and to identify the specifics of the cases.

4.7 Results: Four Single Cases

In the following the results of analysis are presented, substantiated by short excerpts from the interviews (for an overview see Table 4.1). The original language of the interviews is German, implying that the translation of linguistic details is also the result of an interpretation. The main focus in this chapter is on the content level of the interviews, and any fine-tuned linguistic analysis is only hinted at (see Table 4.2).

Table 4.1 Overview of cases for interpretation

Pseudonym, age (interview 1), interview code[a]	Status (interview 1)	Career aspiration (interview 1)	Career aspiration (interview 4)
Sophie, 16 years, S-1	Pupil (one more year of high school)	Something related to physics	Drop out after the second interview
Julia, 18 years, J-1/J-4	Finished high school, heading for a voluntary service	Maybe research in the field of mathematics, chemistry or computer science	Applied mathematics (logistics and information management or technoinformatics)
Karolin, 17 years, K-1/K-4	Pupil (one more year of high school)	Something theoretical (e.g., in quantum chemistry)	Mathematics with physics as minor subject
Emma, 16 years, E-1/E-4	Pupil (one more year of high school)	Something related to chemistry	Computer science in a dual study programme

[a]The interview code consisted of the first letter of the first name and the time of the survey

Table 4.2 Transcription rules (Nohl, 2009)

Rule example	Description
(3) or (.)	Number of seconds a pause lasts or a short pause
no:	Emphasised
mayb-	Termination of a word
no:::	Elongation, the frequency of: corresponds to the length of the elongation.
have=we	Dragging, spoken words merging into each other
(but)	Uncertainty in transcription
( )	Incomprehensible utterance, depending on length
((smiles))	Paraphrased events
@no@	Spoken laughingly
@(.)@	Short laugh
//mmh//	Audible signal of the interviewer
°no°	Very soft spoken

In order to ensure the validity of the interpretations, consistency was sought throughout the interview. In addition, interview extracts were interpreted in 'interpretation workshops' in order to be able to compare and weigh up different readings of the data. Experts from the field of science education and from pedagogy were involved in these workshops in various groupings. Particularly relevant interview excerpts were jointly interpreted sequentially and different readings of the sequences were compared and discussed.

In addition to the selection of participants, who we assume have a comparatively strong interest in STEM, the interview context can also influence the fact that certain topics are discussed to a greater or lesser extent in the interviews. Due to the extracurricular context and the academies' focus on career orientation of women in the STEM field, topics such as future planning, career choice, gender and STEM appear to be a natural topic of conversation for the interviewees. In comparison, topics such as school lessons or physics as a specific science had to be more strongly triggered by the interviewer for some participants.

The interviews were conducted by the second author of this paper, the first onsite at the academy, the second via a video conferencing system. The age difference between the interviewees was small, they were already on first name terms before and after the interviews, and did not use the German formal *Sie* form of address. This contributed to a personal and confidential interview atmosphere at eye level. Of course, the interviewees were assured of confidentiality and anonymisation of the interview data, and they had the option to end their participation in the study at any time. During the interviews, the statements were not evaluated, but attention signals were given to encourage as free a narrative as possible.

In the following, four cases will be presented and analysed along the levels of attention mentioned above: autobiographical self-representation, approach to physics and to physics lessons, perspectives on "physics people", views on women and physics, ideas about their (professional) future. Not all points are equally important in every case, but other relevant aspects – which we did not look at from the beginning – might become apparent.

Occasionally we will highlight differences and similarities between the cases. In the summary we finally relate our observations/results with findings from the wider research literature and ask what can be confirmed across the educational and social systems and where there may be particularities.

4.7.1 The Case Sophie

In the interview with Sophie, hesitant, short answers alternate with phases in which she talks about her own experiences spontaneously and extensively. Many of her statements end with the phrase "No idea" (32 references in S-1), by which she classifies her statements as less reliable.

The keyword 'social' is central to Sophie, but in two different ways that are in contrast to each other. She repeatedly presents herself as 'weak on the social side' and sometimes finds it difficult to deal with other people. She also sees this quality in physicists and perceives it as a similarity between herself and them. On the other hand, it is extremely important for her to pursue activities and a profession that allow her to help other people (for example, by passing on her knowledge in the context of development aid). That is why the academy's focus on the social aspects of STEM subjects was the reason for her participation.

With regard to her interest in physics, Sophie portrays herself as a 'late bloomer'. As already indicated above, in her identity work concerning physics she refers to similarities and differences to other physics people. In the first interview, Sophie first compares herself with schoolmates who were interested in physics earlier than her, to conclude that she is not a typical physics person, because she has not always been enthusiastic about the subject but "Somehow came late to physics" as she puts it repeatedly (e.g., S-1, 308f). Physics for her seems to be a topic for which typically a firm decision is made early on, whether one is interested in it or not.

However, her assessment changes when she thinks of another significant person to her, a physicist she only met by chance and who left a deep impression on her. In the course of the interview, she compares herself directly with this physicist and finds similarities in interests and communication behaviour. As a result, she states that she – Sophie – and physics fit together quite well. So here, in telling 'her story with physics', Sophie negotiates who she actually is and how she has seen herself in the past.

This can be found in other sections of the interview as well. In view of her own trilingualism, it seemed obvious to her earlier that her preoccupation with language was something she was born with. At the same time, this is an idea that seemed incompatible with seeing herself as someone with an affinity and capacity for science. Sophie's relationship, her positioning with respect to the natural sciences, thus seems to be closely linked to her self-image, which in turn undergoes biographical changes and adaptations – as does her image of physics. In addition, Sophie's interest in physics and her experience of competence are inseparably linked.

Sophie initially denies any influence of her gender on her access to physics, but after some seconds of reflection she changes her mind: "[…] That it could be that I came so late to physics because few girls think about doing something with natural sciences and uh you get it from your friends so I'm taking languages now and no idea //yes// that was also back then I think one of the reasons why I took Spanish and not umm [here she names a specific science course]" (S-1, 615–622).

Just as Sophie's interest in physics arose late, there have been changes in her career aspirations. When asked if she already knows what she wants to become, Sophie answers, "No I have a lot of @directions@ but nothing fixed uh (.) I think right now the tendency is really to physics //mhm// er since the fifth or so it was architecture all the time //yes// and then it shocked me a bit that it wasn't anymore. I don't know" (S-1, 550–552). The provisional, open character of the processes on which educational decisions are based becomes evident here – Sophie herself is amazed ("shocked") that her original clear desire to study architecture has changed to an interest in studying physics. A change she describes with "oddly enough" and "totally strange".

For Sophie, physics is synonymous with research and therefore positively linked to change and transformation. To her, physics is the opposite of boring and monotonous, especially since she associates physics with very different occupational fields (which she does not elaborate on) and fields of application. In her perception, other people regard physics as very difficult and almost unattainable, and her use of language indicates that she shares this view to a certain extent: "You have to hang in, it's far away, but not that far away" (S-1, 486). The difficulty of physics is in particular attractive for her as she mentions several times that she likes challenges and needs them in order to keep up with something: "Pretty much the hardest subject I have at school and I find that exciting" (S-1, 103). Her memories of physics lessons are largely positive, mainly because she almost always had teachers (male) who made understanding possible. She particularly likes it when the lessons do not stick to the standard curriculum which, in her view, is far from reality. Sophie seems to distinguish between physics in class and physics as it is "Really done" (S-1, 202) – the latter arousing her interest. Similarly, she distinguishes between the skills that a physics teacher needs ("The pedagogical", S-1, 375) and the skills that a physicist needs.

4.7.2 The Case Julia

Julia finished high school recently, has already participated in several science contests and was due to take up a voluntary service at the time of the first interview. In the first Interview Julia appears insecure when she talks about herself, she stumbles and interrupts her story repeatedly. But at the same time, she clearly speaks from a first-person perspective.

When asked about her special traits, she mentions her excellent memory and her mathematical abilities, which seems very self-assured at first glance: "So I think that my (.) memory could be quite good for me, because it feels like I actually remember everything. //aha// but also my (.) mathematical abilities a bit @(.)@" (J-1, 54–59). At the same time Julia marks her self-description as uncertain by using the subjunctive and formulations such as "It feels like" or "A bit". Similar to Sophie she weakens her statements. This could reflect the social expectation not to present oneself too self-confidently as a woman. Unlike Karolin (see below), however, she does not try to legitimise her statement by drawing on the perspectives of other people.

Though Julia says "I just love the natural sciences" (J-1, 414) she does not talk about physics until she's asked about it. Although this indicates that physics is not very relevant for her, she had decided to take physics as an additional subject at school. She characterises her interest in physics as intrinsic and high in comparison to fellow students who attended the same classes. Her classmates did not understand she was doing physics voluntarily: "And most of them just couldn't understand why I chose physics again (.) although I can't use it for my Abitur anyway" (J-1, 349–351).

Julia says that her gender does not play a role in her relationship to physics. While she sees advantages for women looking for a job in physics because of the promotion of women, she also mentions that women are paid less. However: “As a woman you should still stand by physics just as much as you do as a man” (J-1, 401–402). Julia seems to mix the expressions “to stand one’s ground” and “to stand by someone”, which allows two interpretations: she sees women in the responsibility to assert their position in the domain of physics and/or she feels that women have an obligation towards physics to devote themselves to physics despite any difficulties. Accordingly, it would be the responsibility of women to position themselves in relation to physics and to adapt if necessary and not vice versa.

When she is asked whether she and physics fit together, she answers positively, but at the same time in a distant and impersonal way. “I think it could fit together, (.) but then I think it all depends on what kind of field of physics. If=s is rather based on mathematics I believe rather (.) more” (J-1, 189–191). Here, as at several other points in the interview, she emphasises her particular interest in mathematics. She does not see herself as a typical physicist and compares herself with a stereotypical image whose stereotypical character she is also aware of: “I find it (.) Nothing against physicists but that they are always shown so well-behaved? //mmh//A:nd (.) so clever and mostly with glasses °or so° and I don’t think that @I could convey it like this@” (J-1, 216–219). Although she herself laughs at this statement, it becomes clear that she assumes that the stereotype should be at least partially fulfilled in order to be a physics person.

For Julia, physics is clearly linked to physics lessons, and above all to experiments. This fits in with the fact that physics is primarily an experimental subject for her, while theoretical physics – which would fit her mathematical inclinations – is unfamiliar to her: “So: (.) physics in general (.) is also very much bound (.) to (.) experiments which (.) then (.) (.) explain something resulting from it (.)” (J-1, 264–265).

Julia also raises the issue of the quality of explanations in the classroom: “Even though the teachers may sometimes (.) explain (.) strangely and assume that the students understand this, (.) otherwise I always liked physics lessons” (J-1, 237–240).

Julia’s reaction to the question of how she imagines her own future is noticeable: “So: it would be nice if you could (.) find a man @(.)@ and then a reliable (.) job or a reliable employment (.) and then at the same time if you can also do your hobbies. //mhm//and then everything fits together” (J-1, 368–372). It is noteworthy that she is not speaking in the first person at this point, although she was asked about her personal future. This could possibly indicate that her answer is guided by social norms or expectations, or that Julia has only a vague idea of her future. However, we will also see more clear statements about wanting to be close to family as guiding factors in the educational choice later in the analysis.

The first thing she mentions about her future is to find a partner, secondly, that it is a job that promises security, and thirdly, the opportunity to pursue her hobbies. Julia seems to be deliberately seeking stable structures in her life and a balance between profession and private life is very important to her (“Then everything fits together", see above). The importance of a work-life balance only came to the fore in Julia's last year of high school. The first interview with her already indicates that stable contacts with other people and her hobby, swimming, would probably play an important role in her decisions.

This interpretation is supported by the fourth interview 1 year later: "Yes, so I would like to continue of course in the future swimming//mhm//to find admission to a study programme (.) and then, I like (.) also to find new people, who make my studies easier (.) yes, at the same time I would also like to keep good contact with my family and with my friends, even though (.) my friends then partly will probably also move away (.)" (J-4, 120–126). Again, the first thing Julia mentions for her near future is from the private sphere – her swimming hobby that she would like to continue.

At other parts of this interview one can see that she is preparing to choose between two study programmes, both related to applied mathematics. Julia is struggling between deciding for her favourite study programme which would mean moving away from home, or to continue living at her parents' place with the swimming pool for her hobby in the back yard and with more continuity in friendships, but only with her second-choice study programme. The choice of study location is repeatedly an important topic for her in the interviews; she hopes to get to know new people in her university town, but at the same time she wants to stay close enough to her family. Her final decision is in favour of the study programme with the compromise of going home every weekend.

4.7.3 The Case Karolin

Karolin still has 1 year of high school left, has taken part like Julia in scientific competitions and already has good contacts with doctoral students of physics at a research institute. She presents herself as a self-confident individualist and rebel in general and as a high achiever and insider with regard to physics.

Karolin quickly enters a narrative flow, in which she often emphasises the extremes and which at some points indicates that she is not recounting episodes for the first time. What is striking, however, is that she repeatedly talks about other people's points of view regarding her. What others think about her is a recurring theme and seems to be particularly important and relevant for Karolin to justify statements about herself.

In both interviews Karolin repeatedly highlights her own remarkable achievements and her individuality for which she also desires recognition from others. This gives the impression that she wants to appear extraordinary to others, but at the same time wants to be seen as relaxed and laid-back. Her reaction to the same prompt that Julia received at the beginning of the interview – what is special about you – is quite different: "What else do many people say about °me°? That I (.) have very opposite sides (.) That on the one hand I can be very quiet and very thoughtful, but then again totally excited. //yes// that they find it extreme how many sides I have, or that they did not think that I have so many interests, or that I partly do things out of the blue, something totally unexpected, uhm (.)" (K-1, 54–60). Karolin describes herself as a person of contrasts, who is quiet and thoughtful and then again excited. She even emphasises the extremes with "Very" and "Totally". And even more, her behaviour is unpredictable for others, she "Does things out of the blue", "Totally unexpected". Karolin presents herself as the opposite of boring. Another central characteristic she mentions a few times in this first interview is her interest in theoretical issues, and again to the extreme: purely theoretical things: "Uhm (3) I am very interested in philosophy and purely theoretical things that would be possible or pure theories only //aha// I am a very theoretical person, practice isn't so mine" (K-1, 49–52).

The natural sciences are central to Karolin's life and seem to be more important than other interests. But similar to Sophie, she also stresses that she is interested in many other topics and that she differs from other "physics people" in this respect.

Karolin describes her circle of friends as important for her attitude towards physics and stresses that most of her friends are male and interested in physics: "Which is why I am influenced by this and why people often talk about physical topics" (K-1, 249–250). Other important people in the field of physics for Karolin are the doctoral students of physics who give her an insight into studying and researching in the natural sciences, her physics teacher and finally Stephen Hawking, whom she sees as a role model. The physicists at the research institute are a point of reference for Karolin when she deals with stereotypical ideas of physicists in the interview: "One has for example a certain picture of a physicist //yes// and then the picture was partly fulfilled ((smiles)) and partly not at all and that I found very interesting, because=s just proves again that quasi not only your way of being is decisive, whether you can do physics or not" (K-1, 269–273). As evidence for this statement, she then describes a woman and a man who are both physicists, but at the same time represent typical gender ideas: "The doctoral student [...] she likes such long nails ((smiles)) and she would also like to have children and something like that, //°aha!°// so she is a bit girlie ((smiles)), which I think is really sweet (.) and the other doctoral student for example only walks around in flip-flops or barefoot, no matter where he is, //yes// which I also find very interesting" (K-1, 276–282). How Karolin imagines a typical physicist can only be deduced indirectly from what she says here about physicists who are not stereotypical in her eyes. Regarding the stereotyped picture, she refers to physicists as being uninterested in fashion and not caring about their appearance, instead walking around in flip-flops. It is crucial that Karolin negates or deconstructs the stereotypical ideas based on her own experiences.

Karolin does not want to "Subordinate to her gender" (K-1, 642–644), indicating that she does not want to fulfil the typical social expectations of women's behaviour or is willing to disregard the persistent idea that women and physics do not go together. On the contrary she sees a fit between physics and herself, because physics is something "To bite into" (K-1, 295), using the metaphor of a dog that bites into a leg and will not let go (the English phrase "to get wound up in" is a translation of the content, but uses a different linguistic metaphor). In Karolin's eyes, physics is a subject to be fully engaged with, which suits her self-image as someone who likes and lives the extremes (in the sense of 'all or nothing').

Karolin's opinion about physics lessons is very negative though, and she describes them as uninteresting and incomprehensible. Although she opted out of physics at school, in her free time she is occupied with physics and thus shows a certain resilience towards school experiences. Thus, she distinguishes between the subject of physics she learned at school and the extracurricular physics she has experienced.

At this first interview she mentions that she is undecided between studying physics or chemistry, and it becomes obvious that she already has extensive insights into the subjects and courses of study, even though she is a year away from graduation.

One year later, and in the fourth interview, Karolin still emphasises her preference for theory and this became the striking argument in her choice of mathematics as study programme: "Where I was really fascinated by the chemistry was the theoretical chemistry and this is also quickly just a lot of mathematics //yes// a:nd (.) I'm extremely interested in this topic, but I take just as application physics and not chemistry [she refers here to a minor subject], because that would be too practical for me again" (K-4, 217–222). Now it is not physics and chemistry in the first place, but mathematics with physics "Just as an application" that she wants to study. We assume here that her identity as a theoretical person, as she puts it before, is reflected in her choice of mathematics as a field of study. In contrast to Julia, for Karolin to change residence and to move to a new town is an attractive option when choosing a place of study.

4.7.4 The Case Emma

Emma is in the 11th grade at the time of the first interview, so she will graduate from high school in 1 year. In her spare time, she plays water polo as a competitive athlete. She only hesitantly begins to talk about herself and seems rather restrained.

Emma gives the impression that she likes to go her own 'trial and error' way instead of the given one (E-1, 67f). This is in contrast to Karolin, who plans more strongly and attaches importance to conscious influence on her own actions (cf. K-1, 246f). Emma describes herself as an uncomplicated person who gets along with different people (in contrast to Sophie who does not see herself as socially at ease) and as someone who people come to who need an "Open ear" (E-1, 47–48). In this way Emma stresses a social ability that is stereotypically attributed to women (Briton & Hall, 1995) and which is typically rated positively by others. She mentions another characteristic only after being asked explicitly: she likes to experiment (E-1, 67ff). Somewhat unexpectedly in the context of the interview and in view of her current wish to study something related to science, by 'experimenting' she refers to the example of cooking which is again often perceived as a 'feminine' occupation (Holm et al., 2015).

For Emma, physics is primarily present as a teaching subject (e.g., E-1, 135–163) and not as a scientific discipline (as for Karolin) or as phenomena that can be experienced. Physics is one of her two advanced courses in high school (in Germany, students at most high schools have to choose two courses at advanced level), although she likes physics only "In connection with chemistry" (E-1, 83). She also refers to physics teachers as physicists, which implies that she does not distinguish between physics within and outside of class. In connection with an internship in a crime lab it becomes clear that she likes activities related to the natural sciences and she prefers it if there are social applications to the natural sciences (she uses the word "Helping" twice in this context (E-1, 100–103)). Emma perceives that other people react with incomprehension when someone like her voluntarily chooses physics as a subject and that for these other people physics is the greater challenge compared, for example, with chemistry (E-1, 324).

When it comes to describing physics, Emma's answer is characterised by uncertainty markers (E-1, 264ff), which are in contrast with her other answers. For Emma physics is a mixture of philosophy and maths, and she describes physics as requiring advanced theoretical thinking and only partly as a practical activity. She does not see that physics and herself fit together, "Because //mhm// it becomes very theoretical at some point and I want to do something more practical and (.) so I didn't find anything now, umm (.) where you can do so much practical, respectively the study becomes very theoretical (.) towards the end and that's (.) not my thing" (E-1, 183–188). This conclusion contrasts with the viewpoint of Karolin, who justifies her fit with physics precisely by attesting that physics is theoretical – so both girls perceive physics as a theoretical matter, but their conclusions are opposite in terms of feelings of rejection versus attraction.

In contrast to Emma's description of physics as theoretical is her associating of physics lessons mainly or only with experiments (that do not work) (E-1, 139ff + 155ff). Emma describes the natural sciences as interdependent, but then suggests a hierarchy of sciences with physics at the top, followed by chemistry and finally biology (E-1, 298ff). In a similar way, Karolin also places biology at the bottom of the hierarchy (cf. K-1, 435f).

Even more than Karolin and Julia, Emma emphasises how important physics teachers are for students to enjoy the subject and gain access to it. She previously had a "Wacky teacher" (E-1, 216) in whose class she did not understand physics, whereas it was only with a new teacher that she became interested in physics.

In Emma's ideas for the future, the goal is to find an attractive study programme and she does not think about a specific subsequent profession (E-1, 392). This study programme should in some way be related to the discipline of chemistry, and she currently favours physical chemistry (E-1, 392). Water polo will remain important in Emma's life. Time for the family, however, will depend on the job: "That, for example, one could still fit in the family" (E-1, 415f). Like Julia, Emma is also concerned with the balance between work and family, but gives higher priority to work, to which family members somehow will have to subordinate themselves.

Emma realises that women are more likely to face pressures to justify themselves and to be looked at askance compared to men when they choose a profession related to physics, and she is aware that incomes are different (E-1, 428ff). She hopes that this will change, "That it will become normal" (that is, in her view it is not yet socially recognised as normal that women do physics). But Emma also says that this is no reason for her to reject physics.

At the time of the fourth interview, Emma had graduated from high school and had enjoyed the support of her family during the final months. With her close family she also enjoyed talking about her future (E-4, 187f). Emma does not discuss her place of study in the interview. It is possible that this is not in question for her, but has already been fixed. At least it does not seem to be a relevant criterion for her choice of study programme that needs to be discussed (unlike in the case of Julia) (E-4, 264f) but she is hoping to leave her parental home "In order to become a little bit independent" (cf. E-4, 102f). For Emma the study format is more important. She will begin a so-called 'dual study programme' (parallel to her university studies, she will be working in a company in her future field of work) in computer science (E-4, 19f) and is already preparing for this by teaching herself programming (E-4, 9f), which she "Really enjoys" (E-4, 79).

Emma would have decided to study chemistry instead of computer science only if necessary (E-4, 177f), and she justifies this by mentioning the geographical flexibility that one would have to bring along as a researcher in the field of chemistry (cf. E-2, 183; E-3, 44f). It should be added here that in the second interview Emma said that she had found a research activity "Now quite interesting", and in the third interview she said that she intended to go into research if there were no requirement for geographical flexibility. In the first interview she had not indicated this wish.

4.8 Perspectives of Identity Work in the Context of Educational Path Decisions: Summary and Comparison of the Individual Cases

In the following, the results from the four cases will be summarised, compared and related to the research literature. However, each of the cases analysed is unique and the individual findings should only be generalised with caution. The following thematic bundling of the results according to the research question is an analytical one, which is not laid out in this way in the interviews themselves. Rather, the aspects are closely interwoven.

4.8.1 How Do the Young Women Present Themselves as Individuals, and Which Self-Images Do They Display, Including Their Ideas About Their Future?

In their autobiographical narratives, the four young women reveal diverse negotiations of identity. Their different personalities can be seen in the way they tell their stories and how they present themselves. All four though have a fundamentally positive approach to physics. However, this approach differs greatly in detail and is reflected in the fact that physics has almost no relevance (Emma) to great importance (Sophie) for their professional future.

4.8.2 How Do They Position Themselves Towards Their Idea of Physics and to "Physics People"?

Physics as a Whole and "Physics People" Only Karolin already has a deeper insight into the everyday life of a physics institute and its structure and she perceives a diversity there. The other interviewees partly have an undifferentiated picture of physics and physics-related professions as well as of the natural sciences in general. This can be seen, for example, in the fact that physics is equated with theoretical work (Emma), while areas of application of physics are not discussed. It is noteworthy that the assessment of physics as theoretical leads to the avoidance of physics in one case and is evaluated as a positive attractor in the other. This fits with earlier results that a "lack of knowledge about science-related work has been the broad finding of a number of studies" (Tytler, 2014, p. 93).

All four girls in our survey refer to stereotyped culturally influenced images of physics and physicists (Kessels et al., 2006), to which they (critically) relate in the interview. It becomes apparent that the overarching discourses ("Discourses with a big D" in line with Gee, 2010, p. 34) on gender and physics play a role for all of the girls, albeit in different ways and that these discourses still seem to be viewed in a rather clichéd way. It seems that even today to relate oneself to these discourses and to position themselves in relation to stereotypes is unavoidable for young women who are thinking about a career in the natural sciences (see below). Nevertheless, it is evident from the interviews that, similarly to what Lykkegaard and Ulriksen (2016) observed, the comparison with prototypes reaches its limits when the young women encounter real people – significant others (see below) – and individual sympathies or distancing occur. Stereotypical images of people working in physics thus remain a pertinent, but by no means dominant, aspect in identity negotiations.

School Physics That the girls make a distinction between school physics and physics in general can only be gathered from one of the interviews (Karolin). This result contrasts with other studies that have observed differences in attitudes to school science and science in general (Osborne et al., 2003), which presupposes that these are perceived as different areas. In our cases *physics lessons* seem very important for the idea of what physics is, but tend not to be well thought of. It is not recognisable that the young women's physics lessons promote interest in physics or that they have a noticeable orienting effect, which is a rather unfortunate diagnosis for physics lessons in Germany.

Other studies have highlighted the *role of the teacher* and the "quality of teaching as a major determinant of student engagement with and success in school subjects" (Tytler, 2014, p. 93). In telling about their physics lessons, three of the girls (all except Sophie) refer to the quality of teacher explanations which are perceived rather negatively. The girls have positive memories of doing experiments (even if they present themselves as tending to be more interested in the theoretical side), which is in line with "students often nominating these activities as positive experiences in science" (Tytler, 2014, p. 93).

In two cases (Sophie and Julia) we can see how the *institutional framework* of high school restricts access to physics. In the German *Gymnasium*, courses on two different levels (basic and advanced) are chosen after the tenth grade, which are binding for the following 2 or 3 years. In most cases only two natural sciences can be chosen. Two interviews show that these conditions prevented the choice to study (or return to) physics in school.

4.8.3 In What Ways Do Gender and Their Own Gender Identity Play a Role in Relation to Physics?

Overall, gender is an issue in the young women's negotiations. Sophie reflects that gender may have played a role in her approach to physics. Although Julia denies that it has any significance for her, it is clear from her comments that women have to "confess". Karolin (as well as Emma) mentions that there are certain stereotypical ideas about women and physics, but that she does not want to subordinate herself to them. Emma also sees lower income opportunities for women and hopes that these will be equalised in future.

4.8.4 Who Are Significant Others and What Role Do They Play?

The influence of significant others on positioning towards physics can be seen in most of the cases. In Julia's case these are a factor in her decision-making, while for Karolin other people are rather a source of information, though she claims that she would not base her decision on their opinions. Sophie's case shows the peculiarity that a random acquaintance – a physicist – became a significant person with whom Sophie compares herself and in whom she finds confirmation for her own personality.

4.8.5 What Other Aspects Are Visible as Relevant to the Identity Negotiations in the Context of Educational Choices?

The place of study is included in the decision for a course of study. A move to another city seems rather unsettling for Julia, while it is attractive to Karolin, who associates a move with further development of her own personality in the sense of greater independence. In Julia's case the place of study even influences the choice of the study programme, while in Karolin's case the decision for a study programme comes first and is supplemented by the additional wish to find a university with a reputation in the field of mathematics.

Considerations on reconciliation of work and family have extremely different effects on the interviewed girls. While for Julia the balance of her future private and professional life is very important, Karolin focuses more on her professional career and for Emma the family has to somehow fit into her working life and has to be subordinate to work.

Additionally, the societal usefulness and relevance of contents or the respective subject seems to be important for almost all girls, most clearly visible with Emma and Sophie.

The process-oriented and preliminary nature of educational decisions are evident in all four cases – for instance, during the survey period Karolin changed from chemistry and physics to "pure mathematics". We observed the complexity of potential influences on the decisions in which coincidences in terms of unpredictable encounters with people who influence decision-making also play a role. This finding underpins the unpredictability of educational choices at the level of individuals as opposed to supporting a mechanistic notion of decision-making. In all this, educational choices appear to be closely interwoven with the identity work of young women. For example, when they relate to significant others, reflect on where they study or on the role of family and work in their lives, or position themselves in relation to physics, they are also negotiating how they see themselves as people now and in the future.

4.9 Discussion

We initially argued that a qualitative approach that draws on the construct of identity allows access to the process-oriented and complex nature of educational choices at the level of the individual. Indeed, our data reveal a large heterogeneity within our cases, which at first glance could have been regarded as homogeneous with regard to access to physics.

Quantitative data from a questionnaire survey, conducted during the academies, indicate that all four interviewees can be classified as "interested in physics" (interest in various subject areas was rated using a four-point Likert scale). The diversity of their attitudes to physics and to the natural sciences in general is not captured (cf. Avraamidou, 2021, p. 9) and consequently all four young women appear as likely candidates to study physics.

In the interviews, our participants were encouraged to speak about experiences and episodes from their own lives that were linked to the natural sciences and especially physics. This allowed them to start from their own personal perspective and to draw on their own ideas about what they found relevant. This specific kind of data and using identity as a lens in analysis allowed us to observe differentiated considerations within this supposedly homogeneous group of physics-interested women with regard to career choices. As we have seen, their choices have led them (so far) to other courses of study, but not physics.

The focus on identity enables a reconstruction of individually distinct superimpositions of influences on choices of study and career that are negotiated and balanced very differently by each person. This relates well to the intersectionality approach to identity, proposed by Avraamidou (2020), and underlines the need for "recognition of the myriad dynamic factors that shape decisions over time in multiple contexts" (Tytler, 2014, p. 94). Empirical survey methods intended for larger cohorts are likely to reach their limits there, because they are incapable of grasping this complexity. With the help of the interviews, however, a more in-depth analysis of the negotiation processes was possible.

The process character (Holmegaard et al., 2012) of the educational path decisions and its link to identity work also became clear. Continuities and stable identity negotiations (for example Karolin's consistent interest in theoretical approaches to the subject) can be observed as well as incidental influences (Sophie's chance acquaintance with a physicist). It is also remarkable that justifications for path decisions are not always consistently formulated at the time, so that their retrospective justifications can appear contradictory and suggest that decisions are being reinterpreted (as we discovered in Emma's interview). In this sense we see identity negotiations in the context of path decisions that are consistent yet can be contradictory over time, as has been reported before (Holmegaard et al., 2012).

To summarise, we agree that a focus on identity "provides a basis for understanding students' long-term personal connection to physics and is a more meaningful measure than a general assessment of students' attitudes" (Hazari et al., 2010, p. 979). To avoid misunderstandings, we should stress that we do not think that quantitative studies should be replaced by studies of this provenance, but rather that the approaches are complementary.

In addition, there are some implications on the methodological level. Our findings indicate that a differentiated examination of identities specific for subjects or domains is worthwhile, because the girls position themselves specifically in relation to particular subjects and do not simply view the various natural sciences in the same way. The study provides evidence that it makes sense to specifically examine the physics identity of young people and not only a general science identity. It is not easy, however, to decide at which points a separation of physics identity and science identity is possible or meaningful.

We also faced the methodological difficulty of identifying negotiation processes relating to physics, because a basic idea of narrative interviews is their openness to the relevance systems of the interviewees. To a certain extent, this contradicts the specific research interest as physics is not necessarily relevant to the young women. This point is closely linked to the question we raised above, whether physics identity always means a positive relationship to physics or whether the demarcation of a person from physics also says something about their 'physics identity'. At the same time the negotiation of study choices in physics or other subjects is strongly overlaid by or embedded in subject-unspecific topics. For this reason, it is methodologically difficult to focus on physics-specific considerations on the one hand, but on the other hand not to overestimate their influence in relation to subject-unspecific influences. The challenge is to focus on physics identity in the interviews without overemphasising the importance it has for the interviewees. Similarly, another difficulty is that the interviews themselves can trigger reflections among the girls (Svašek & Domecka, 2012), which in turn influence identity work and educational decisions.

From our findings, we conclude that, given the complexity of the negotiation processes, the expectations of (short) interventions aiming at recruiting more young women for science careers should not be too high. This also applies to the career orientation academies organised in the project. As the identity negotiation process starts early on and does not follow simple and always rational pathways, the probability of influencing decisions sustainably with a single intervention is low, particularly because this study shows that the social discourse on gender and physics in Germany might still refer to stereotypes to a certain extent (Kessels et al., 2006). As we saw in the data, such stereotyped images do not leave our interviewees unscathed, but require positioning towards this discourse.

Physics education at school – at least theoretically – allows for reaching all children and adolescents regardless of their parents' home situation, social setting or cultural background. It should hence be examined more closely whether and how the physics identity of teachers is related to the identity work of students. This leads to limitations of this study with regard to further overarching discourses that are important in the German education system and elsewhere. We have not investigated how social and cultural backgrounds of the girls are incorporated into the negotiation of their physics identity and educational choices, although this is of particular importance in a society to which people have migrated, as in Germany. This points to the relevance of intersectionality. As Avraamidou (2020) argues, "researchers ought to turn their attention to gain a more comprehensive and explicit understanding of the nexus and complexity of how intersecting identities form social positioning and how this intertwines with forming a science identity" (p. 331). Likewise, the possible influence of discourses in media and social networks on positioning in relation to physics has not been taken into account, although these probably grow in importance compared to school influences (cf. Höttecke & Allchin, 2020, highlighting the gatekeeper function of social media). Future research projects should consider these aspects.

Compared to short-term and isolated interventions, in light of our results we consider physics lessons at school in particular to be responsible for improving access to physics. Although any invitation to STEM outside of school is of course welcome, it should remain an important goal to improve physics teaching in such a way that it promotes interest and provides orientation about physics professions. In our interviews physics is often seen as a rather theoretical subject and the distinction between experimental and theoretical physics common in German study programmes is not evident to the young women. A conclusion could be that in Germany it should already be highlighted in school that there is not 'the *one* physics' and that there are many fields in the spectrum, from (industrial) applications and basic research to more theoretical or more experimental orientations as well. This includes highlighting possible links between physics and social or ethical issues, which was relevant for some of the girls. The conditions for reconciling work and family life are also, as the interviews show, still significant for young women and should be addressed further. These issues could lower the barriers discouraging young women from choosing a physics-related course or profession and thus help to reduce the shortage of female STEM professionals in Germany, an issue which has been a key driver of our research.

References

Allegrini, A. (2015). Gender, STEM studies and educational choices. Insights from feminist perspectives. In E. K. Henriksen, J. Dillon, & J. Ryder (Eds.), *Understanding student participation and choice in science and technology education* (pp. 43–59). Springer.

Andersen, H. M., Krogh, L. B., & Lykkegaard, E. (2014). Identity matching to scientists: Differences that make a difference? *Research in Science Education, 44*(3), 439–460. https://doi.org/10.1007/s11165-013-9391-9

Anger, C., Koppel, O., & Plünnecke, A. (2011). *MINT – Report 2011.* Zehn gute Gründe für ein MINT-Studium. http://eee.lizzynet.de/ww3ee/bin/34470180-34473252-1-studie_mint-report_2011.pdf

Archer, L., & DeWitt, J. (2015). Science aspirations and gender identity: Lessons from the ASPIRES project. In E. K. Henriksen, J. Dillon, & J. Ryder (Eds.), *Understanding student participation and choice in science and technology education* (pp. 89–102). Springer.

Archer, L., DeWitt, J., Osborne, J., Dillon, J., Willis, B., & Wong, B. (2010). "Doing" science versus "being" a scientist: Examining 10/11-year-old schoolchildren's constructions of science through the lens of identity. *Science Education, 94*(4), 617–639. https://doi.org/10.1002/sce.20399

Archer, L., DeWitt, J., Osborne, J., Dillon, J., Willis, B., & Wong, B. (2013). "Not girly, not sexy, not glamorous": Primary school girls' and parents' constructions of science aspirations. *Pedagogy, Culture & Society, 21*(1), 171–194. https://doi.org/10.1080/14681366.2012.748676

Avraamidou, L. (2020). Science identity as a landscape of becoming: Rethinking recognition and emotions through an intersectionality lens. *Cultural Studies of Science Education, 15*(2), 323–345. https://doi.org/10.1007/s11422-019-09954-7

Avraamidou, L. (2021). Identities in/out of physics and the politics of recognition. *Journal of Research in Science Teaching, 59*(1), 1–37. https://doi.org/10.1002/tea.21721

Bennett, J., Lubben, F., & Hampden-Thompson, G. (2013). Schools that make a difference to post-compulsory uptake of physical science subjects: Some comparative case studies in England. *International Journal of Science Education, 35*(4), 663–689. https://doi.org/10.1080/09500693.2011.641131

Bøe, M. V., Henriksen, E. K., Lyons, T., & Schreiner, C. (2011). Participation in science and technology: Young people's achievement-related choices in late-modern societies. *Studies in Science Education, 47*(1), 37–72. https://doi.org/10.1080/03057267.2011.549621

Briton, N. J., & Hall, J. A. (1995). Beliefs about female and male nonverbal communication. *Sex Roles, 32*(1–2), 79–90. https://doi.org/10.1007/BF01544758

Carlone, H. B. (2004). The cultural production of science in reform-based physics: Girls' access, participation, and resistance. *Journal of Research in Science Teaching, 41*(4), 392–414. https://doi.org/10.1002/tea.20006

Carlone, H. B. (2012). Methodological considerations for studying identities in school science: An anthropological approach. In M. Varelas (Ed.), *Identity construction and science education research. Learning, teaching, and being in multiple contexts* (pp. 9–25). Sense Publishers.

Carlone, H. B., & Johnson, A. (2007). Understanding the science experiences of successful women of color: Science identity as an analytic lens. *Journal of Research in Science Teaching, 44*(8), 1187–1218. https://doi.org/10.1002/tea

Carlone, H. B., Scott, C. M., & Lowder, C. (2014). Becoming (less) scientific: A longitudinal study of students' identity work from elementary to middle school science. *Journal of Research in Science Teaching, 51*(7), 836–869. https://doi.org/10.1002/tea.21150

Cleaves, A. (2005). The formation of science choices in secondary school. *International Journal of Science Education, 27*(4), 471–486. https://doi.org/10.1080/0950069042000323746

DeWitt, J., Osborne, J., Archer, L., Dillon, J., Willis, B., & Wong, B. (2013). Young children's aspirations in science: The unequivocal, the uncertain and the unthinkable. *International Journal of Science Education, 35*(6), 1037–1063. https://doi.org/10.1080/09500693.2011.608197

Gee, J. P. (2000). Identity as an analytic lens for research in education. *Review of Research in Education, 25*(1), 99–125. https://doi.org/10.3102/0091732X025001099

Gee, J. P. (2010). *An introduction to discourse analyses. Theory and method* (3rd ed.). Routledge. https://doi.org/10.1016/0346-251x(88)90022-x

Gee, J. P. (2011). *How to do discourse analysis: A toolkit.* Routledge. https://doi.org/10.4324/9781315819662

Hannover, B., & Kessels, U. (2004). Self-to-prototype matching as a strategy for making academic choices. Why high school students do not like math and science. *Learning and Instruction, 14*(1), 51–67. https://doi.org/10.1016/j.learninstruc.2003.10.002

Hazari, Z., Sonnert, G., Sadler, P. M., & Shanahan, M.-C. (2010). Connecting high school physics experiences, outcome expectations, physics identity, and physics career choice: A gender study. *Journal of Research in Science Teaching, 47*(8), 978–1003. https://doi.org/10.1002/tea.20363

Helfferich, C. (2011). Die Qualität qualitativer Daten. *In Manual für die Durchführung qualitativer Interviews.* Wiesbaden: VS Verlag für Sozialwissenschaften. https://doi.org/10.1007/978-3-531-92076-4.

Heublein, U., & Schmelzer, R. (2018). *Die Entwicklung der Studienabbruchquoten an den deutschen Hochschulen. Berechnungen auf Basis des Absolventenjahrgangs* 2016. DZHW-Projektbericht, German Centre for Higher Education Research and Science Studies.

Holm, L., Ekström, M. P., Hach, S., & Lund, T. B. (2015). Who is cooking dinner? Changes in the gendering of cooking from 1997 to 2012 in four Nordic countries. *Food, Culture & Society, 18*(4), 589–610. https://doi.org/10.1080/15528014.2015.1088191

Holmegaard, H. T., Madsen, L. M., & Ulriksen, L. (2012). To choose or not to choose science: Constructions of desirable identities among young people considering a STEM higher education programme. *International Journal of Science Education, 36*(2), 1–30. https://doi.org/10.1080/09500693.2012.749362

Holmegaard, H. T., Ulriksen, L. M., & Madsen, L. M. (2014). The process of choosing what to study: A longitudinal study of upper secondary students' identity work when choosing higher education. *Scandinavian Journal of Educational Research, 58*(1), 1–20. https://doi.org/10.1080/00313831.2012.696212

Höttecke, D., & Allchin, D. (2020). Reconceptualizing nature-of-science education in the age of social media. *Science Education, 104*(4), 641–666. https://doi.org/10.1002/sce.21575

Hsu, P.-L. L., Roth, W.-M. M., Marshall, A., & Guenette, F. (2009). To be or not to be? Discursive resources for (dis-)identifying with science-related careers. *Journal of Research in Science Teaching, 46*(10), 1114–1136. https://doi.org/10.1002/tea.20352

Kane, J. M. (2012). Multiple identities and the science self: How a young African American boy positions himself in multiple worlds. In M. Varelas (Ed.), *Identity construction and science education research. Learning, teaching, and being in multiple contexts* (pp. 27–42). Sense Publishers.

Kessels, U., Rau, M., & Hannover, B. (2006). What goes well with physics? Measuring and altering the image of science. *The British Journal of Educational Psychology, 76*, 761–780. https://doi.org/10.1348/000709905X59961

Kruse, J. (2015). *Qualitative Interviewforschung. Ein integrativer Ansatz.* Beltz Juventa.

Lee, Y.-L. (2012). Identity-based research in science education. In B. J. Fraser, K. Tobin, & C. J. McRobbie (Eds.), *Second international handbook of science education* (pp. 35–45). Springer International Handbook of Science Education. https://doi.org/10.1007/978-1-4020-9041-7

Loeken, M. (2015). When research challenges gender stereotypes: Exploring narratives of girls' educational choices. In E. K. Henriksen, J. Dillon, & J. Ryder (Eds.), *Understanding student participation and choice in science and technology education* (pp. 277–295). Springer.

Lucius-Hoene, G., & Deppermann, A. (2002). *Rekonstruktion narrativer Identität.* VS Verlag für Sozialwissenschaften.

Lykkegaard, E., & Ulriksen, L. (2016). Role model and prototype matching: Upper-secondary school students' meetings with tertiary STEM students. *Nordina, 12*(1), 73–89. https://doi.org/10.5617/nordina.1209

Lyons, T., & Quinn, F. (2010). *Understanding the declines in senior high school science enrolments.* University of New England. http://www.une.edu.au/simerr/pages/projects/131choosingscience.pdf

Morf, C. C., & Koole, S. L. (2014). Das Selbst. In K. Jonas, W. Stroebe, & M. Hewstone (Eds.), *Sozialpsychologie* (pp. 141–195). Springer.

Nohl, A.-M. (2009). *Interview und dokumentarische Methode.* Verlag für Sozialwissenschaften.

Osborne, J., Simon, S., & Collins, S. (2003). Attitudes towards science: A review of the literature and its implications. *International Journal of Science Education, 25*(9), 1049–1079. https://doi.org/10.1080/0950069032000032199

Pohlmann, B., & Möller, J. (2010). Fragebogen zur Erfassung der Motivation für die Wahl des Lehramtsstudiums (FEMOLA). *Zeitschrift für Pädagogische Psychologie, 24*(1), 73–84. https://doi.org/10.1024/1010-0652/a000005

Przyborski, A., & Wohlrab-Sahr, M. (2014). *Qualitative Sozialforschung. Ein Arbeitsbuch.* Oldenbourg Verlag.

Rabe, T., & Krey, O. (2018). Identitätskonstruktionen von Kindern und Jugendlichen in Bezug auf Physik – Das Identitätskonstrukt als Analyseperspektive für die Physikdidaktik? Identity constructions related to physics – The identity construct as an analytic lens for physics education. *Zeitschrift Für Didaktik Der Naturwissenschaften, 24*(1), 201–216. https://doi.org/10.1007/s40573-018-0083-x

Regan, E., & DeWitt, J. (2015). Attitudes, interest and factors influencing STEM enrolment behaviour: An overview of relevant literature. In E. K. Henriksen, J. Dillon, & J. Ryder (Eds.), *Understanding student participation and choice in science and technology education* (pp. 63–88). Springer.

Shanahan, M.-C. (2009). Identity in science learning: Exploring the attention given to agency and structure in studies of identity. *Studies in Science Education, 45*(1), 43–64. https://doi.org/10.1080/03057260802681847

Sjaastad, J. (2012). Sources of inspiration: The role of significant persons in young people's choice of science in higher education. *International Journal of Science Education, 34*(10), 1615–1636. https://doi.org/10.1080/09500693.2011.617935

Smith, B., & Sparkes, A. C. (2008). Contrasting perspectives on narrating selves and identities: An invitation to dialogue. *Qualitative Research, 8*(1), 5–35. https://doi.org/10.1177/1468794107085221

Statistisches Bundesamt. (2019). Studienanfänger WS 2018/19: Deutschland, Semester, Nationalität, Geschlecht, Studienfach. Retrieved from https://www-genesis.destatis.de/genesis/online

Stokking, K. M. (2000). Predicting the choice of physics in secondary education. *International Journal of Science Education, 22*(12), 1261–1283. https://doi.org/10.1080/095006900750036253

Svašek, M., & Domecka, M. (2012). The autobiographical narrative interview: A potential arena of emotional remembering, performance and reflection. In J. Skinner (Ed.), *The interview. An ethnographic approach* (pp. 107–126). Bloomsbury.

Tytler, R. (2014). Attitudes, identity, and aspirations toward science. In N. G. Lederman & S. K. Abell (Eds.), *Handbook of research on science education* (Vol. 2, pp. 82–103). Routledge. https://doi.org/10.4324/9780203097267

Whitehead, J. M. (1996). Sex stereotypes, gender identity and subject choice at A-level. *Educational Research, 38*(2), 147–160. https://doi.org/10.1080/0013188960380203

Chapter 5
Student Identity, Aspiration and the Exchange-Value of Physics

Billy Wong

In English schools, the subject physics is the least popular of the traditional sciences. Despite an upward trend in uptake in the last 10 years, biology and chemistry typically have 50% more students than physics at A level, the main post-16 qualification used for university entry (Joint Council for Qualifications [JCQ], 2020a). The comparative lack of physics students has been a national concern in the past decade, contributing to wider debates of the 'leaky science pipeline' (and the 'rigged bingo game', see Archer, 2019), especially for students from underrepresented backgrounds (e.g., girls and some ethnic minorities, see Lykkegaard & Ulriksen, 2019).

In this chapter, the notion of science identity is explored through the concept of *symbolic exchange*, which recognises that our decisions and choice of consumptions are bounded up with status and power. Drawing on sociological theories of social reproduction, the chapter highlights the added value of studying science for school students, especially in physics and amongst high achievers. Empirical data from 42 A level Physics students (aged 16–18) in England inform the discussion, with the focus on student identity and aspirations in physics. This chapter highlights examples of student decisions to study advanced-level physics due to its exchange-value, symbolic identity and the pull factors of physics. The chapter concludes by discussing the potentials of the concept of symbolic exchange for research into young people's science and physics trajectory, especially in the context of identity.

B. Wong (✉)
University of Reading, Reading, UK
e-mail: b.wong@reading.ac.uk

H. T. Holmegaard, L. Archer (eds.), *Science Identities*, Contributions from Science Education Research 12, https://doi.org/10.1007/978-3-031-17642-5_5

5.1 Studying A Level Physics

In England, the study of A level (officially known as the General Certificate of Education Advanced Level) is the most common qualification for students aged between 16 and 18. Students typically study for three subjects and their outcomes can be used for university admission. High-tier universities typically require students to have an A level grade set of AAB or above. Some degrees also specify or have preferences in particular A level qualifications, such as A level mathematics in the study of a physics degree (although A level physics is rarely a requirement to study for a mathematics degree). As such, A level qualifications are similar to university entrance exams in other parts of the world (e.g., Gaokao in China; SAT or ACT in the US), albeit students' degree options may already be restricted by the type of subject qualifications acquired in school, as explained below.

It is compulsory for students in England to study science in primary and secondary schools (between ages 5- to 16), which tend to include the sub-disciplines of biology, chemistry and physics. At GCSE – a qualification normally taken by students aged 14–16 – these sciences can be studied as three individual subjects, through what is commonly known as Triple Science. The most popular science at GCSE is Double Award, which is taken by c. 880,000 students in 2020 (compared to c. 170,000 for Triple Award, see JCQ, 2020b), where the sub-disciplines of science are taught together.

Students who take Triple Science are typically high-achieving students and are more likely to be from socially advantaged backgrounds, which raises concerns about the diversity, equality and inclusivity of the subject (Reiss, 2000). Furthermore, school students often have little or no choice in their science education pathways, which are bounded by school structures, policies and priorities (Archer et al., 2017a). For example, some schools – due to external factors such as league tables and school rankings – may restrict which of their students are 'eligible' to study advanced sciences; most likely based on students' academic profile, such as their existing subject choices and attainments (e.g., at GCSE, typically at age 14). An appreciation of the GCSE system is therefore necessary to understand the challenges of students to study post-compulsory physics, especially at A level (ages 16–18).

Most A level subjects, especially the sciences, demand students to have prerequisite knowledge, as evidenced in their relevant grades at GCSE or equivalent. It is common for schools to only recommend (and in practice, limit) students with certain grades to continue at A level (to strengthen the success rate).

Given that physics (and chemistry and biology) only exist as an individual subject at GCSE under Triple Science, the pool of potential A level science students is already smaller, although students with a strong outcome in GCSE Double Science may also be eligible. In any case, A level physics seems to attract the fewest of students in the traditional sciences. In 1976, over 40,000 students took A level physics and the numbers peaked at over 53,000 in 1982. By 2007, there were fewer than 24,000 students and a gradual upward pattern then emerged and by 2014, the figures were just over 32,000 (Wong, 2016b). In 2020, student numbers have risen to 38,000 (compared to 65,000 for biology, 56,000 for chemistry and 94,000 for mathematics, see JCQ, 2020a), which is still the least popular traditional sciences at A level.

Over the years, researchers have investigated the reasons for the low participation rate of physics in post-compulsory education, especially for females (Hazari et al., 2010; Sax et al., 2016). Girls constitute just one in five A level physics student, whilst there are often balanced if not more girls in the study of biology and chemistry than boys (JCQ, 2020a). Concerns about lack of girls in physics has been studied extensively, especially by the UK's Institute of Physics (IOP) and their reports under the 'Gender Balance' research theme (e.g., IOP, 2013, 2015; Murphy & Whitelegg, 2006). These studies and others highlight a range of challenges experienced by girls in physics, including the environment of physics as overly masculine, gender stereotypes and the subject being seen as difficult, unrelatable and 'not for me'. Girls in single-sex schools, however, are two-and-a-half times more likely to study A level physics than girls in co-educational schools (IOP, 2011), which further highlights the issue of gender in school structures and systems as barriers for further physics education.

Research from the ASPIRES2 project reported that A level physics student are most likely to have done Triple Science at GCSE, be a high achiever/in the 'top set' in science (and mathematics), have high levels of cultural and science capital and have family members working in science (Archer et al., 2017b; Francis et al., 2017). Existing studies suggest that the key reasons for students to choose to study A level physics was due to their experience of enjoyment of the subject, perceived usefulness of the subject for future aspirations and the assumed intellectual identity afforded by the difficulty of physics (DeWitt et al., 2018; Gill & Bell, 2013; Mujtaba & Reiss, 2016).

As with most subjects, an intrinsic interest in the discipline is central to support and reinforce motivation and enthusiasm for study and learning (Mujtaba & Reiss, 2014). For physics, such interest can extend to extrinsic factors, especially the added value of the discipline. For example, the inclusion of physics as one of 'facilitating subjects' for university application (Russell Group, 2018) has meant that the qualification of A level physics has an added value for entry into higher education (although a broader range of subjects were recommended in their relaunched 'informed choices' website, see informedchoices.ac.uk). Popular stereotypes and perceptions of science and scientists as people of high intelligence can also appeal to particular individuals (DeWitt et al., 2013; Wong, 2012), even though such identities can also promote discourses of science as 'not for people like me'.

At A level, students tend to study fewer subjects as preparation for specialism, especially at university. This chapter focuses on students who are studying A level physics. The decision to study A-level physics is likely to reflect students' aspirations and identities, especially their intrinsic and extrinsic motivations. More specifically, I am interested in the symbolic/exchange values afforded by physics, which I argue is important for understanding the concept of science identity.

5.2 Science Identity, Capital and Symbolic Exchange

My approach to science identity is informed by sociological theories and literature, where identity as a concept is understood as a social construction. Informed by key writers such as Stuart Hall (1990) and Judith Butler (1990), identity is conceptualised as something fluid, relational, 'always in process' and entangled within complex relations of power (e.g., structural inequalities of gender, class and 'race'/ethnicity). In other words, identity can be thought of as a continuous project that constitutes an ongoing process of negotiation within multifaceted structural and agentic relationships. For individuals, our identities are therefore developed and performed in negotiations with dominant identity discourses (Lawler, 2007). In science education research, the lens of identity can offer us an insight into the nuances of science participations and science aspirations, especially the symbolic values and identities for studying advanced sciences such as physics.

The concept of science identity, including identifications with science, has gain popularity in the last decade or so. Notably, Carlone and Johnson (2007) posited that a sustainable science identity within the classroom would require recognition by self as well as by others, such as teachers and peers (Hazari et al., 2017; Hazari et al., 2015). Consistent with Hall (1990) and Butler (1990), an identity in science – or science identity – would therefore require approval, recognition and acknowledgement from members of the immediate environment, otherwise the claim to such identity is unsustainable and temporary. With a stronger focus on individual agency, Calabrese Barton et al. (2013) also explored the concept of *identity work* to tease out the complex ways in which students navigate their identities across different science learning contexts (e.g., inside or outside the classroom, in pairs, groups or whole-class discussions). Here, identity work focuses on the processes of identity formation and how individuals construct their sense of selves within different learning spaces (Carlone et al., 2011).

In understanding young people's science aspirations, the concept of science identity is also prominent, especially in appreciating the extent to which careers in or from science are considered as feasible, plausible and desirable for 'people like me' (Archer & DeWitt, 2017; Wong, 2015). For instance, people in science are often stereotyped as intelligent but socially inept in popular media (Chimba & Kitzinger, 2010), as exemplified in CBS's popular American sitcom The Big Bang Theory. The work of scientists is typically presented as intensive and laborious, with long working hours and minimal social life (Masnick et al., 2010). The portrayal of scientists in science-fiction movies such as the classic trilogy, Back to the Future, strikes a high resemblance to Albert Einstein, one of the most recognisable scientists of our time. Einstein's image as a white old man with a distinctive hairstyle dressed in long laboratory coat is arguably synonymous with being a scientist. Such ingrained images and ideas have, for a long time, reinforced a particular (and mostly exclusive) identity of science for prospective scientists and science students.

An identity in science can therefore be quite particular and peculiar, in the sense that people who associate with science, such as a science student, a science enthusiast or a science professional, can also attract stereotypes about who they are (i.e., their identity), be it positive and negative. In schools, students who study or aspire to study more advanced level sciences may find the study of science attractive because it symbolises intelligence (Wong, 2012), especially since students who are good at science are popularly regarded by their peers as clever and brainy (DeWitt et al., 2013). Here, an identity in science can be paralleled with a symbolic identity because of what associations with science can also represent, such as intelligence. These students, which I have previously termed as *science extrinsic* students (Wong, 2016a), would be interested in science such as GCSE Triple Science and A-level physics, biology and chemistry because these qualifications are recognised to have an added value, especially for university entry applications, even if their own interests or aspirations in science are minimal. Similarly, students may avoid or reject an identification with science due to the very same intellectual stereotype, which can include descriptors such as nerds, geeks and boffins, and consider these characteristics to be inconsistent with the identities that students wish to embody or be recognised (e.g., 'interesting, but not for me', see Archer et al., 2010). It is therefore important to consider the symbolic and additional meanings of what associations with science can mean for different students as part of our understandings of their science identities.

Underlying these student choices and identifications with sciences, there seems to be value in delving into the semiotics of science education, and more specifically, the symbolic/exchange value of studying science. I therefore suggest that in thinking about science identity, or an identity in science, it is important to consider the exchange value of science, especially in the context of science aspirations. Jean Baudrillard (1981), who extends Marx's (1867) writing in the sociology of capitalism and consumption, extensively discusses the concept of symbolic exchange (or *exchange value*, as advocated by Bourdieu, 1977, see below). In short, Baudrillard argued for the importance of recognising the exchange value of a commodity as much as the use value, which is the directly applicable or obvious purpose of the commodity. Exchange value constitutes the symbolic and added value of the commodity, especially what it means to have acquired such commodity. An example could be buying a t-shirt – the use value is to provide warmth and something to wear, and the exchange value refers to what is means to buy/wear that t-shirt, be it the brand, the design, the materials or the production process (e.g., ethical or sustainable). An expensive or branded t-shirt may signify that the owner is wealthy or fashionable, although it could also be received negatively, such as pretentious or materialistic. Individuals, with their own values and resources, will interact with brands in ways that reflect their own identity. The idea of exchange value is applicable for most commodities, such as cars, smartphones and various lifestyle choices (and therefore *taste*), which invites us to consider the thinking tools of Pierre Bourdieu (1977, 1984) and his theory of practice.

In brief, Bourdieu's work (1977, 1984, 1986) arguably builds and refines on the ideas of Marx, with a focus on the social reproduction of inequalities, by socioeconomic background, through the intergenerational transmission and embodiment of dispositions (habitus) and resources (capital) that ensures the privileged maintains their privileges across societies (field) and generations. In particular, Bourdieu's (1986) theory of capital is in alignment with Baudrillard's symbolic exchange, although Bourdieu also conceptualised capital with respect to economic, social and cultural dimensions. Most importantly, Bourdieu considered capital as exchangeable resources, where possession of one capital can lead to another, including symbolic capital, which is understood as 'the acquisition of a reputation for competence and an image of respectability and honourability' (Bourdieu, 1984, p. 291).

In science education research, Bourdieu's concepts are increasingly popular, especially with the concept of *science capital* (Archer et al., 2015; Wong, 2016b) that considers the breadth and depth of science-related resources that individuals, especially young people, can access and deploy in support of their science trajectories. Students with high levels of science capital are more likely to identify with science (through expressions of science careers aspirations and high attainment) than those with medium or low science capital (see also Du & Wong, 2019 in the Chinese context). Recalling that capital is all exchangeable resources, capital that are related to science (i.e., science capital) will also have symbolic values.

With this in mind, the consideration of symbolic exchange within the lens of science identity has the potential to make an important contribution to the understanding of the aspirations and decisions that young people make about their science pathways, especially in the study of advanced sciences such as A level physics. For instance, what is it about A level physics that appeals to students? Why kind of science identities do A level physics students embody?

5.3 The Study

This chapter draws on data collected in 2018 for an evaluation research interested in the ways that contemporary A level physics students would like to be supported in their learning of physics, in and out of school. Supported by IOP – a UK-based professional body and learned society for physics, the project explored the views, experiences and ideas that these young physicists have about their science pathways, including aspirations toward advanced science education and future careers. The evaluation itself focused on the IOP's membership system and engagement strategy to support A level physics students.

Empirical data comprised of 42 A level physics students from two major English cities (London and Manchester), who participated in seven focus groups. The method of focus group was utilised to promote discussions and interactions between participants to highlight, debate and reflect on similar and different views, expectations and aspirations about A level physics and beyond (Krueger & Casey, 2014). In other words, the emphasis here is less on the individual, but their collective views and opinions. Through purposeful sampling, students were recruited using the IOP's local network of schools, where a call for participant was emailed and distributed to A level physics students through physics teachers or the school. Interested students were then informed of the date, time and location of the research. The research was conducted at local community centres that are reachable by public transport. Student participants travelled to these respective sites with their travel costs reimbursed.

Whilst recruitment was passive in that participants would actively sign up to take part, the recruitment process was mindful of the importance to include a diverse range of students to appreciate different views and perspectives, rather than a proportional representation of existing figures for A level students, given it is a male-dominated subject. Of the 42 focus group students, an equal number of girls ($n = 21$) and boys ($n = 21$) were participants, which was coincidental despite an awareness on gender balance. Students came from 11 different schools, including one private girls-only school to boost the number of girls in the study ($n = 12$). Our students are ethnically diverse, with 12 self-identified as White, six as Black, nine as Asian, five as Middle Eastern, four as Chinese and six as mixed or with multiple ethnicities. Based on their GCSE or equivalent grades, the A level physics students recruited in this study are considered as high achievers. The vast majority of our students had studied GCSE Triple Science, with a handful who completed the more popular (but less demanding) Double Science.

Each focus group lasted an hour and a half on average and was audio-recorded, with the data transcribed verbatim and more sensitive details anonymised. All names of students are pseudonyms. The size of each group was between five to seven students and I conducted two girls-only group (FG1, FG2), two boys-only group (FG3, FG7) and three mixed-sex groups (FG4, FG5, FG6). The composition of each focus group was pragmatic and primarily based on the availability of students. Most groups included students from at least two schools, although the two girls-only focus groups were comprised of students from the private school for girls. Participants were prompted to discuss their views and experiences about science, especially their reasons for studying physics at A level and the support available. There were further questions designed to evaluate the IOP's membership system for young physicists, which is not part of this analysis. As an observation, most students in the focus groups were open to share their thoughts about A level physics and their trajectories so far in post-compulsory science education.

Data analysis is informed by a social constructionist perspective which recognises social phenomena as socially constructed and discursively produced (Burr, 2003). Focus group data were managed and organised using the software NVivo. Initial codes were created through the identification of relevant themes that emerged in the initial stages of data analysis where there is a 'back and forth' movement between the data and analyses in an iterative process through which the dimensions of concepts and themes were refined or expanded through the comparison of data (Corbin & Strauss, 2014). These codes were subject to an iterative process of gradual coding refinement, with the themes being revised with emerging research data and further coding.

In this chapter, the lens of science identity is used to interpret the data, including the concepts of exchange value and symbolic identity. Here, focus group discussions and narratives of students' views and experiences of science and physics were analysed as potential expressions, indications and performances of identities in science. Furthermore, these interpretations also considered the possibilities of symbolic exchange in students' identity and aspirations in science, which shed further light into the extrinsic value of physics. In other words, what are the added meanings of being a physics student?

5.4 The Exchange Value of Physics

In English schools, science is usually taught as an overarching subject that encompasses a range of sub-disciplines. Typically, it is not until at GCSE (age 14), via Triple Science, where the individual sciences of biology, chemistry and physics can be studied as distinctive science subjects. Arguably, this stage is where students begin to explore more specialised science identities in school, which is further developed at A level. To appreciate the symbolic exchange and complexities of a physics science identity, it is important to explore why students study A level, what aspirations these students have and the extent to which they identify with physics.

From the data, it is no surprise to report that students who chose to study A level physics said they did so because of their interest in the discipline, contrary to other research (Mujtaba & Reiss, 2016) which noted the higher importance of extrinsic motivation. As also reported by DeWitt et al. (2018), fascinations with the scientific discoveries, concepts and applications of physical science are echoed by many students when asked about the reasons for their choice of study:

> I really enjoy physics because everything that I've ever done and, it sounds cheesy … Has been a direct influence of physics. So, I'm drinking a cup of coffee now, and when I poured it, it was too hot. But thanks to thermodynamics, now it's fantastic to drink. (Nathan, 17, White male, FG7)

> I feel like physics explains life, in a sense. I feel like it's more than just a subject. Physics is literally everything around us, and that's the best part. You look at something and, oh my gosh, this is staying down or using Newton's Laws, all that stuff. It's just so binding, and follow everything, and it somehow connects to everything. (Lydia, 17, Black female, FG5)

Existing studies agree that the capacity of young people to appreciate science and physics in their everyday lives is likely to increase and strengthen their scientific interest, aspirations and even school attainment (Archer & DeWitt, 2017; Archer et al., 2020; Godec et al., 2017). Whilst these intrinsic interests are critical, the appeal of A level physics extends beyond the content of the discipline, with ample evidence to infer that the study of physics is also recognised by students to have added, or exchange, value.

5.4.1 Careers 'From' Physics

Our students clearly recognise that A level physics is a 'facilitating subject' (Russell Group, 2018) that would be particularly useful for university application and future careers, even though few have expressed an interest in a physics career (DeWitt

Table 5.1 A level physics students' primary career aspiration

Careers mentioned	Girls	Boys	Total
Engineering	8	12	20
Medicine	3	0	3
General science	3	0	3
Physics	2	1	3
Finance/Banking	0	2	2
Architecture	1	0	1
Geology	1	0	1
Sports science	0	1	1
Undecided	3	5	8
Total	21	21	42

et al., 2018). Instead, most wanted a career 'from' physics (Wong, 2015), especially in engineering.

Table 5.1 presents the primary career aspirations that were mentioned by our students. Most expressed an interest in a STEM career and just under half (20 out of 42, 48%) explicitly stated engineering as their main career goal (which includes careers in aerospace, bio, civil, mechanical and structural engineering). While our recruitment strategy certainly contributed to the type of students (i.e., A level physics) who self-selected to participate, it is rather concerning that only three students (out of 42, or 7%) appear interested to pursue a career 'in' physics, and these students appear to be driven by their intrinsic interests and motivations. Six students also stated a secondary career aspiration, which include working the fields of physics ($n = 2$), general science ($n = 1$), neuro-technology ($n = 1$), radiotherapy ($n = 1$) and economics ($n = 1$). For most of our A level students, their study of physics was understood to be a process or gateway into their desired career pathway:

> I want to go into engineering or something like that, so obviously you need to do physics. (Abby, 16, White female, FG1)

> The basic principles of engineering are based around physics and maths, so physics will be central to everything I want to do in engineering. (Larry, 17, White male, FG7)

Students with an engineering career aspiration clearly recognised the relevance of A level physics, as well as mathematics, to be essential subjects for a career in engineering, which is unavailable as a school subject. In fact, all but one student (41 out of 42, or 98%) were also studying A level mathematics, which was mentioned by some to be a requirement in their schools in order to study physics (see also Kemp et al., 2018, p. 131). The other popular subjects are A level chemistry and A level further mathematics, which are studied by 19 (or 42%) and 15 (or 36%) of the 42 A level physics students, respectively.

For students with career aspirations in medicine, general science, finance/banking, architecture, geology and sports science, the subject physics is generally considered as a useful or complementary subject. For those with undecided or unknown career aspirations, their reasons for studying A-level physics were due to personal interest, but also because they believed they were 'good' at the subject, especially at GCSE (age 14–16). I return to the symbolic identity of being good, or intelligent, in the next section.

When probed to discuss why few young people appear to find a physics career attractive, it was suggested that the subject itself lacked a strong and distinctive identity. For instance, physics is seen, by some students at least, to be a 'supporting' subject, rather than as a 'mainstream' discipline such as mathematics and engineering. Hannah (16, White British female, FG2) explains that:

> Even if you do like a physics question … So much of it is sometimes just maths. [Physics] doesn't seem like it has much of an identity. It is just like, "Oh, this is sort of like maths with a few extra bits"

In a related vein, a handful of students was also uncertain about the range of career options available for those with a physics specialism:

> I don't really know what else I'm going to do with physics, other than the subjects I chose … I looked for other professions and there's not that many linked with physics and other subjects. (Mark, 16, Black British male, FG3)

However, when asked to envision the role of physics in their futures, such as in five or ten years, most students were positive that their knowledge of physics would play a useful part in their professional or personal lives. Here, most students acknowledge that physics is a versatile subject, especially for employment:

> Physics has a lot, it can open a lot of doors because it can, you can use it in a lot of different kind of areas. (Stephen, 17, White male, FG3)

> It opens many doors to you because if you get good grades in physics people see that you have high abilities, like you can cope with anything really. (Maha, 16, Middle Eastern female, FG4)

Our students appear to appreciate that the study of A level physics has symbolic exchange (Baudrillard, 1981), or added value, as the subject can be a pathway into other careers, especially engineering. The study of physics also infers a status or perception as a competent person, which is explored below.

5.4.2 An Intelligent Identity

Several students, including those with science career aspirations, attributed their impressive GCSE grade in science as a key validation of their academic ability. To put it more bluntly, it appears that some students took A level physics because they can, as evidenced by their previous attainment. These students may be strategic in their subject choices, with rational rather than sentimental decisions, to ensure that their academic trajectories are guided by the probabilities of best possible outcomes:

> I would never pick a subject that I'd enjoy but I'm not good at, if that makes sense. There's a lot of subjects which I might enjoy, but I just don't get good grades in so I wouldn't pick

> it. When you pick subjects, it's not really what you enjoy. It's what you're good at. (Luke, 16, Black male, FG3)

> I did it [A-level physics] because in GCSE [Triple Science] I liked it, and I was good at it, and that's the only reason. (Onnika, 17, Middle Eastern female, FG5)

For students such as Luke, the decision to study A level physics is based on perceptions of their academic ability to perform and achieve ("It's not really what you enjoy. It's what you're good at"). This reason can also be used to explain the comparative lack of A level physics students, who may be dissuaded by the belief that they are not good enough to do physics, especially since physics is popularly regarded as a hard and difficult subject. As such, being good at or achieving high grades in physics can entail a symbolic identity that is unavailable in most other subjects, namely as academically competent or even as clever (DeWitt et al., 2013), especially from peers and even family or community members (Wong, 2012). For example:

Yasmine: When we tell people out of school, like family or something they're like, "Wow, you must be really clever because you take maths and physics".
Lorraine: Like, everyone thinks it must be such a hard subject to take.
Yasmine: Yeah. (FG1)

For other students, this (mis)recognition from others on physics being a difficult subject can also have a positive influence on their identity, especially as a smart or intelligent person:

Researcher: And how does that [physics seen as a hard subject] make you feel?
Gary: Yeah. It makes me feel better.
Leah: Definitely.
Researcher: In what ways?
Edward: Makes you feel smart, like you're doing something that other people really can't do.
Researcher: Does it make everyone else feel smart as well?
All: Yeah. (FG5)

As can be seen, some students recognise and even feed off recognitions from others that the study of A level physics entails intelligence and cleverness. Studying physics offers student a plausible identity as a genius, at least from peers ("If you take physics, they'll assume you're a genius", Markus, 16, White male, FG7), especially from those who have struggled with science and physics themselves. While A level physics students felt the subject physics is less intimidating than it appears, they believe it is one of the harder subjects, and our students generally agree that the symbolic identity afforded by physics as people of high intelligence is an appropriate and attractive identity to embody.

5.4.3 Girls in Physics: Has the Tide Turned?

Across the seven focus groups, our students generally acknowledge the added challenge for girls to persevere in science and physics, citing the socialisations and expectations derived from traditional gendered discourses, including the lack of female role models (Archer et al., 2012, 2013).

The boys-only groups (FG3, FG7) reasoned that fewer girls study physics because "Girls are more into" biology and chemistry, with science-related career aspirations more likely to be in medicine than in engineering or physics, as the latter is more of "A boyish thing to do" and physics "Nobel prize winners are mostly men".

The girls-only groups (FG1, FG2), all from a private school for girls, also recognised the barriers from traditional gendered discourses, although these were interpreted as more historical. Instead, the girls suggest that "The tide has now turned" for women in science, driven by a number of local/national initiatives, as well as the international #metoo movement against sexual harassment and assault. The girls said that their school has been actively promoting physics, with an attractive package of support from field visits to attendance of external lectures to specialised resources in school.

Some girls saw the recent emphasis of the STEM sector as a whole to promote and encourage women to participate in the field as a unique opportunity on which they ought to capitalise, with ever more and even exclusive science-related opportunities (e.g., placements/internships) that are only available to girls. It seems that these girls are highly aware of their current (even if temporary) advantage to maximise the opportunities afforded by this 'gender turn' which celebrates and promotes women in STEM, especially the appreciation of female in traditionally male-dominated fields such as physics.

Some of the state-school girls in the mixed focus groups (FG4, FG5 & FG6) also articulated similar views, with one speaking about "Being a girl who is kind of successful in physics… [Is a] kind of empowering because it's like you want to prove people wrong" (Alison, 16, Middle Eastern girl, FG4). Here, the symbolic meaning for girls to be successful in STEM seems to offer these A level physics girls an added sense of pride, through their supported challenge of the dominant and gendered discourses of physics (see also Avraamidou, 2020). The study of A level physics therefore provided some girls with a powerful, meaningful and even noble identity in science/physics.

5.5 The Symbols of an Identity in Science

In this chapter, I have explored the reasons behind why students study A level physics, with a focus on science identity and the concept of symbolic exchange (Baudrillard, 1981). The student identities of those who study advanced physics are not just those of 'any' student, but specifically physics students, which, when interpreted through the lens of symbolic exchange, encompasses extrinsic and external meanings and implications. Students who study A level physics are mostly aware of the high exchange value afford by the subject for university applications (Russell Group, 2018), future career options and even a heightened sense of self-belief. Young people may make study choices as a result of intrinsic interest and motivation, but it is evident from the data that these decisions, especially for post-compulsory education, can also be strategic and symbolic. Referring to one of the five 'types' of science participation I identified in a previous study with minority ethnic young people (blinded), students who are science extrinsic will continue with their science education as long as the subject is considered to be useful for their future educational and career plans.

The lens of science identity enables us to take a deeper understanding of young people's trajectories in science, by considering the type of person that one wants to become. Studying A level physics (and beyond) is therefore more than just the learning and understanding of physics content and concepts; it is also the recognition and gradual embodiment of the cultures of science and physics. This brings us back to the question of what it means to be a scientist, especially a physicist, for those who are studying post-compulsory physics. Is their choice driven only by scientific curiosity and fascination? Or, is the symbolic exchange of A level physics, with the various added valued as already discussed, an important factor of consideration? The evidence suggests both, which means our understanding of science identities, especially for young people, ought to consider what the perceived added value and extrinsic meaning of doing, studying or working in/from science means for the individual, as well as perceptions of them from others (Carlone & Johnson, 2007).

If we wish to promote further access and participation in advanced level physics, either for study or as a career, then perhaps the added values of physics – be it financial or personal prestige (i.e., symbolic) – ought to be marketed more strongly by educators and employers, rather than to wait for future physicists to emerge purely from their own intrinsic interests. Whilst this approach may not be as appealing as those where participation is nurtured and supported through students' scientific interests, it is a pragmatic method that should not be completely dismissed as we must try and explore a range of different ways to initiate the shift in the existing imbalances of physics.

It is also important to appreciate that whilst the symbolic meanings and identities of studying physics can be positive, with various added values, these symbolic identities can be a negative, especially when associated with undesirable traits or values that appear inconsistent with the self-identities of individuals. These could be stereotypes of scientists/physicists as socially inept and eccentric, or the 'chilly climate' of physics (perceived or actual) for girls and/or minority ethnic groups in a traditionally masculine and white field (Aikenhead, 1996; Blickenstaff, 2005). As such, an identity in physics can certainly be exclusive and exclusionary, and so the concept of symbolic exchange can be positive with added value, as well as negative and 'subtracted' value, when we consider the symbolic meanings and cultural understandings/interpretations of what an identity in physics can potentially mean.

So, what does this mean for science identity research? By highlighting the importance of exchange value and symbolic identity, future studies informed by the lens of science identity need to ensure that external and extrinsic factors are thoroughly considered when attempts are made to interpret and make sense of the science educational trajectories and career aspirations of young people. Thus, the concept of symbolic exchange should be considered as a part of our understandings into how individuals identify (by self or by others) with science, in recognition of the added meanings of studying or working in science, especially physics.

References

Aikenhead, G. S. (1996). Science education: Border crossing into the subculture of science. *Studies in Science Education, 27*(1), 1–52. https://doi.org/10.1080/03057269608560077

Archer, L. (2019, May 16). *Identities, inequalities, education: Sociology for social justice with youth in (and out) of school* (UCL-IOE Professorial Public Lecture Series).

Archer, L., & DeWitt, J. (2017). *Understanding young people's science aspirations*. Routledge.

Archer, L., DeWitt, J., Osborne, J., Dillon, J., Willis, B., & Wong, B. (2010). 'Doing' science versus 'being' a scientist: Examining 10/11-year-old schoolchildren's constructions of science through the lens of identity. *Science Education, 94*(4), 617–639. https://doi.org/10.1080/09500690902759053

Archer, L., DeWitt, J., Osborne, J., Dillon, J., Willis, B., & Wong, B. (2012). 'Balancing Acts': Elementary school girls' negotiations of femininity, achievement and science. *Science Education, 96*(6), 967–998. https://doi.org/10.1002/sce.21031

Archer, L., DeWitt, J., Osborne, J., Dillon, J., Willis, B., & Wong, B. (2013). Not girly, not sexy, not glamorous': Primary school girls' and parents' constructions of science aspirations. *Pedagogy, Culture and Society, 21*(1), 171–194. https://doi.org/10.1080/14681366.2012.748676

Archer, L., Dawson, E., DeWitt, J., Seakins, A., & Wong, B. (2015). 'Science capital': A conceptual and empirical argument for extending Bourdieusian notions of capital beyond the arts. *Journal of Research in Science Teaching, 52*(7), 922–948. https://doi.org/10.1002/tea.21227

Archer, L., Moote, J., Francis, B., DeWitt, J., & Yeomans, L. (2017a). Stratifying science: A Bourdieusian analysis of student views and experiences of school selective practices in relation to 'Triple Science' at KS4 in England. *Research Papers in Education, 32*(3), 296–315. https://doi.org/10.1080/02671522.2016.1219382

Archer, L., Moote, J., Francis, B., DeWitt, J., & Yeomans, L. (2017b). The "exceptional" physics girl: A sociological analysis of multimethod data from young women aged 10–16 to explore gendered patterns of post-16 participation. *American Educational Research Journal, 54*(1), 88–126. https://doi.org/10.3102/0002831216678379

Archer, L., Moote, J., MacLeod, E., Francis, B., & DeWitt, J. (2020). *ASPIRES 2: Young people's science and career aspirations, age 10–19*. UCL Institute of Education.

Avraamidou, L. (2020). 'I am a young immigrant woman doing physics and on top of that I am Muslim': Identities, intersections, and negotiations. *Journal of Research in Science Teaching, 57*(3), 311–341. https://doi.org/10.1002/tea.21593

Baudrillard, J. (1981). *For a critique of the political economy of the sign* (C. Levin, Trans.). Telos Press.

Blickenstaff, J. C. (2005). Women and science careers: Leaky pipeline or gender filter? *Gender and Education, 17*(4), 369–386. https://doi.org/10.1080/09540250500145072

Bourdieu, P. (1977). *Outline of a theory of practice*. Cambridge University Press.

Bourdieu, P. (1984). *Distinction: A social critique of the judgement of taste*. Routledge.

Bourdieu, P. (1986). The forms of capital. In J. Richardson (Ed.), *Handbook of theory and research for the sociology of education* (pp. 241–258). Greenwood.
Burr, V. (2003). *Social constructionism*. Routledge.
Butler, J. (1990). *Gender trouble*. Routledge.
Calabrese Barton, A., Kang, H., Tan, E., O'Neill, T. B., Bautista-Guerra, J., & Brecklin, C. (2013). Crafting a future in science tracing middle school girls' identity work over time and space. *American Education Research Journal, 50*(1), 37–75. https://doi.org/10.3102/0002831212458142
Carlone, H. B., & Johnson, A. C. (2007). Understanding the science experiences of women of color: Science identity as an analytic lens. *Journal of Research in Science Teaching, 44*(8), 1187–1218. https://doi.org/10.1002/tea.20237
Carlone, H. B., Haun-Frank, J., & Webb, A. (2011). Assessing equity beyond knowledge- and skills-based outcomes: A comparative ethnography of two fourth-grade reform-based science classrooms. *Journal of Research in Science Teaching, 48*(5), 459–485. https://doi.org/10.1002/tea.20413
Chimba, M., & Kitzinger, J. (2010). Bimbo or boffin? Women in science: An analysis of media representations and how female scientists negotiate cultural contradictions. *Public Understanding of Science, 19*(5), 609–624. https://doi.org/10.1177/0963662508098580
Corbin, J., & Strauss, A. (2014). *Basics of qualitative research: Techniques and procedures for developing grounded theory* (4th ed.). Sage.
DeWitt, J., Archer, L., & Osborne, J. (2013). Nerdy, brainy and normal: Children's and parents' constructions of those who are highly engaged with science. *Research in Science Education, 43*(4), 1455–1476. https://doi.org/10.1007/s11165-012-9315-0
DeWitt, J., Archer, L., & Moote, J. (2018). 15/16-year-old students' reasons for choosing and not choosing physics at A level. *International Journal of Science and Mathematics Education., 17*(6), 1071–1087. https://doi.org/10.1007/s10763-018-9900-4
Du, X., & Wong, B. (2019). Science career aspiration and science capital in China and UK: A comparative study using PISA data. *International Journal of Science Education, 41*(15), 2136–2155. https://doi.org/10.1080/09500693.2019.1662135
Francis, B., Archer, L., Moote, J., DeWitt, J., MacLeod, E., & Yeomans, L. (2017). The construction of physics as a quintessentially masculine subject: Young people's perceptions of gender issues in access to physics. *Sex Roles, 76*(3), 156–174. https://doi.org/10.1007/s11199-016-0669-z
Gill, T., & Bell, J. F. (2013). What factors determine the uptake of A-level physics? *International Journal of Science Education, 35*(5), 753–772. https://doi.org/10.1080/09500693.2011.577843
Godec, S., King, H., & Archer, L. (2017). *The science capital teaching approach: Engaging students with science, promoting social justice*. UCL Institute of Education.
Hall, S. (1990). Cultural identity and diaspora. In J. Rutherford (Ed.), *Identity: Community, culture, difference* (pp. 222–237). Lawrence & Wishart.
Hazari, Z., Sonnert, G., Sadler, P. M., & Shanahan, M.-C. (2010). Connecting high school physics experiences, outcome expectations, physics identity, and physics career choice: A gender study. *Journal of Research in Science Teaching, 47*(8), 978–1003. https://doi.org/10.1002/tea.20363
Hazari, Z., Cass, C., & Beattie, C. (2015). Obscuring power structures in the physics classroom: Linking teacher positioning, student engagement, and physics identity development. *Journal of Research in Science Teaching, 52*(6), 735–762. https://doi.org/10.1002/tea.21214
Hazari, Z., Brewe, E., Goertzen, R. M., & Hodapp, T. (2017). The importance of high school physics teachers for female students' physics identity and persistence. *The Physics Teacher, 55*(2), 96–99. https://doi.org/10.1119/1.4974122
Institute of Physics. (2011). *It's different for girls: The influence of schools*. Institute of Physics.
Institute of Physics. (2013). *Closing doors: Exploring gender and subject choice in schools*. Institute of Physics.
Institute of Physics. (2015). *Opening doors: A guide to good practice in countering gender stereotyping in schools*. Institute of Physics.

Joint Council for Qualifications. (2020a). *GCE A level & GCE AS level: Results summer 2020*. Joint Council for Qualifications. https://www.jcq.org.uk/wp-content/uploads/2020/09/A--Level-and-AS-Results-Summer-2020.pdf

Joint Council for Qualifications. (2020b). *GCSE (full course) outcomes for all grade sets and age breakdowns for UK candidates: Results summer 2020*. Joint Council for Qualifications. https://www.jcq.org.uk/wp-content/uploads/2020/09/GCSE-Full-Course-results-Summer-2020.pdf

Kemp, P. E. J., Berry, M., & Wong, B. (2018). *The Roehampton computing education report: Data from 2017*. University of Roehampton.

Krueger, R. A., & Casey, M. A. (2014). *Focus group: A practical guide for applied research* (5th ed.) Sage.

Lawler, S. (2007). *Identity: Sociological perspectives*. Polity Press.

Lykkegaard, E., & Ulriksen, L. (2019). In and out of the STEM pipeline – A longitudinal study of a misleading metaphor. *International Journal of Science Education, 41*(12), 1600–1625. https://doi.org/10.1080/09500693.2019.1622054

Marx, K. H. (1867). Das Kapital. Kritik der politischen Oekonomie von Karl Marx. Erster band. *Buch I: Der Produktionsprocess des Kapitals*. Hamburg: O. Meissner.

Masnick, A. M., Valenti, S. S., Cox, B. D., & Osman, C. J. (2010). A multidimensional scaling analysis of students' attitudes about science careers. *International Journal of Science Education, 32*(5), 653–667. https://doi.org/10.1080/09500690902759053

Mujtaba, T., & Reiss, M. (2014). A survey of psychological, motivational, family and perceptions of physics education factors that explain 15-year-old students' aspirations to study post-compulsory physics in English schools. *International Journal of Science and Mathematics Education, 12*(2), 371–393. https://doi.org/10.1007/s10763-013-9404-1

Mujtaba, T., & Reiss, M. J. (2016). 'I fall asleep in class … But physics is fascinating': The use of large-scale longitudinal data to explore the educational experiences of aspiring girls in mathematics and physics. *Canadian Journal of Science, Mathematics and Technology Education, 16*(4), 313–330. https://doi.org/10.1080/14926156.2016.1235743

Murphy, P., & Whitelegg, E. (2006). *Girls in the physics classroom: A review of the research on the participation of girls in physics*. Institute of Physics.

Reiss, M. J. (2000). *Understanding science lessons: Five years of science teaching*. Open University Press.

Russell Group. (2018). *A Russell Group guide to making decisions about post-16 education 2017/18* (6th ed.). Russell Group.

Sax, L. J., Lehman, K. J., Barthelemy, R. S., & Lim, G. (2016). Women in physics: A comparison to science, technology, engineering, and math education over four decades. *Physical Review Physics Education Research, 12*(2). https://doi.org/10.1103/PhysRevPhysEducRes.12.020108

Wong, B. (2012). Identifying with science: A case study of two 13yearold 'high achieving working class' British Asian girls. *International Journal of Science Education, 34*(1), 43–65. https://doi.org/10.1080/09500693.2010.551671

Wong, B. (2015). Careers 'from' but not 'in' science: Why aspirations to be a scientist are challenging for minority ethnic students? *Journal of Research in Science Teaching, 52*(7), 979–1002. https://doi.org/10.1002/tea.21231

Wong, B. (2016a). Minority ethnic students and science participation: A qualitative mapping of achievement, aspiration, interest and capital. *Research in Science Education, 46*(1), 113–127. https://doi.org/10.1007/s11165-015-9466-x

Wong, B. (2016b). *Science education, career aspirations and minority ethnic students*. Palgrave.

Part III
Student Science Identities in Higher Education

Chapter 6
Science Talent and Unlimited Devotion: An Investigation of the Dynamics of University Students' Science Identities Through the Lens of Gendered Conceptualisations of Talent

Henriette Tolstrup Holmegaard and Bjørn Friis Johannsen

Historically, western universities have served as the exclusive final step of the educational system for the privileged, select few. Traditionally, and still now, a university education was a key factor in shaping the minds and values of young members of the elite. This university-educated elite comprised primarily white, upper-class men, and during the 1960s and 1970s, also working- and middleclass men and – more cautiously – women too (Trow, 2010). Changing the student population by admitting an increasingly diverse range of students is why some talk about massified and diversified higher education (Kogan, 1997; Scott, 1995) which in turn can be a political goal in and by itself. In Denmark, for example, the current goal is that 50% of the population have finished a higher education degree by the time they turn 30 (Ministry of Higher Education and Science, 2018). Despite the increased accessibility of higher education, inequality persists (e.g., Liu et al., 2016). Rather than moving from an elite to a mass higher education system, it is suggested that a parallel system developed that comprises an elite *and* a 'mass system', while still, the myth of meritocracy persists (Hartsmar et al., 2021; Leathwood & O'Connell, 2003; Scott, 2019). These parallel systems can be observed in various ways. For example, higher education is increasingly socially stratified globally, where universities in the anglophone countries make up the top of university rankings (Marginson, 2016), and there is social stratification to be observed both across type of institutions and types of study-programmes (Reimer & Thomsen, 2019). These parallel systems are also installed at a student level. As institutions appear massified, less selective and elite discussions about how the system recognizes extraordinary individuals

H. T. Holmegaard (✉)
University of Copenhagen, Copenhagen, Denmark
e-mail: htholmegaard@ind.ku.dk

B. F. Johannsen
University College Copenhagen, Copenhagen, Denmark

H. T. Holmegaard, L. Archer (eds.), *Science Identities*, Contributions from Science Education Research 12, https://doi.org/10.1007/978-3-031-17642-5_6

followed. Gale and Tranter (2011, p. 42) for example, worried that "the inclusion of more people from disadvantaged backgrounds may be seen to undermine the talent and hard work of 'deserving individuals' and traditional notions of merit and standards". With the masses entering universities new sorting-mechanisms found their ways into universities. In this chapter we will explore talent as such a mechanism that works to identify and privilege a certain kind of student performance maybe to preserve the idea of higher education as elite.

Study programmes for the talented, honours programmes, talent camps and competitions and fast tracks have been established to provide special students with special benefits (e.g., time with teachers, access to lab-equipment, exchanges, tailored educational activities) (Rasmussen & Rasmussen 2015; Wolfensberger, 2012). Common to many of these initiatives is that they treat talent as a personal trait, something that derives from individual characteristics, abilities, and experiences, as a way to legitimize this privileging of the participants (Scager et al., 2012). The underlining idea is thus (as per, for example, Gagné's, 2011 writing on gifts, talent and equity) that society deserves to have these talented and gifted people develop their talents for the good of all. Talent is thus configured as particular personal traits that match societal needs (Csikszentmihalyi et al., 1997). However, fifty years of research have shown how education tend to treat social gifts as being natural ones and in consequence act to mask inequalities (Bourdieu, 1974, p. 32, in Radnor et al., 2007, p. 296).

This is the platform that this chapter will build on. We do not attempt to define talent. Instead, we investigate how teachers and students ascribe meaning to notions about what talent is and who the talents in higher education science practices are. We do so, referring to the common theme of this book on science identities, because we believe that thinking about doing and being in science with reference to the notion of talent, is an interesting way to unmask the norms and values that feed into how some positions appear attainable for some and not others.

6.1 The Ideal and Celebrated Student in Higher Education Science

Much of science education research is premised on needs to attract, retain and support talented students throughout the education system (e.g., Taber, 2007). Science identity research has provided valuable analytic lenses to unpack, in part, the social practices of who are recognised as someone who do or do not belong in science, and in part, how these social practices intersect with different students across social, gendered, and racial categories. Science identity researchers have demonstrated through a large body of literature how science education activities are historically and culturally constituted around preferences that determines how a science student is expected to be and what a 'good' science student looks like (see for example the work of Barton, 1998; Carlone, 2004). Thus, students who can navigate the system and gain recognition are often, not surprisingly, also students who have relevant social resources to draw on (Archer et al., 2012b).

Archer and colleagues found how the ideal student in English schools were associated with being active and engaged – which is not as straightforward as it may sound. Working class girls, for example, engage actively in class by 'speaking their mind' and thus risk being recognised as contesting the teaching:

> Trying to 'change' and achieve 'success' is a difficult and costly endeavour. In many ways, the project appears 'impossible' because the girls inhabit social positionings and embodied identities that are always already read as 'wrong' within dominant educational discourses (Archer et al., 2007, p. 565).

And Archer and Francis (2006) show how the 'ideal' student is ascribed attributives associated with white, male and middle class why girls risk being dismissed as 'diligent, conformist, culture-bound, followers, passive, unassertive' (p. 66).

Wong and colleagues conceptualise the notion of the ideal student further and shows how it varies between national and cultural contexts (Wong & Chiu, 2020; Wong et al., 2021). What is recognised as ideal by students and teachers produces a spectrum of attractive and desirable attributes. This is why Wong and colleagues discuss how students who are not recognised or do not recognise themselves within the spectrum risk feeling 'not good enough' or feel they must work harder than those who are recognised, in order to achieve the same recognition from teachers or peers.

From this perspective, notions of the ideal student or talented students can be thought of as a category that works to keep power and privilege with the already powerful and privileged.

This category resonates with a study by Jackson and Nyström (2015) which investigates the notion and celebration of 'effortless achievers' in Sweden and the UK. They document how effortless achievers are often thought of as extra intelligent. Moreover, the category ensures that, rather than being poisoned as 'not good enough', effortless achievers who failed in their achievement continue to gain recognition as ideal students, they just need to adjust their effort the next time. The category also associates with prestige and social popularity and the authors conclude that the category interacts with gender, social class, ethnicity, and institutional settings, which explains why it is available only to some students (Jackson & Nyström, 2015).

Effortless achievers as ideal students have similarities with what Carlone (2004) describes as 'raw talent'. Carlone et al. (2014) use the notion of celebrated identities in science teaching (identities being recognised, highlighted, and valued) as a lens to approach the ways of acting that were celebrated, marginalized and/or condemned (p. 843). They find that these identities set the scene for students' understanding of who science is for and how one should navigate it to gain recognition.

Taken together, we find that the above authors bridge the notion of talent. Where Wong and colleagues' (2020, 2021) ideal student are suitable to explore what teachers and students value and recognise as desirable and good student characteristics, Archer and colleagues (2006, 2007) along with Jackson and Nyström (2015) and

Carlone (2004) bridge how ideal students' imaginations are played out on a subject level. With this as our empirical perspective, we will outline our theoretical approach.

6.2 Theoretical Lenses

Our purpose in this chapter is to investigate how the idea of talent in higher education science is produced, recognised, and rendered meaningful at the level of the individual. We will do so by interviewing both teachers and students and by applying the notions of power, positioning, recognition (mechanisms of inclusion and exclusion), and gender.

6.2.1 Norms and Meaning Making

Our starting point is an interest in how power is exercised at subject level. Power is the vehicle that shapes students' and teachers' norms and practices. Power is not about someone (such as a teacher) dictating thoughts and acts. Rather, power is continuously being produced and maintained by the subjects who inhabit it (Foucault, 1997). Teaching is a powerful space. It provides a platform for deciphering the norms and values of academia and thus communicates the expectations of how and who to become (Danielsson et al., 2018). Teachers have a central role in recognizing who and what belong within or outside science and thus in shaping students' science identities. Inherent in this recognition lies a risk of pushing students into taking on identities they feel violate their sense of selves in order to gain a sense of belonging in science, or because doing otherwise, they would risk being excluded by or from the teaching-regime (Avraamidou, 2020). However, teaching is just one aspect of science culture. Other aspects tie to the research that forms the basis of teaching and the students' study culture (e.g., Hasse, 2016), and as our data are produced at a research-intensive university, this context set situates our analysis.

Foucault (1997) links power and the ongoing construction of knowledge with social practices, and this means that power is exercised as norms, ideas, or notions of what and who is recognised as legitimately belonging and what and who is not. Both are historically shaped and continuously negotiated in social relations. Thus, science culture entails ideas historically embedded through decades, and this informs what ideal science participation looks like (Carlone, 2004). These ideas, however, are continuously (re)produced and find new ways of exercising themselves in various settings. To examine the ongoing negotiations of what is celebrated and recognised as desirable and what is excluded, we are interested in teachers' and students' narratives about ideal students and how they produce and reproduce meaning as Staunæs (2003) phrases it: "Social categories [such as talent] are tools of selecting and ordering. They are tools of inclusion and exclusion and they are tools of positioning and making hierarchies" (p. 104). These mechanisms, however, are subtle and therefore researchers must investigate how they are played out in everyday practices and how knowledge and meaning making are produced. We are inspired by Søndergaard (2005) to investigate the negotiation of meaning-making practices by asking "who would talk in what ways about which potential positioning?" (p. 193). Thus, by analysing interviews in line with Søndergaard (2005), our purpose is to explore the patterns in the meaning-making processes.

6.2.2 *In- and Ex-clusion Through Available Gendered Subject Positions*

Davies and Harré (1990) have theorised how identities are fluently and continuously negotiated through social practices. They provide the tools to investigate how identities can be perceived as a net of subject positions being offered, recognised, obtained, and neglected. Identities are in this understanding more dynamic than describing different sides of a person, offering lenses to see the subtle inclusion and exclusion practices at stake in social practices and who one is allowed to become through the social practices one participates in. This also means that the positions we obtain are not necessarily coherent, as we can be positioned in contradictory ways through different contexts or even within the same interaction entailing different forms of meaning. What is relevant to highlight is thus how different positions allow for different viewpoints and ways of making meaning:

> Recognition of oneself as having the characteristics that locate oneself as a member of various sub classes of dichotomous categories and not of others i.e., the development of a sense of oneself as belonging in the world in certain ways and thus seeing the world from the perspective of one so positioned. This recognition entails an emotional commitment to the category membership and the development of a moral system organised around the belonging (Davies & Harré, 1990, p. 48).

Being recognized as a talented science student thus form an individual's sense of self, while at the same time acts to emphasise a certain set of positions which are available in the particular context of higher education science. Recognition may signal or demonstrate disciplinary norms and may demonstrate power. Here we are not interested in seeing how inclusion and exclusion play out in daily interactions, however. Instead, we explore how they are embedded with the meaning that students and teachers create to justify or explain their engagement with higher education science programmes. We place talent as a social and cultural practice that acts to recognize or reject student positions within science culture at higher education institutions. Thus, the talented student is both produced in and by the cultural-historical setting of higher education, through ongoing acts of recognition, that also works to maintain and solidify an ideal and desirable subject position that students and teachers can see and evaluate themselves and each other through and draw meaning from.

Carlone et al. (2015) argue that a gender perspective is inescapable if we wish to make sense of structure or agency (and everything in between) in settings such as science teaching and learning "where inequities are blindingly pronounced" (p. 474). For example, studies show that schooling teaches students that they are understood through their gender and that their gender means they are expected to behave in certain ways in relation to science (Archer et al., 2012a). One effect is that female students who engage with science are being recognised as performing good student identities while some of the male students are recognised as possessing 'raw' talent (Carlone, 2004).

We are inspired by gender as performed as in line with the widely cited work of Butler (1990). Here the idea of gender as biology and naturally embedded in bodies is abandoned. Butler (1990) proposes an analytic lens to understand the dynamic and continuously communicated expectations that set the scene for how one can perform oneself. Gender is not stable, nor inner characteristics but shaped through the available discourses. This also means that different discourses allow for different opportunities to negotiate gender. This provides the platform to study how gender is dynamic, negotiated yet inescapable as gender is a powerful category dominating western societies (see e.g., Acker, 1990; Ridgeway & Correll, 2004; Rottenberg, 2014).

Consequently, this chapter investigates how meanings of the ideal or talented student are produced and gendered, because in science, ideals most often are (e.g., Keller, 1985; Thomas, 1990). Our aim is on the one side to explore the inclusion and exclusion mechanism entailed in recognition, and on the other to understand the consequences of recognizing, enacting, and producing talent as a gendered subject position.

6.3 Method

6.3.1 Background

The chapter brings together two work packages and datasets produced within the same project with the aim to investigate some of the challenges evoked by the massified higher STEM education. The data was produced at different sites at the same university, and conceptual and analytical work was developed in cooperation. One work package centred around the transition from master's degree program to professional employment and consisted of workshops with master's students, as well as individual and group interviews conducted three times in the transition phase with students at three selected programmes, one at a biomedical study programme which will be central to this chapter. The other work package centred around teachers' meaning-making of talent and was based on individual interviews with teachers of undergraduate science courses. While these informants speak from different subject positions and experiences, we were increasingly struck by the force with which gender emerged as a common theme in data. We thus decided to use this opportunity to bring the two datasets into dialogue to unfold ideas about talent and gender that appear to pervade higher education science in this context.

In the country where this study was performed it is not a requirement, and only a recent practice, that research involving qualitative research on adult humans using methods such as interviews and participant observation undergoes ethical reviews. This study adheres to common ethical practices, recommendations and norms set by the national Data Protection Agency and social science research council, that mainly requires the researcher to be considerate of the people and groups that are directly and indirectly objects of the research. This is done by informing subjects of the purpose of the research, and by keeping participants anonymous and identifiable information at a minimum.

6.3.2 Study Context

As the context is central when approaching an understanding of science identities, we find it important to share some background of the study. The empirical data is produced at a research-intensive university with over 40,000 students in an urban setting in Scandinavia. All students can access and complete higher education without paying tuition, and they also receive a monthly government grant that cover essentials such as books, food and rent. Students who inform this chapter are at the end of their master's (MS) part of the biomedical programme (BioMed), which has an undergraduate (BS) and a master's part (MS) that qualifies for a Ph.D. programme in BioMed if students secure the necessary funding and/or support from faculty. It is a characteristic of the programme that it emphasises activities that are close to and relevant for biomedical research practices. This research focus is reflected in the teaching, some of which takes place through projects in research laboratories. In practice students seek out research groups to discuss opportunities for doing project work. When good matches and projects are identified, the MS students join the research groups as student members, who work on their MS thesis projects typically under the supervision of a senior and junior member of the group. The thesis-projects are formulated to contribute to the research otherwise carried out in the lab, and interviewees had been told that if everything went well, they should expect to contribute data and co-authorship to a first international peer-reviewed publication before they finish the programme.

In contrast to most other STEM-programmes in the region, admission to BioMed requires one of the highest GPAs of all courses of study. In addition, the university webpage states that three quarters of BioMed students are women. Thus, that the gender composition in this science programme is different from courses of study that have difficulties attracting and retaining women such as physics and computer science. Most students in BioMed continue from the BS programme. It also means that the composition of students in this programme often comprise an unusually large group of women who have earned and are good at earning high grades. A recurring theme in the dataset is students who appear preoccupied with earning top grades are referred to as "Straight-A girls". The term is derogatory and reserved for young women who excel at making top grades, who appear to care about their grades, who have developed strategies to maintain high grade point averages. The women are referred to as "Girls" which could work to emphasize that this grade seeking behaviour is immature or not congruent with adult behaviour, or it could work to denigrate the grade and achievement by feminising it. Furthermore, it reserves the category and connotations for women. To our knowledge, there is no similar name for boys or men who care about their grades. One might speculate that they are thought of as talented, nerds or merely ambitious and clever or just schooled.

6.3.3 Interviews with Teachers

This chapter thus draws on two studies. One study, carried out by the second author, investigated teacher practices and teachers' understandings of what talent in higher education science is, how it is recognized and produced during teaching. The data used in this chapter draws on a larger data-set collected through interviews with four undergraduate science and mathematics teachers, interviews with groups of their students and participant observations during several of the teaching activities associated with their courses. The teachers were asked to participate because they taught courses that are more advanced than the introductory courses but still sequenced somewhat early in the undergraduate program. This choice was made to be able to discuss the types of negotiations of meaning, purpose and values that students depend on when they enter into an academic field, but also to see how these negotiations tied to contents that might have more bearing to the teachers' thinking on their research disciplines than standard introductions do. Typically, the courses identified as meeting these criteria were offered during the end of the first year to mid-second year of the undergraduate programmes. The teachers were contacted and informed about the purpose of the study, but needed some persuasion to be interested in informing a study about talent. The argument that was persuasive to them was that the investigation would be close to their practice and reflection on teaching, learning and their research disciplines, and would not depart from preconceived, decontextualized ideas about what talent is.

The interviews were conducted as a semi-structured approach with few themes allowing for the teacher to elaborate on questions such as 'What is a good student to you?', 'How do you recognize talent in your teaching?', 'What opportunities do you plan for students to show their talents?' One interview was carried out before teaching started and a follow-up interview was carried out at the end of the course in order to compare and contrast intentions and experiences. Observations of teaching activities and interviews with students provided context for the interviews, but also for shared critical discussion and reflection on what was being said during interviews in a more agonistic or confronting way (e.g., Kvale, 2006). That is, in effect, talent would be investigated as a reflection of their expectations of and experiences with students during their own teaching – a topic they were all interested in or did not mind discussing. This is important, also for ethical reasons, because higher education in this context is considered a matter of equal access which does conflict somewhat with conversations about elitism and talent. Allowing participants to define talent themselves was key to their willingness to participate. It happened during the interviews that participants asked not to be quoted on specific aspects of their sentiments. However, this never happened when they reflected on how they felt gender being relevant to talent. This means that the teachers unwittingly have added to an analysis they did not expect, anticipate or would have agreed to inform. However, the interviewer never designed the interview to be about gender – this was an aspect that appeared in the co-constructed interview.

In this text we use interview quotes outside of the context they were discussed in, but do so, because these quotes address themes that span all datasets and contexts. Consequently, they should be viewed not as sentiments expressed by particular individuals, but as perspectives produced by a larger system that span disciplinary settings and individuals, and which are produced in an interview-setting that ask about talent and the 'ideal' student.

6.3.4 Workshops and Interviews with Students

The other study, carried out by the first author, explored the transition of MS students from their studies and into their first job. All MS students at the start of their final theses were invited to participate in a workshop. Two two-hour workshops were conducted, where students were engaged in different activities, for example, writing an essay reflecting on their choice of MS programme and thesis topic, and their expectations for the future, or making a portrait of themselves three years hence.

The workshops made it possible to obtain various empirical data and provided insights into the culture of the study programmes. Based on the workshop, seven students were asked to participate in narrative interviews (see Holmegaard, 2020a, b). The first interview was conducted at the beginning of their MS programme, and the final interview was one year after graduation. The students were selected to enable maximum variation of participants (Flyvbjerg, 2011). The interviews centred on the following themes: (1) current interests, challenges, and aspirations; (2) sense of belonging at the study programme and thoughts on student practices; (3) competences they expected to gain from their MS programme; (4) expectations of the future; and (5) interaction with the world outside their studies. The interviews lasted about one hour. All were transcribed. All names in this chapter are anonymized, randomly assigned, and are not reused from previous or in later papers that treat this dataset, meaning that the same students are given different names to prevent recognition. Also, experiences that could be identified were changed so they retain the intended meaning, but obscure possibilities for recognition.

6.3.5 Positionality and Production of Data

The interviewers' own positionality informed a reflexive analysis. The first author is a white woman with working-class background and with an educational background outside science. In particular gender seem to interplay with the participants, who were willing to share both very personal narratives of being sick with stress, and of fearing inadequacy. Not having a background in science allowed for what has been described as the 'naïve' interviewer position, which allows for asking questions to commonalities and allowing for an insider (same race and gender as the participants)/outsider (not from within science) positionality (Adriansen & Madsen, 2009). The second author is a white man of working-class origins with a background in science who could thus to some extent interview white, male scientists from an insider-position. This allowed for in-depth discussion of teaching practices and to probe interviewees' perspectives on learning. However, our positionality also made us blind to white hegemony, and the second author to the more implicit aspects of masculine hegemony left uncontested in the interviews. The result appears to be very open, honest and maybe even revealing interviews. This appearance must not be confused with truthfulness, as our blind-spots might just as well result in a form of co-constructed naivety that shapes how the interview appears. And here lies maybe the most accurate characterisation of the data that is analysed here: it is our co-construction with science students and teachers in higher education, respectively. We could not have created the data without the participants; nor could the participants have created similar data without us to guide discussions, focus, interests, and attention.

6.3.6 Analytic Approach

Through student' and teachers' narratives about performing talent and being talented, our focus in this chapter is on how talent is ascribed meaning. We are interested in the subject positions that are made available and included in the cultural context of higher education science, and the ones that are rejected or not recognized. Based on the transcriptions the empirical material was thematically analysed from an inductive approach jointly by both authors (see Braun & Clarke, 2006). Using our theoretical background as lenses we formulated a set of analytic questions to approach the transcriptions by applying the following analytic questions. Analytic questioning is in line with Søndergaard (2006), questions that act to bridge the aim and theory with a set of supportive questions to approach the analysis:

1. How are ideal students ascribed meaning by teachers?
2. How are ideal students ascribed meaning by students?
3. How do students relate themselves to and position themselves in these meanings (inclusion and exclusion mechanisms at stake)? How are ideal students recognised, enacted and negotiated?

4. And how does it more generally set the scene for the identity negotiations taking place in science?

Based on these questions we constructed themes that were reframed, collapsed and re-categorised. Analysis of the two data sets was carried out treating the sets as a dialectic, moving back and forth between teachers' and students' narratives as described in the introduction to this section. The result is presented in the analysis.

6.4 Theme 1: What Talent Is and Is Not

Theme 1 explores how talent is recognized and enacted as an aspect of positioning and being positioned as a good student. First, the scene is set, and sets of rules for engagement among students are located. Second, the playing field that these students believe they engage with is viewed from the teachers' perspective. Third, a typology comprising two kinds of good student is explored as described by teachers. Only one kind is allowed status as 'talented'.

6.4.1 Complicity in Silencing

From the interviews with BioMed students, it is clear they understand themselves as members of an extremely hardworking group, characteristically comprised of high ability students pursuing a professional future in a field considered up-and-coming in both academia and industry. In the interviews the students explain how throughout their studies they hear teachers express sentiments like: "You are the flower of our country's youth", "You are special students" and "By the end of the programme, 85% of you will enter a Ph.D. programme, but only 20% can get funding from the university." Thus, BioMed students know they need to compete for the attractive positions on offer, and they know that competition is intense, and mutual expectations are high. This discourse dominates how students describe engagement in BioMed studies. Anna explains the possibilities as well as challenges this kind of culture brings:

> It's just really a fiercely competitive environment. Ehm, that, but both for good and bad. I mean, you are pressured to achieve, right? And if sometimes you think 'I don't want to study for tomorrow,' you know there'll be 20 others who did, and then you are like 'You didn't study?! Don't you get your reading done? Are you sick?' 'No, I'll read it for tomorrow then.' It's a fierce, fiercely competitive environment (Anna, student).

This competitive culture produces certain positions for students to take on to gain recognition, through producing oneself as 'engaged'. Above, Anna tells us that you definitely need to be well prepared, but below Cecilie adds how preparation does not necessarily legitimate participation:

> It's this atmosphere, you quickly sense in a course if someone's there who believes they know everything and don't want to listen to people who doesn't know 100% what they ought to know. It gets a bit gloomy sometimes. If teachers ask a question, no one says anything because they sit there, feeling a bit watched over. You don't want to say anything wrong. It's not like anyone is booing. I think, maybe, it's got something to do with, there's really, really many girls, at least in my year, so it kind of turns into infighting and slander (Cecilie, student).

Above, Cecilie describes an atmosphere of gloomy watchfulness. Participation in class requires that you believe in your own abilities and are sure that your inputs are relevant, both by being based on thorough preparation, and on knowing the content absolutely. This on the one hand produces the position of knowing and participating with the risk of being someone who then at the same time 'Don't want to listen to people who doesn't'. But it moreover on the other hand produces the position of not participating and hence not knowing. Silence or being in class without actively participating appears as a viable form of engagement during teaching for students who are not comfortable with participating or who did not prepare sufficiently to feel their voices are legitimate to share. Silencing however holds the consequence of positioning oneself as someone who do not 'Know'. The quote shows how the students govern themselves and each other, producing positions as quiet bystanders to their own education. Power is, as we outlined in the theory, proceeded and excised in daily practices. This is an example of how students govern each other and how competitive behaviour encouraged by the programme shape student-student interactions.

Together these quotes illustrate an unhealthy learning environment, where student learning becomes an individual matter, and where being afraid of making mistakes prevents students from sharing questions, insecurities and challenges. This contradict higher education pedagogy where elements as active participation, student ownership and sharing questions and challenges are pointed out as central in order to achieve learning (Rienecker et al., 2015). While Cecilie explains how the silencing of each other are due to the fact that there are a majority of women at the study program, the literature suggests something else, namely that the competitive study culture can take a form where silencing of each other is the exact purpose. In this case the competition is not only about getting good grades, because this, as shown in the next quote, is considered a precondition. Rather we can understand student participation as a game of outperforming each other so the advantage of some become the disadvantages of others (Becker, 2002). Similar conclusions are reached in settings where women are the minority, here other women can be experienced as a threat because as all are aware that only few will make it all the way (Wright, 2016). As a consequence, not performing in line with these expectations produces insecurities in their own abilities:

> They're all really good and ambitious and you can only tell people what you got if you got an A. Many get an A, but there are also many who don't. Sometimes you can't help thinking 'Gee, everybody's just so smart and good and why do I need to sit and read it three times to understand?' (Katarina, student).

The quote shows how there is a risk of exposing yourself as not being as smart as your peers, but moreover how students' challenges are experienced as a sign of not being as good as the other students. Tina explains how students' insecurities are also reinforced in teaching and how the concept of talent pervades the programme, including teaching:

> I remember these classes when no one could answer. People thought it was hard and someone said like 'But it's super hard, this, we haven't met it before at all and you need to explain it with some other words.' Then I remember that [the teacher said] kind of 'Really? That's unbelievable! Aren't you? Weren't you supposed to be the best in this country? That can't be right' (Tina, student).

Summing up the above quotes, the ideal student at BioMed entails students who are nationally eminent, get top grades, are expected to continue into research, believe in their abilities and always are well-prepared. These expectations are communicated clearly in teaching exemplified in Tina's quote – when students express insecurities and challenges, they risk their eminent positioning. However, the expectations are also produced and maintained in student practices as we have also shown. While the students themselves are complicit in the silencing of their uncertainty and doubts regarding their own abilities, it seems from the quote that the teacher reinforce such notions by silencing the students when finally speaking up. Done intentionally, this would be a vicious and irresponsible move on the teachers' part. However, as we will show in the next section, hearing teachers reflect on the silent high-performing women in their classrooms reveals a much more structurally embedded pattern that draws direct lines to hegemonic masculine beliefs concerning women and academic performance.

6.4.2 Invisibility in Teaching

In interviews with educators who teach undergraduate science, we found it a habitual practice of disassociating talent with high-achievement if it involved women. Instead, their roles in narratives about student behaviour are as unobtrusive, almost diffident bystanders. Below, Tom complains about students' general indolence, that sometimes, and especially during early mornings, manifest in total absence:

> I came in at 8 in the morning. It was pouring outside. The only ones present were female students. I mean, they [i.e., students in general] don't take it seriously, right? And they were supposed to work 20-hours a week on this. And they don't (Tom, teacher).

In this quote, the teacher does not quite ignore the female students. It seems as if their presence in his class emphasizes the absence of the others; those he identifies as his students when he is recounting an experience that is useful for lamenting their lack of engagement. But once the point is made, it seems, the women disappear from his awareness. Below is another example from interviews with Tom that shows this invisibility:

> I almost left Wednesday, that one class. There was simply no response. Almost none. Except from those two girls on first row, and who had been at the study-café to finish everything. But they are typical female students. They aren't the kinds who show off (Tom, teacher).

What the teacher picks up on here, when he discounts the two women on his first row as engaged with typical female student behaviour, may be similar to Cecilie's description of their mutual silencing; but also, a reflection of the expectations revealed by his way of talking about female students. However, the interviews with teachers show that to be recognized as talented it is not sufficient to meet the course requirements. Supposedly, you must be good in a particular kind of way to gain recognition. First, the above quote suggests that you cannot be silent in class. If you are, and you are a woman, you are erased from active awareness. Instead, it is crucial for students to show an ability to move beyond the scope of the course and curriculum to pose questions that might even challenge the teacher, here explained by Henry:

> The straight-A girls, we all have them in our courses and as a teacher it's delightful to have them, because they show up, they've read their stuff, they've done their problems, they only pose relevant questions. Of course, they all get A's. They become tremendously capable – no doubt about it – but it's just, they are not the ones you, like, notice. They aren't extroverted, necessarily. But then, statistically speaking, there are a few boys. They ask questions the first day where you don't know the answer and as a teacher you just go 'Pff, what's happening here?!' (Henry, teacher).

Suggesting that the questions these young men come up with for this teacher to notice them are beyond course and curriculum may be a friendly reading of the interview transcript. Digging deeper, it would be what Lucas (2001) describes as 'effectively maintained inequality'. Even if struggles for resources and privileges are settled, for example, through redistribution, norms shift and goal posts are moved. In the case of educational access "Universality … may be largely irrelevant to the intensity of class conflict, as the focus of conflict may simply change once access is universal" (Lucas, 2001, p. 1680). Put differently: when women start outperforming men, what counts as performance is simply changed – power and knowledge are ongoing produced and find new ways and form. In this case this is an example of masculine hegemony that is recognized and propagated by the male educators and maintained and enacted by the students. Henry, who spoke above, continues in the same breath:

> I mean, it's two different ways to be good at it, and it's just different kinds of jobs they need to be good at, right? I mean, I think you should be aware that now you are choosing teachers to talk to, you probably need to remember that they…what they think is good is a bit like what they themselves aspire to or want to be, right? (Henry, teacher).

Talent, it seems, is a cultural reproduction of masculine dominance in science. Across the interviews with male science teachers of undergraduates, we found similar sentiments regarding good, yet silenced and thus invisible women. Tonso (1998), Faulkner (2007) and Gonsalves (2014) have used the term in/visible as a lens to understand the experiences of women in engineering and science. Tonso (1998) shows how power mechanisms embedded in engineering education act to maintain the invisibility of women and erase them. Faulkner (2011) recounts: "Even when faced with evidence to the contrary, there can remain lingering doubts about women's ability" (p. 281). There are at least two important points here. One is how teachers across different sciences, whether or not women make up a majority, share similar sentiments that render women invisible with respect to talent. It is the gendered construction of active versus passive; 'natural ability' versus 'plodding diligence' that Archer and Francis (2006) find separates the production of the 'ideal' and the 'pathologized' student. Another point, which is treated in the discussion, is the complicity of the BioMed students in producing their invisibility by silencing each other in the classroom. If students can be good, silent, invisible and erased through gendered practices, it still leaves the question: How does a student position herself to be ascribed as good or talented student and to gain recognition as such? That is the focus of the next section.

6.4.3 Good Students: The Organisers and the Artists

The intention behind interviewing the teachers was to understand how they expected students to perform in order to recognize them as talented. A related aspect was to ask them how this recognition tied to students' ways to participate in their teaching. What came out of these conversations shows what we have categorised as different ways to perform 'good student'.

A teacher, Simon, describes how many young women in his course perform 'good student' identities by being structured, prepared, and delivering perfect results through meticulous work:

> Many of the girls keep from going for the Ph.D. because it overwhelms them. I mean, they become paralyzed with stress because they're so dutiful. But Christ they are a bunch of good girls to bring on a fieldtrip to make databases and such. But they refrain from thinking new and exciting thoughts and from writing papers because they say 'But it stresses me out right away'. But to put in numbers, check them, they look fine and dandy, run the database, go do labs and measure analyses – 200 analyses. And then go home, kick back, my head is empty afterwards. We need that kind. But we also need [the other kind] (Simon, teacher).

Based on this characterisation, the good student performativity is about being well structured and organized. However, students in this category are considered fragile and easily stressed out. They are useful to perform data processing but cannot be relied upon to challenge assumptions or for the daunting tasks of writing up papers which he considers a central trait of the talented, necessary to venture into Ph.D. studies.

The other group characteristically comprises primarily male students. They are described as less organized, they go their own way, even if it might bring on extra work. The aim is to have fun, and the attitude is playful and curious, and contrary to the 'organizers', almost fearless. And often the teachers draw out the best examples from their own past when describing their attributes:

> Back when I was a Ph.D. student, we needed some mathematics that didn't exist. I mean, maybe it did, but I didn't know about it and my advisor definitely didn't. Well, then we had to develop that kind of mathematics. And I'm sure standard text books exists about it, but we hadn't learned it so we just had to develop it now we needed it. You know, that's the way you do real research, often (Henry, teacher).

In the example above the organisers might have searched the literature to find the math needed. It might even have been a shortcut that had allowed the students to spend time on the intended aim. However, this other group of 'good students' make an effort to develop their own solutions, a kind of hard-headed do-it-yourself attitude, engaging in work without being sure that it leads anywhere. This is therefore not the same as Jackson and Nyström's (2015) finding that 'good students' are effortless achievers. In our data, effort is made however not the one outlined in teaching. Rather, it assumes a willingness on the students' part to take risks, to be able to invest time and energy on trying out approaches that do not necessarily lead to solutions. However, taking risks or playing with the content matter, is not a question about engagement as pointed out by the teacher above, it presumes that all students have equal access to sharing their ideas and that these will be recognised for their substance rather than by who shares them. Some students may hold back their views and opinions not due to lack of confidence, but because they have experienced lack of recognition or even that their views have been silenced – this is a gendered pattern (Read et al., 2001). Taking risks includes risking that your voice is not heard, but also it means risking your sense of self. Risk-taking is not an individual endeavour, rather we saw examples in our data of how risk is indeed socially embedded and a practice that is shared amongst peers. Again, a teacher recounts his own past in describing good student practices:

> And thank God they exist, we've got a few of them. In my time we went to a summerhouse – red wine on the table, dataloggers and planes and all kinds of stuff and spend the weekend on that. And we enjoyed ourselves. Some of it turned out well. And sure, some of it was a bloody catastrophe and nothing sensible came out of it. But in the long run something did. And that is because we bothered, dared, had opportunities to play. There bloody well needs to be a need for creative play in this life. Like, I mean, otherwise you might as well ask what the hell are artists good for (Simon, teacher).

Both 'organisers' and 'artists' perform good student identities. However, it is clear from the interviews that the teachers experience an asymmetry between the two groups. The preference for structure and efficiency exhibited by the first group are perceived as feminized and superficial. In the interview above we see a mechanism that both rewards this surfaced engagement because it is so effective, but also discourages access to those aspects of doing science that are more open-ended, risky and valued. The masculine performativity of the good student is passionate, engaged, possibly reckless and hence truly talented. In the next section we investigate the consequences of this perspective at the individual level, in part as recounted and recollected by the male science teachers and in part derived from the interviews with women students.

6.5 Theme 2: The Price of Talent

Theme 2 takes up the characterisation of talent from Theme 1 as an inherently masculine performative aspect of the 'good student'. In this theme we will investigate how these meanings of good student practices set the scene for the identity-negotiations played out at subject level amongst students and teachers at higher education science programmes. The theme is ended by exploring the challenges that students who attempt to combine talent and feminine performativity face, when talent as a mechanism for producing inequity favours masculine performativity and renders the feminine good student performativity invisible by dismissing it as insincere or inauthentic.

6.5.1 To Invest the Whole Self to Science

A crosscutting theme in the interviews with the teachers is descriptions about how, in order to become a researcher (which is pointed out by students as the expected aspiration when being a student at BioMed), a student must be willing to devote their whole self to science. One teacher, Henry, explained how he cannot help himself from reading scientific papers while watching movies with his partner: "It has to come naturally because if you feel it as a pressure then it's no good." If you are really passionate about your field of research, the presumption is, that it is almost unavoidable for you to invest the whole of yourself in it. Talent is essentialized. Commitment exists as a natural condition for the talented person, it cannot be developed. A few examples describe this condition, centred around getting absorbed in the content matter in a way that make you forget about time and space. Again, a teacher draws on own experiences to describe what he is looking for in talented student practices:

> I remember my own advisor came in to me. It was Friday and it was seven in the evening and he just had this and this idea, we just needed to calculate and write these codes, just do this and that. Suddenly he realized that I was twenty something and I was going out dancing. Then he said 'No, you don't need to do it now, you'll do it early tomorrow.' He was being serious. In no way a provocation. It was so natural – of course we'll do it Saturday morning, what else would we do? And I think this is what I'm looking for with the students. Those where it comes naturally. Of course, they've got a life on the side, but where they just 'When it's exciting, we'll be here all the time' (Henry, teacher).

Interestingly, Henry allows for the students to have 'a life on the side'; but only when nothing else interesting is going on at work. When looking for talented students in teaching it seems as it is also a matter of the students' willingness to invest all their hours and give the whole self to their studies – for whenever they do not, one may infer, it may be read as an indication they do not find the work sufficiently interesting.

Another sign of talent is to be able to engage to the extent you forget about space and time. In the interviews Henry remembered a particular student he has advised, who he describes as "The best doctoral student ever, *ever* produced in this country." Rhetorically he asks, "What was it, that made us chose him?" and ventures into a story about a trip to a conference and a group of colleagues who went for a walk on the beach. A stone is kicked, it breaks open to reveal a fossil and the group searches for more:

> When two to three hours had passed I looked up and noticed that everyone had gone. The only one I could see was this guy. He was sitting down the beach like some kind of caveman cracking these rocks open. Completely consumed by it and I'm sure the others had left. And I'm sure that if I hadn't brought him along, he'd still be sitting there two hours later. And it's simply that ability he's got. He can be completely consumed by whatever he does and it is this flow. He has that flow and he can get into flow quick and he can just stay there for a long time. It's the exact same thing with science, no matter what he does when he takes something up (Henry, teacher).

Above, Henry describes how he knows that a doctoral student has talent as a way to explain what he looks for when teaching students. Note how the experience is reproduced: Henry discovers his student losing himself to the fossils the same way he did himself. To have talent, good grades are not enough. And if students want to do their MS in his group, he explains how he tests them with an exercise that is almost impossible to solve. What matters is the time and effort they have put into solving it:

> Some of them return four days later and it's obvious from what they've done, that they haven't done anything else and haven't thought about anything else, right. Then you know, now I've got something that is right (Henry, teaching).

But in order to read papers in the evening, and spent all your time and energy another teacher points out that you will need a partner that supports that kind of life. This includes having a partner working part-time who can be the primary caretaker, someone who will not mind being alone during summers for ten years straight because it is when fieldtrips take place. Simon reflects:

> So, there are some fierce consequences you must make sure that you've, or else it'll cost a divorce. We see this. Riiight, so then they switch to a newer wife. It's usually the men who switch. It's rarely the girls who switch to a younger man. Instead they leave. Or become single (Simon, teaching).

The system works when (heterosexual) men find new younger partners as years go by and women become single if they want to keep up the 'good work'. The system is not questioned – rather it is something you are urged to submit yourself to. However, the students in BioMed have been brought up as women with what Rottenberg (2014) would characterize as neoliberal feminism, and face quandaries about how they balance demands across their spheres of life.

6.5.2 Balancing Different Life-Spheres

Rottenberg (2014) describes how the neoliberal "Feminist subject accepts full responsibility for her own well-being and self-care ... crafted on felicitous work-family balance based on a cost-benefit calculus" (p. 1). Similarly, a general concern expressed by the students in BioMed is how they feel the study programme may absorb all their time and energy. Thus, as a self-care measure it is important for them to find time for sports and leisure such as dancing, drawing, cooking, and being with friends and partners. Still, there is a sense of that studies invade all their spheres of life. Here Sandra recounts:

> I guess it's okay if there's something you absolutely need to come in and do in the weekend, but I don't think it should be the norm that I need to work every day. I want my weekend, and I also want to be in time to shop for groceries. If I'm off at eight every day I won't have time to prepare dinner. Stuff like that. I like drawing. I'd like to have time for that (Sandra, student).

As a student who is also associated with a lab, Sandra reacts to perceived expectations that she works evenings and weekends. She has discussed these expectations with friends in similar situations:

> I've got this impression that many, also my friends and so, they've experienced the same pressure that, well 'I can't go now even if I'm practically done so now, I have to sit here and just pretend I'm doing something because obviously I can't leave at three' (Sandra, student).

It seems that this high-workload environment is sometimes just pretending, or alternatively, very difficult to gauge and find ways to legitimately participate in. Still, the environment is invasive of what is Sandra perceives as rightly earned free-time. Even after having finished work students like Sandra still feel that they are expected to stay in order to position themselves as engaged. In the previous section Henry, a male teacher, reflected that one is allowed to have "A life on the side" as long as nothing interesting is going on at work. Clearly these women are not in a position that allows them to know whether what is going on in the lab is considered sufficiently interesting to merit the invasion of their leisure time. Instead, they feel forced to stay at their bench till someone else signals it is time to leave. They can do this for a while, but are increasingly concerned for how their future life will be shaped by this trajectory:

> I'd like a Ph.D. where I can be ambitious because I think, right now there is no partner, husband or children, so in no way, I mean, all-in on science. But after that I'd like to have a family at some point and then I'd like to have an option for a job where, it's just a job and not a life's work or anything (Maria, student).

In the quote, Maria describes how forming a family eventually will have consequences for her possibility to position herself as engaged. The students have definitely picked up on the masculine sentiment on ambition and engagement and they feel the pressure to submit themselves to spending all their time on a career that can be prioritized only until other priorities take over. Life in science cannot be combined with other aspects of womanhood. Thus, a Ph.D. is perceived as a means to a safe well-paid job in the industry, as the drawing below illustrates. It is titled 'Where I am in 3 years' and was made by one of the students during an interview (Fig. 6.1).

The strong emphasis that the BioMed students put into balancing life in science with other aspects of life is in line with are a range of studies showing how non-traditional students, such as women in science, face a cultural gap between academia and their values. The feeling of alienation and of being required to distance oneself from the values of academia is particularly salient in studies of working-class students (Bryan & Simmons, 2009). For some, this means balancing multiple identities. Moreover, studies focusing on underrepresented ethnic minorities in higher education show how some students are required to establish a sense of biculturalism in order to gain membership in the predominately white higher education setting (Rodgers & Summers, 2008, p. 183). Thus, academic success is contingent on students' abilities to break with non-academic backgrounds, practices, and home communities (Reay et al., 2010, p. 121). These researchers argue that instead, students can be supported in their identity work affiliated with negotiating different social contexts but conclude that continued racialised and classed inequalities means for "The very few working-class students who make it' that they risk ending up with what will later turn out to be a 'devalued degree'" (Reay et al., 2010, p. 121). To render science education equitable, it is necessary to consider

Fig. 6.1 Where I am in 3 years: Illustration by student drawn during interview. (*Note.* Ph.D. building says 'National Hospital'; Job building says 'Industry')

fundamental changes to pedagogy, changes that aim to go beyond standard inclusive practices, to instead introduce and use critical perspectives on the science that is taught (e.g., Mansour 2020; Mensah & Jackson 2018).

In our case, the sense of bi-culturalism has to do with performing successful student identity and talent in a way that is essentially masculine, white and middle-class on the one hand, but in opposition to, unaligned with other identities that are expected outside university. When female students attempt to combine these identities, they are interpreted as inauthentic or insincere in their commitment to science. The result is that the feminine identity of the 'good student' leaves the individual struggling to position themself as intelligibly authentic science learners within the dominant discourse of science. This is a challenge for the female informants. They describe their engagement as a materially limited supply – if you use it all while studying, there is nothing left for family and other interests. None of the narratives pointed out life outside university as a resource the students could draw on within academia. This is essentially an untenable situation for the women, whereas men can and do derive status from being able to prioritize work over family – even, or maybe especially, when signs point to health hazards. It is a perspective that teachers use to inform their pedagogy, and it in turn feeds into the mechanisms that ensures that the sciences are restricted to "all but a privileged few" (Archer et al., 2020, p. 373).

6.5.3 The Production of Stressed Students

The teachers also describe stress in their work life, but contrary to the students, they seem to perceive this as a condition for their work. Simon recounts his experience realizing that he had developed stress-induced migraines:

> Suddenly I saw a spot, a grey spot that just popped up. So now I had to go to the doctor. 'You are stressed,' she said. 'You need to cut down and that's got to be now! Otherwise, you'll risk losing your sight'. Management said, 'Oh well, we've just been waiting for you to come and ask to reduce your hours, because you've got far too much to do here'. Afterwards, I've been talking to some of the old geezers at the other departments and they are like 'We've got colleagues who've experienced the same,' simply had the same problems with their eyes, and they continue on and on and on and on and suddenly they get retinal migraines. That's something you must beware of. That it's a sign of stress. I just thought I'd been looking at a screen too long or something (Simon, teacher).

Interestingly, he describes this physical reaction as the first indication ever, that his engagement with work is excessive. However, colleagues and management alike, seems aware that the phenomenon exists why this pretence of disinterestedness with one's health could be considered another masculine performativity associated with science in general and 'talent' in particular. Being engaged with science has since ancient Greek philosophy been perceived as a matter of the mind, articulated as something detached from and superior to the body. Mind in opposition to body has been associated with rationality, intellect and masculinity, and femininity with body, emotions, and irrationality (Allegrini, 2015). Henry describes how his work "Doesn't cost blood, but surely sweat and surely tears". While it might be true that working with this teacher does not actually draw blood, figuratively speaking however, it might. In our data stress and consequences were articulated by teachers and students alike. But where the teachers as Simon seem to suppress the stress-signals, these are perceived different by the students. A student, Maya, describes how she has developed strategies for avoiding research environments that are too ambitious to allow for its members to take care of their mental health:

> I've seen people in this programme who just break along the way because they were at a place that was extremely ambitious. So, I think it's important not to just go for the most ambitious you can find but something that suits you. If you are the type who just loves to be there around the clock, then okay. I've met them, but they are few, I think. Before I started, I also thought it was important to choose something where you were sure to have something published. But my experience is perhaps more that what's important, is to find something you are comfortable with. I know there are places that are very ambitious. I keep far away from those places (Maya, student).

Maya's reasoning is a perfect example of resistance to submitting herself to an ambitious environment, where breaking down seem to be unavoidable. Maya is fully aware of her own well-being, and that some people may thrive working in such an environment – but that they are not for someone like her. In the quote she articulates how she is sensitive to masculine norms regarding talent, but how getting into such an environment would require her to compromise her own sense of self. This is why she refrains from following the path. This is in line with Holmegaard (2020b) who illustrates how exclusion mechanisms work on subject level when potential future challenges are imagined and coped with in present time. While some male students tended to think of potential future challenges as something they would deal with in the future, female students did not perceive having as much agency in their future, so they felt a need to avoid potential challenges rather than planning to deal with them. Maya's present resistance prevents her from pursuing prestigious research environments, and at the same time support the idea of female students as not interested in following such career tracks as her strategy is to avoid. Simon described the costs of being part of such environments by risking his health, and while he succeeded in entering the elitist culture, his strategy of avoiding rather than challenging the discourse is similar.

6.6 Discussion

With higher education moving from being exclusively for the elite to admitting the majority of a youth cohort as is the case in this study, a discussion on how the educational system can identify talented students has emerged. In this chapter we have investigated how teachers and students respectively ascribe meaning to what talent is and who talented students are in the cultural setting of higher education science. By applying science identity to investigate the notion of talent, we aim to examine values that feed into inequitable practices and produce some positions as desirable or as unattainable because they imply certain privilege. In particular, our aim was to explore the gendered inclusion and exclusion mechanisms that determines what and who were recognised in science, and to understand which positions were made available (and which were not) and the consequences of these recognition-practices as played out on subject level. To do so, we analysed two datasets produced in distinct work packages within the same project. One dataset drew on workshops and interviews with female MS students in a biomedical study programme that requires top GPA and is majority women. The other dataset was based on interviews with men who teach undergraduate science. A joint thematic analysis was carried out by both authors.

Theoretically, we approach talent as a social and cultural practice that is enacted everyday practices. We are interested in how students and teachers produce norms and ideas about what talent is and who it is for, which gendered consequences these ideas produce and how female students who are positioned as talented recognise and negotiate these positions. We conceptualise science identities as a net of subject positions being offered, obtained, and neglected, which also work to maintain and solidify talent as a subject position that students and teachers can see and evaluate themselves and each other through and draw meaning from.

The analysis showed how students from BioMed were competing against each other. They explained how early on in their studies they were positioned as a special group, the national best and thus natural aspirants to the limited number of Ph.D. positions in prestigious research environments. This competitive culture produced narrow positions for students to take on to obtain recognition as ‘engaged’. Participation in class required students to position themselves as well prepared, knowing the content matter, and totally believing in own abilities. As a result, the competitive environment produced a study culture where students governed themselves and each other and practiced a silencing of their own uncertainties and doubts by engaging passively and refrained from risking exposing themselves and potentially their lack of abilities.

The teachers in the study shared a perspective of talent as involving a willingness to take risks, to be able to invest time and energy without having a sense of what and where you are going or what may happen, because interest and fascination are intrinsic rewards. At the same time, talented students were expected to hold confidence in own abilities, and they were recognised in class by posing questions beyond course aim. Rather, students were expected to perform engagement that was rooted in a pure interest in the subject matter rather than being instrumental.

These practices of both students and teachers help maintain what is pointed to in the literature as the invisibility of women in STEM (e.g., Faulkner, 2007; Gonsalves, 2014; Tonso, 1998), as the ideals require certain levels of preestablished privileges. In our analysis, talent appeared to be associated closely with hegemonic masculine ideals of being active and taking risks which inherent the idea that all students have equally access to speaking their mind and gaining recognition for substance rather than by who shares them. Female students engaging in silence literally rendered the physical presence of the female body invisible to the teacher.

Moreover, the analysis showed how students and teachers seem stuck in a framing of talent that produces highly stressful and unhealthy study and work environments. To be recognized as talented during teaching, students were presumed to desirably invest their whole self and time in science. Teachers frame this way of engagement as a natural desire that students must feel the need to free, while inescapably reducing their lives outside university as less important. The female students resisted taking up this identity, as they experienced lack of coherence between their sense of self, and the expected science identities available in higher education. In more general terms, experiences with prestigious research laboratories taught the students that aspiring towards a celebrated career in research risked that their hard work would go unnoticed or that they would be met with scepticism if they were not willing to invest everything. From the interviews with the four teachers who were asked to relay their perception of talent based on their experience teaching science students, we understand that prioritizing life outside university and working conditions over science must result in being met with deep suspicion from established scientists like themselves. It suggests that femininely performed identities, within or outside of science, will never be associated with scientific talent.

Our study shows how higher education science continuously find new ways to keep and reproduce inequities in everyday practices (Lucas, 2001) by pretending to reward students who are talented. Talent work as a mechanism to position some students as naturally right no matter their grade-performance, but solely based on their ability to perform science identities that resonate with those of their teachers. It is time to stop talking about talent in science as anything but dangerous and harmful hegemonic masculinity and sexism shrouded by myths of meritocracy because it reproduces the narrative of the privileged white male, coming-of-age academic science-experience romance.

The analysis is notable in several ways. First of all, the lack of coherence between the students' identities is similar to what is found in studies of minoritized students in higher education (e.g., Rodgers & Summers, 2008). However, in this case the female students are the majority group – but still, they are kept from taking up dominant positions. This highlights the fact, that even if more women enter science, they are still required to submit themselves to masculine ideals and rules for doing science (as it has always been done), leaving limited space to negotiate identities and position oneself for recognition. Secondly, the ideas of investing your whole self in science is in line with what Rose (1998) explained as the expectation that modern individuals manage and invest themselves to maximize quality of life, which Rottenberg (2014) identifies as prime-motives driving neoliberal feminism. Explanations for why women are not present in research careers often points to priorities concerning reproduction and family-life. However, as Doerr (2021) argues with their study of scientists stuck in eternal temporary teaching positions, these explanations, tied in with meritocratic logic "Act as a guise to perpetuate inequity and maintain glass ceilings" and produce women as second-class academics who can be "Denied or deprived of basic elements integral to academic labor such as … appropriate pay and credit for intellectual property, recognition for teaching, a fundamental mission of the university, and normative respect needed to operate in civil society" (pp. 125–127). This bears repetition: women must be a fundamental mission of the university (as men and their work have always been), if we want to ensure them just conditions for doing academic labour – that is to say, taking up identities as scientists inside the university setting.

Changing culture is never easy. It is necessary to include *people* as a mission of the university and consequently also research science. This requires management support, teacher training and the willingness to actually not only admit a diverse student body to university, but also to support different ways of doing and being a student and doing and being within science. Thus, it is necessary to continue this line of research into how notions of science talent (or ideal student, effortless achiever, natural ability, genius and other myths or normative positions emphasising desirable science identities) produce women (and other nondominant groups) as 'second-class', 'unnatural', and 'inauthentic' practitioners of science. Here, critical ethnographies, for example, using participant observation or other immersive methodologies that allow a focus at the inter-personal and institutional level on interaction during the initiation and implementation of, for example, talent-initiatives should be of special interest. Such studies could wisely be accompanied and contrasted with a similar focus on anti-essentialist and gender-inclusive science initiatives (e.g., Hughes, 2001). Besides aiding equitable development of higher education, studies and actions are needed that will help policy-makers and teachers understand how and why *talent* is yet another way to advance white male middle-class at the expense of *the other*. Such understanding may also aid equity work in schools to make science identities more readily available to people of all ages.

References

Acker, J. (1990). Hierarchies, jobs, bodies: A theory of gendered organizations. *Gender & Society, 4*(2), 139–158. https://doi.org/10.1177/2F089124390004002002

Adriansen, H. K., & Madsen, L. M. (2009). Studying the making of geographical knowledge: The implications of insider interviews. *Norsk geografisk tidsskrift - Norwegian journal of geography, 63*(3), 145–153. https://doi.org/10.1080/00291950903238966

Allegrini, A. (2015). Italian students' ideas about gender and science in late-modern societies: Interpretations from a feminist perspective. In E. Henriksen, J. Dillon, & J. Ryder (Eds.), *Understanding student participation and choice in science and technology education* (pp. 331–347). Springer.

Archer, L., DeWitt, J., Osborne, J., Dillon, J., Willis, B., & Wong, B. (2012a). "Balancing acts": Elementary school girls' negotiations of femininity, achievement, and science. *Science Education, 96*(6), 967–989. https://doi.org/10.1002/sce.21031

Archer, L., DeWitt, J., Osborne, J., Dillon, J., Willis, B., & Wong, B. (2012b). Science aspirations, capital, and family habitus: How families shape children's engagement and identification with science. *American Educational Research Journal, 49*(5), 881–908. https://doi.org/10.3102/2F0002831211433290

Archer, L., & Francis, B. (2006). *Understanding minority ethnic achievement: Race, gender, class and 'success'*. Routledge.

Archer, L., Halsall, A., & Hollingworth, S. (2007). Inner-city femininities and education: 'Race', class, gender and schooling in young women's lives. *Gender and Education, 19*(5), 549–568. https://doi.org/10.1080/09540250701535568

Archer, L., Moote, J., & MacLeod, E. (2020). Learning that physics is 'not for me': Pedagogic work and the cultivation of habitus among advanced level physics students. *Journal of the Learning Sciences, 29*(3), 347–384. https://doi.org/10.1080/10508406.2019.1707679

Avraamidou, L. (2020). Science identity as a landscape of becoming: Rethinking recognition and emotions through an intersectionality lens. *Cultural Studies of Science Education, 15*(2), 323–345. https://doi.org/10.1007/s11422-019-09954-7

Barton, A. C. (1998). Reframing "science for all" through the politics of poverty. *Educational Policy, 12*(5), 525–541. https://doi.org/10.1177/2F0895904898012005004

Becker, H. S. (Ed.). (2002). *Boys in white: Student culture in medical school*. Transaction Publishers.

Bourdieu, P. (1974). The school as a conservative force: Scholastic and cultural inequalities. In J. Eggleston (Ed.), *Contemporary research in the sociology of education* (pp. 32–46). Methuen. https://doi.org/10.4324/9780203128374

Braun, V., & Clarke, V. (2006). Using thematic analysis in psychology. *Qualitative Research in Psychology, 3*(2), 77–101.

Bryan, E., & Simmons, L. A. (2009). Family involvement: Impacts on post-secondary educational success for first-generation Appalachian college students. *Journal of College Student Development, 50*(4), 391–406.

Butler, J. (1990). *Gender trouble: Feminism and the subversion of identity*. Routledge.

Carlone, H. B. (2004). The cultural production of science in reform-based physics: Girls' access, participation, and resistance. *Journal of Research in Science Teaching, 41*(4), 392–414. https://doi.org/10.1002/tea.20006

Carlone, H. B., Johnson, A., & Scott, C. M. (2015). Agency amidst formidable structures: How girls perform gender in science class. *Journal of Research in Science Teaching, 52*(4), 474–488. https://doi.org/10.1002/tea.21224

Carlone, H. B., Scott, C. M., & Lowder, C. (2014). Becoming (less) scientific: A longitudinal study of students' identity work from elementary to middle school science. *Journal of Research in Science Teaching, 51*(7), 836–869. https://doi.org/10.1002/tea.21150

Csikszentmihalyi, M., Rathunde, K., & Whalen, S. (1997). *Talented teenagers*. Cambridge University Press.

Danielsson, A. T., Berge, M., & Lidar, M. (2018). Knowledge and power in the technology classroom: A framework for studying teachers and students in action. *Cultural Studies of Science Education, 13*(1), 163–184. https://doi.org/10.1007/s11422-016-9782-0

Davies, B., & Harré, R. (1990). Positioning: The discursive production of selves. *Journal for the Theory of Social Behaviour, 20*(1), 43–63. https://doi.org/10.1177/2F0959353505051728

Doerr, K. (2021). *Is college science teaching women's work? Gender inequity in the physical sciences* [Doctoral dissertation]. The University of Texas at Austin.

Faulkner, W. (2007). 'Nuts and bolts and people': Gender-troubled engineering identities. *Social Studies of Science, 37*(3), 331–356. https://doi.org/10.1177/2F0306312706072175

Faulkner, W. (2011). Gender (in)authenticity, belonging and identity work in engineering. *Brussels Economic Review, 54*(2/3), 277–293.

Flyvbjerg, B. (2011). Case study. In N. K. Denzin & Y. S. Lincoln (Eds.), *The sage handbook of qualitative research* (4th ed., pp. 301–316). Sage.

Foucault, M. (1997). *Ethics: Subjectivity and truth (essential works 1954–1984)* (P. Rabinow, Ed.). Penguin.

Gagné, F. (2011). Academic talent development and the equity issue in gifted education. *Talent Development & Excellence, 3*(1), 3–22.

Gale, T., & Tranter, D. (2011). Social justice in Australian higher education policy: An historical and conceptual account of student participation. *Critical Studies in Education, 52*(1), 29–46. https://doi.org/10.1080/17508487.2011.536511

Gonsalves, A. J. (2014). "Physics and the girly girl - there is a contradiction somewhere": Doctoral students' positioning around discourses of gender and competence in physics. *Cultural Studies of Science Education, 9*(2), 503–521. https://doi.org/10.1007/s11422-012-9447-6

Hartsmar, N., Leathwood, C., Ross, A., & Spinthourakis, J. (2021). Can educational programmes address social inequity? Some examples from Europe. In A. Ross (Ed.), *Educational research for social justice* (pp. 271–297). Springer.

Hasse, C. (2016). *An anthropology of learning*. Springer.

Holmegaard, H. T. (2020a). Complexity, negotiations, and processes: A longitudinal qualitative, narrative approach to young people's transition to and from university. In N. E. Fenton & W. Ross (Eds.), *Critical reflection on research in teaching and learning* (pp. 107–130). Brill Sense.

Holmegaard, H. T. (2020b). Master students' imagined futures. The interaction of students' resources, narrative repertoires and their thoughts about postgraduate futures within selected STEM master programmes. *Scandinavian Journal of Educational Research, 65*(7), 1217–1237. https://doi.org/10.1080/00313831.2020.1789213

Hughes, G. (2001). Exploring the availability of student scientist identities within curriculum discourse: An anti-essentialist approach to gender-inclusive science. *Gender and Education, 13*(3), 275–290. https://doi.org/10.1080/09540250120063562

Jackson, C., & Nyström, A. S. (2015). 'Smart students get perfect scores in tests without studying much': Why is an effortless achiever identity attractive, and for whom is it possible? *Research Papers in Education, 30*(4), 393–410. https://doi.org/10.1080/02671522.2014.970226

Keller, E. F. (1985). *Reflections on gender and science*. Yale University Press.

Kogan, M. (1997). Diversification in higher education: Differences and commonalities. *Minerva, 35*(1), 47–62.

Kvale, S. (2006). Dominance through interviews and dialogues. *Qualitative Inquiry, 12*(3), 480–500. https://doi.org/10.1177/2F1077800406286235

Leathwood, C., & O'Connell, P. (2003). 'It's a struggle': The construction of the 'new student' in higher education. *Journal of Education Policy, 18*(6), 597–615. https://doi.org/10.1080/0268093032000145863

Liu, Y., Green, A., & Pensiero, N. (2016). Expansion of higher education and inequality of opportunities: A cross-national analysis. *Journal of Higher Education Policy and Management, 38*(3), 242–263. https://doi.org/10.1080/1360080X.2016.1174407

Lucas, S. R. (2001). Effectively maintained inequality: Education transitions, track mobility, and social background effects. *American Journal of Sociology, 106*(6), 1642–1690. https://doi.org/10.1111/gwao.12074

Mansour, N. (2020). The dissonance between scientific evidence, diversity and dialogic pedagogy in the science classroom. *International Journal of Science Education, 42*(2), 190–217. https://doi.org/10.1080/09500693.2019.1706114

Marginson, S. (2016). Global stratification in higher education. In S. Slaughter & B. Taylor (Eds.), *Higher education, stratification, and workforce development* (pp. 13–34). Springer. https://doi.org/10.1007/978-3-319-21512-9_2

Mensah, F. M., & Jackson, I. (2018). Whiteness as property in science teacher education. *Teachers College Record, 120*(1), 1–38. https://doi.org/10.1177/2F016146811812000108

Ministry of Higher Education and Science. (2018, 9 March). *Nye målsætninger for de videregående uddannelser [Press release] [New goals for higher education]*. https://ufm.dk/aktuelt/pressemeddelelser/2018/nye-malsaetninger-for-de-videregaende-uddannelser-skal-geare-os-til-fremtiden/nye-malsaetninger-for-de-videregaende-uddannelser

Radnor, H., Koshy, V., & Taylor, A. (2007). Gifts, talents and meritocracy. *Journal of Education Policy, 22*(3), 283–299. https://doi.org/10.1080/02680930701269186

Rasmussen, A., & Rasmussen, P. (2015). Conceptions of student talent in the context of talent development. *International Journal of Qualitative Studies in Education, 28*(4), 476–495. https://doi.org/10.1080/09518398.2014.916013

Read, B., Francis, B., & Robson, J. (2001). Playing safe': Undergraduate essay writing and the presentation of the student 'voice. *British Journal of Sociology of Education, 22*(3), 387–399. https://doi.org/10.1080/01425690124289

Reay, D., Crozier, G., & Clayton, J. (2010). 'Fitting in' or 'standing out': Working-class students in UK higher education. *British Educational Research Journal, 36*(1), 107–124. https://doi.org/10.1080/01411920902878925

Reimer, D., & Thomsen, J. P. (2019). Vertical and horizontal stratification in higher education. In R. Becker (Ed.), *Research handbook on the sociology of education* (pp. 308–328). Edward Elgar Publishing.

Ridgeway, C. L., & Correll, S. J. (2004). Unpacking the gender system: A theoretical perspective on gender beliefs and social relations. *Gender & Society, 18*(4), 510–531. https://doi.org/10.1177/2F0891243204265269

Rienecker, L., Jørgensen, P. S., Dolin, J., & Ingerslev, G. H. (2015). *University teaching and learning*. Samfundslitteratur.

Rodgers, K. A., & Summers, J. J. (2008). African American students at predominantly white institutions: A motivational and self-systems approach to understanding retention. *Educational Psychology Review, 20*(2), 171–190. https://doi.org/10.1007/s10648-008-9072-9

Rose, N. (1998). *Inventing our selves: Psychology, power, and personhood.* Cambridge University Press.

Rottenberg, C. (2014). The rise of neoliberal feminism. *Cultural Studies, 28*(3), 418–437. https://doi.org/10.1080/09502386.2013.857361

Scager, K., Akkerman, S. F., Keesen, F., Mainhard, M. T., Pilot, A., & Wubbels, T. (2012). Do honors students have more potential for excellence in their professional lives? *Higher Education, 64*(1), 19–39. https://doi.org/10.1007/s10734-011-9478-z

Scott, P. (1995). *The meanings of mass higher education*. Open University Press.

Scott, P. (2019). Trow's elite-mass-universal triptych: Conceptualising higher education development. *Higher Education Quarterly, 73*(4), 496–506. https://doi.org/10.1111/hequ.12224

Søndergaard, D. M. (2005). Making sense of gender, age, power and disciplinary position: Intersecting discourses in the academy. *Feminism & Psychology, 15*(2), 189–208. https://doi.org/10.1177/2F0959353505051728

Søndergaard, D. M. (2006). *Tegnet på kroppen. Køn: koder og konstruktioner blandt unge voksne i akademia*. Museum Tusculanum.

Staunæs, D. (2003). Where have all the subjects gone? Bringing together the concepts of intersectionality and subjectification. *NORA: Nordic Journal of Women's Studies, 11*(2), 101–110. https://doi.org/10.1080/08038740310002950

Taber, K. S. (Ed.). (2007). *Science education for gifted learners*. Routledge.

Thomas, K. (1990). *Gender and subject in higher education*. Society for Research into Higher Education.

Tonso, K. L. (1998, April 13–17). *Engineering gender – gendering engineering: What about women in nerd-dom?* [Conference presentation]. American Educational Research Association, San Diego, CA, US.

Trow, M. (2010). *Twentieth-century higher education: Elite to mass to universal.* John Hopkins University Press.

Wolfensberger, M. V. (2012). *Teaching for excellence. Honors pedagogies revealed.* Waxmann.

Wong, B., & Chiu, Y. L. T. (2020). University lecturers' construction of the 'ideal'undergraduate student. *Journal of Further and Higher Education, 44*(1), 54–68. https://doi.org/10.1080/0309877X.2018.1504010

Wong, B., DeWitt, J., & Chiu, Y. L. T. (2021). Mapping the eight dimensions of the ideal student in higher education. *Educational Review, 1-19*. https://doi.org/10.1080/00131911.2021.1909538

Wright, T. (2016). Women's experience of workplace interactions in male-dominated work: The intersections of gender, sexuality and occupational group. *Gender, Work and Organization, 23*(3), 348–362. https://doi.org/10.1111/gwao.12074

Chapter 7
Doing Geoscience: Negotiations of Science Identity Among University Students When Learning in the Field

Lene Møller Madsen and Rie Hjørnegaard Malm

7.1 Introduction

Waiting for the bus to take us on a field trip, a first-year geology student wearing sneakers said: 'I can see that I don't have the right shoes today' as an opening phrase when joining a group of fellow students and me [the first author] at the parking lot. The other students smiled; they were all wearing various types of hiking boots. I participated in the field trip as a pedagogical supervisor of the teacher taking the students on the field trip. I had also been with the students at lectures and in the classroom, observing them in learning situations, for example, classifying rock samples and producing geological profiles. Unlike those situations, it was striking that from the first instance I joined the students in the field, issues of how to look and perform as a geology student came to the foreground (Fieldnotes, first author).

This observation points to interesting considerations about disciplinary culture, belonging and tacit knowledge if we are to understand how students engage in identity work and negotiate their belonging in relation to the geoscience cultural community. How do you know what shoes to wear on a geology field trip? What happens if you have the 'wrong' shoes on? Does the type of shoes you wear influence how you are perceived in the geology community? Can you succeed in geology with sneakers?

In this chapter, we unfold the relations and intersections between science identity, disciplinary culture, belonging and tacit knowledge as they are produced and negotiated in learning situations within the discipline of geology in higher education. With this approach, we aim to explore what is at stake when becoming a geologist and negotiating a geoscience identity in relation to the process of establishing disciplinary knowledge. Science identity, disciplinary culture,

L. M. Madsen (✉) · R. H. Malm
University of Copenhagen, Copenhagen, Denmark
e-mail: lmmadsen@ind.ku.dk

H. T. Holmegaard, L. Archer (eds.), *Science Identities*, Contributions from Science Education Research 12, https://doi.org/10.1007/978-3-031-17642-5_7

belonging and tacit knowledge are not easily separated entities when analysing students' negotiation processes. Our intention is not to use the empirical data to make such a distinction or to provide a framework for analysis. Instead, we draw attention to how analysis of the relations and intersections between these, in concrete learning situations, can unfold the process of becoming a geoscience person. We do this by using our empirical material to show how students continuously engage in identity work and negotiate their belonging within the disciplinary culture of geology in relation to their experiences in the field. Hence, we add to the literature in science identity by providing thick descriptions of how disciplinary knowledge occurs, is negotiated and contested.

7.2 Science Identity and the Doing of Geoscience

We place this chapter within the research field of science identity and use the construct of science identity to investigate learning processes and negotiations in the becoming of a geoscience person. Carlone and Johnson (2007) presented the now seminal model of science identity, with three dimensions of science identity: competence, performance and recognition as well as their intersections. The construct of science identity has been used and developed in a range of studies of both students', teachers' and faculty's science identities. For a review, see, for example, Lee (2012), Varelas (2012), and Hazari et al. (2010). We situate our work in line with studies of science identity that draw on identity not as something one has but as something that one negotiates and performs in cultural settings (Avraamidou, 2020; Carlone & Johnson, 2007; Holmegaard et al., 2014; Malm, 2020).

By conducting science identity research in a disciplinary setting, we wish to contribute with knowledge on both "how science identity might serve in making science learning meaningful and purposeful" (Avraamidou, 2020, p. 4) and explore which "Kinds of people [are] promoted and marginalized by science teaching and learning practices" (Carlone & Johnson, 2007, p. 1189). In the conceptual paper 'Science identity as a Landscape of Becoming', Avraamidou (2020), as others before her, departs from science identity as an individual and isolated project. Instead, she argues for seeing identity as part of a social practice as she links the construction of identities to 'the process of becoming' developed in the theory of communities of practice (Lave & Wenger, 1991; Wenger, 1998). We adapt this approach to science identity, but place the discipline in the foreground by investigating 'the process of becoming a geoscience person' in the context of disciplinary culture. As such, the chapter does not adopt an explicit and critical intersectionality lens to science identity as advocated and discussed by Avraamidou (2020) and Gonsalves (2020), but situates itself within the studies that Avraamidou (2020) describes as studies of science engagement through the construct of identity. Examples are Gonsalves (2014), Johansson et al. (2018) and Danielsson (2012), who study the doing of physics and its intersection with gender and gender performance.

In Danielsson's (2012) study of four women studying physics at university, she argues that: "A more complete understanding of students' learning of physics must expand the meaning of learning (...) [and] needs to be understood in a broader personal and societal context" (p. 26). In line with this approach, we argue for a discipline-specific investigation of science identity in higher education, focusing on how the tacit and embodied practices within a discipline as well as the forms of knowledge intersect with science identity and the process of becoming in specific learning situations. By taking the disciplinary context into account, we follow up on the call by Enyedy et al. (2006) to foreground practice as a missing component in the construct of identity. In their study of middle school teachers, Enyedy et al. (2006) argue that: "A person's identity is shaped and negotiated through everyday activities" (p. 71). To address this, we focus our study on fieldwork as an important practice for learning and developing a disciplinary science identity within geology.

In geology, fieldwork is historically linked to the exploration of the Earth and scientific knowledge production. The practice of fieldwork is perceived to be integrated in a geologist's identity and thereby important when students negotiate their belonging in a higher education geoscience programme (Malm, 2020). Working in the field include a series of embodied, implied notions and practices (Raab & Frodeman, 2002) such as observations, measurements, drawings as well as creating interpretations in the field. The embodied practice of fieldwork includes developing an intuitive selection strategy and learn to make judgements based on both knowledge and direct experiences in the field. Stories of fieldwork are produced in the context of geology as an exploration-centred discipline (Nielsen et al., 2012), where both historical and current discoveries are told as personal stories of overcoming difficulties or dangerous situations in order to retrieve data from the field. These stories are central in the disciplinary culture, as they reproduce fieldwork as central to the identity performance. Thus, the practice of fieldwork ties together the idea about 'who' a geologist is and 'how' geology is performed within a strong disciplinary culture focusing on performance in the field (Malm, 2020).

A cultural practice includes a learning community with shared practices that are accepted through shared language, signs and recognition (Hastrup, 2004; Lave & Wenger, 1991). As the disciplinary culture and the fieldwork practice of geology include a physical element and the use of instruments, we expand our exploration of students' identity work by paying attention to tacit and embodied knowledge in fieldwork. As described by Polanyi (1966), knowledge is both practical and theoretical: "I shall always speak of "knowing", therefore, to cover both practical and theoretical knowledge" (Polanyi, 1966, p. 7). In addition, he argues that to know something includes more than you are capable of expressing, that some of our knowledge within disciplines is tacit and often strongly linked to the practices within the discipline.

The use of instruments is central in understanding tacit knowledge as it is related to "The art of knowing" (Polanyi, 1966, p. 7) with an embodied dimension to knowing. Hence, one also understands and knows things through one's use of the instrument through a bodily knowledge. Particularly the tacit use of instruments is important when collecting geological data in nature. Here, a bodily physical task becomes interwoven with an intellectual reasoning task. Competence is therefore both shown by being able to physically perform the task or handle an instrument and being able to scientifically reason and argue for the choices made in the field. Both tasks are perceived as including a tacit and embodied dimension.

Gaining access to the fieldwork practice, being and feeling competent and receiving recognition in relation to fieldwork are important for students to become part of the disciplinary culture, and at the same time problematic when structural or physical barriers hinder participation (Malm et al., 2020; Núñez et al., 2020; Posselt & Nuñez, 2021). Students in higher education geoscience programmes in Scandinavia often participate in field courses, excursions and fieldwork throughout their studies to learn the practice of fieldwork, as it is believed that these experiences strengthen their engagement and retention in the geosciences (Boyle et al., 2007; Streule & Craig, 2016). The students' engagement with the field and development of a fieldwork practice to be included in the disciplinary culture all start with entering the space of the field, and entering with the wrong shoes is not insignificant when one's appearance and performance in the field is linked to recognition, culture and identity.

7.3 Methodology

The methodology applied in this chapter is eclectic in the sense that we draw on a range of empirical material produced in different research projects with different purposes and teaching situations. The empirical material includes a variety of aspects of fieldwork within the discipline of geology: from looking at minerals in the classroom to practising measurements and discussing geological structures in the field. The multitude of the different types of empirical material as well as various learning situations allow for studying different facets of becoming a geologist.

We organised the analysis in three portraits of learning situations named 'how to make sense of disciplinary knowledge', 'tacit and embodied practice with instruments' and 'being a geologist'. We use the term portraits to denote that we unfold the analysis more extensively in selected parts of the large empirical material. In all three portraits, the point of departure is analysing how science identity intersects with the concepts of sense of belonging, disciplinary culture and tacit knowledge. Hence the main concept in all three portraits is science identity and then the additional concepts are foregrounded in different ways in the different portraits, as shown in Table 7.1. We perceive science identity as a dynamic and social practice that is negotiated and re-negotiated in a continuous process in relation to the learning situations the students are engaged in (Pozzer & Jackson, 2015).

The interviews used in all three portraits were conducted in either Danish or Norwegian, the interactions observed (when verbal) were likewise in these two languages. For this chapter, we have made the translations into English as close to the meaning in the original language as possible. Throughout the chapter we have used they/them, and when using names in portrait 1 and 3, we have used gender-neutral names approved for official use in Denmark.

Table 7.1 The three portraits of learning situations

Portraits of learning situations	Main concept	Additional concepts	Empirical material
How to make sense of disciplinary knowledge	Science identity: Competence, performance and recognition	Belonging Disciplinary culture	First-year student doing coursework and fieldwork
Tacit and embodied practice with instruments		Tacit knowledge Disciplinary culture	Two BSc students doing fieldwork
Being a geologist		Belonging Disciplinary culture Tacit knowledge	MSc student doing fieldwork as part of an MSc thesis

Note. The table shows how the analytical constructs are foregrounded in the portraits and outlines the empirical material used

We provide an outline of the methodologies and materials used in each of the portraits in the following sub-sections.

7.3.1 Methodologies and Materials for Portrait 1: How to Make Sense of Disciplinary Knowledge

In the academic year 2012–2013, the second author researched how first-year students in a BSc study programme of geology negotiated and made sense of geoscience knowledge and developed a geological identity (Malm, 2014). Six students were interviewed individually five times during the first year of their studies (one student was interviewed only twice). The first interviews took place in August, 2 weeks before semester start. The next interviews took place in October and one in December at the end of the first semester. Another round of interviews was held in March and the final interviews were held in June just before the summer break. The interviews lasted between 40-min and 2.5 h and were transcribed verbatim and anonymised. A narrative approach was applied where the students' personal stories were explored (Polkinghorne, 1988; Webster & Mertova, 2007). The analysis of the interviews used a thematic analysis approach (Braun & Clarke, 2006) to find the main themes for each student at each point in time. Then, each interview was analysed using the analytical questions: (1) What influences the students' understanding of the content? and; (2) How do the students negotiate and make sense of the content? Hence, both the development of an understanding of the content as well as the students' negotiations during this process were unfolded. Additionally, the students' narratives were analysed across time using the concept of turning points (Holmegaard et al., 2014). This allowed for analysis of changes in the narratives and the students' negotiations and re-negotiations in relation to their making sense of the content.

In this chapter, we focus on one of the students, Ada, and how they negotiate knowledge production in different learning situations during their first year of studies and how the lens of identity brings new knowledge to our understanding of these processes.

7.3.2 Methodologies and Materials for Portrait 2: Tacit and Embodied Practice with Instruments

In 2019, the first author had been assigned as a pedagogical supervisor of an assistant professor teaching BSc geology students. As a pedagogical supervisor, she discussed and observed teaching practice with the assistant professor in various teaching settings. The teaching settings included lectures, classroom teaching and a field trip to a number of geological sites. At the beginning of the supervision, there were no plans to make any parts of the supervision into research. However, as the observations and interactions with the students intensified during the field trip, it became clear that insights for research could be made by analysing the observations. For purposes of formative feedback to the assistant professor, comprehensive notes were taken by the first author during all observations of teaching in the lecture hall, the classroom and in the field. Notes were taken in relation to students' engagement with the teaching activities and mastering of the geoscience content. These notes serve as the empirical material for the analysis.

In this portrait, we focus on one of the many practises that students are urged to master in order be part of the discipline of geology and become a geologist – the learning of tacit and embodied practices with instruments, in this case using a compass to take measurements.

7.3.3 Methodologies and Materials for Portrait 3: Being a Geologist

From 2016 to 2020, the second author conducted several studies of geology students working in the field as part of her PhD research in Higher Education Earth Science. The studies used ethnographic methods, consisting of participatory observations and interviews of individual student, to study students' practices of fieldwork and associated identity work in the field. In one study, the second author stayed with a group of MSc students and their supervisors for 12-days in the field while the students' collected data for their MSc theses (Malm, 2020).

The ethnographic observations of the fieldwork included both the students' individual work, their interactions with the supervisors and with the group of students working in the field. During the first 2 days everyone worked together as a group covering a large geographical area, after which the students worked alone for 2 days in smaller areas. The observations of one of the students, Ehm, in the field aimed at understanding how the student solved problems, used previous knowledge, collected data and engaged in the field, thus documenting the lived experiences of the student during a period of fieldwork (Feig, 2010). Extensive field notes were recorded before, during and after the fieldwork (Emerson et al., 2011; Walford, 2009) in order to document as many impressions as possible, for example the cooperation and dialogue between participants, the supervision of the student, the physical environment and how the participants engaged in it.

One year later, an interview was conducted with Ehm. By this time, Ehm had graduated from university. The interview aimed at exploring how the student had used the data from the field and how their thesis developed from fieldwork to finished product. The interview was performed as a timeline interview (Adriansen, 2012) to learn how the student made sense of the different events during the year.

In this portrait, we analyse the observations from the field and explore, in depth, how Ehm's practices of doing geology in the field intersect with a continuing process of developing a geoscience identity. An analysis of the timeline interview is used to illustrate how the identity negotiations performed in the field are negotiated and re-negotiated by the student when working with the data in the laboratory, writing up the thesis and after graduating.

7.3.4 *Issues of Positionality*

Empirically, we draw on our own involvement and research in geoscience programmes at two different research-intensive universities in Scandinavia – the University of Copenhagen and the University of Oslo. In this chapter, we use our positions, insights and access to knowledge within the field of geosciences as empirical material, hence, issues of positionality are important (Edwards, 2002; Neal & Gordon, 2001; Rose, 1997). We have worked together during the last 10-years both in a supervisor-student relation and as colleagues in a number of research and development projects (Madsen et al., 2021; Madsen & Malm 2011, 2017; Malm & Madsen 2014, 2015a, b; Malm et al., 2015, 2020). We interact with the people we analyse, and we have different professional and personal relations to them. This creates a wide range of ethical considerations including being both insiders and outsiders in our research. As pointed out by Adriansen and Madsen (2009), "When doing insider research it is necessary to address the insider relationships explicitly in order to reveal the complexity of research relations" (p. 146) both in relation to the research matter and in relation to one's interviewees. To address this, we have tried to make our insider and outsider roles as transparent as possible in the above section without hampering the ethical issues of reporting from our studies and our interaction and relations with the people involved. In addition, we are both in various degrees and positions insiders in relation to the research matter as we are trained within the disciplines of geography (first author) and geology (second author) and have both participated in numerous fieldwork settings. This means that in the analysis we draw on our knowledge of the content, practices and perspectives shared in the community of researchers doing geoscience fieldwork.

7.4 Portraits of Negotiating a Geoscience Identity in Learning Situations

In this analytical part of the Chapter, we provide thick descriptions of how disciplinary knowledge is negotiated and contested in different learning situations. We explore this by showing how students interact with geoscience knowledge and the doing of geoscience in the field. Hence, we show how the three dimensions of science identity (competence, performance and recognition) intersect and are negotiated in specific geoscience fieldwork learning situations.

7.4.1 Portrait 1: How to Make Sense of Disciplinary Knowledge

The first time we met Ada, we talked about their way into the geology programme. After graduating from high school, Ada thought about becoming a geologist but decided to complete an MSc within a different field; becoming a geologist felt like "Too geeky a choice" at the time. However, after completing an MSc degree in another field and not finding a job, Ada applied for the geology programme because "I want to do something that makes me happy" (1st interview, August).

When Ada started attending courses, reflections on the reasoning processes in geology began. The two first courses were a course in Earth Systems (introduction to geological processes, systems and materials) and a course on geophysics (introduction to geophysical exploration and groundwater systems). In the Earth Systems course, Ada was introduced to the rock cycle, how sediments are deposited and how that can provide insights into prehistoric environments, Ada said:

> It seems like they [the teachers] know very little, and then they make a huge story out of it, without really having the evidence, that bothers me (…) It seems like this is how it's done, to make a hypothesis and then I would think, that one should disprove it, but it doesn't really happen here, they just prove the hypothesis (2nd interview, October).

In this quote, Ada is struggling with the interpretive nature of geology (Dodick et al., 2009; Frodeman, 1995) as an epistemology that conflicts with their understanding of science. The reasoning in geology uses a narrative form of logic in contrast to the hypothetical deductive method (Cleland, 2001; Watson, 1969). Within the natural sciences, it has been argued that we use six different types of reasoning; the type described above is referred to as historically based evolutionary reasoning (Kind & Osborne, 2017). Here, evidence relies "On constructing theories about what might have happened in the past" and "Such theories have succeeded because they have been the best possible inferences for what exists" (Kind & Osborne, 2017, pp. 12–13). In the discipline of geology, this means formulating multiple hypotheses based on field observations as part of the research process (Chamberlin, 1890; Cleland, 2001). When deciding between the hypotheses, the geologist constantly re-evaluates observations in the search of that piece of evidence that can make one hypothesis more likely than others (Cleland, 2001). In the quote, Ada reflects on the reasoning process in geology in relation to the hypothetical deductive reasoning process, which is a familiar epistemology for them from other natural sciences. Ada is surprised at not meeting this method when learning to interpret sediments, structures and fossils in geology.

Looking at Ada's negotiations of science identity through the whole year, it seems in hindsight that already from this point, Ada started negotiating a part of the discipline of geology as not 'real' natural science.

The next two courses Ada attended were a palaeontology course (the study of prehistoric species and the development of life on Earth) and a course on mineralogy and metamorphic petrology (the study of the mineralogical and chemical composition of rocks in Earth systems). Later in the first year, Ada reflected on the previous semester:

> I was often really frustrated [during the palaeontology course], but [the teacher] understood it, did not get mad at me and was just trying to answer my questions … [The teacher] understood the frustration and said: "I understand, it's not very smart, but you know there are a lot of other animals, who are also stupid" (4th interview, March).

In the interview, Ada explains how learning palaeontology was challenging when needing to learn details (e.g., how ancient creatures have moved on the ocean floor or eat) and not learning the overall picture (e.g., why these details matter in the geological story). Ada questions the production of knowledge when presented with full interpretations of the ancient species' ways of living and the connection to geological history, knowing that it is based on few observations or small details. When constructing a geological story, the geologist applies a narrative form of logic, where the observed details start to make sense as they are placed in the overall structure of an area or in Earth's history (Frodeman, 1995; Watson, 1969). Ada is therefore on the right track in trying to make sense of the details and the 'full story', here the evolution of living organisms. However, the concerns and difficulties experienced can still be linked to understanding the interpretive type of reasoning, as it includes a judgement of which details are important. The complex nature of this type of reasoning and the challenges in learning become evident in Ada's learning process. However, the quote also shows how Ada is recognised by the teacher for adapting a critical learning approach. The teacher explains that some species are not that complex, they are just simple and 'stupid' and understanding their ways is just a small part of understanding the whole system. This recognition plays an important part in Ada's negotiations of belonging in the study programme, as Ada finds a way to make sense of palaeontology with the recognition and help of the teacher. Ada's need to put the details into a larger picture is acknowledged as an appropriate way of thinking, though not all details are equally important. Learning the difference and making these judgements is important when developing a selection strategy as part of the geological interpretation.

In the second part of the academic year, the students attended a course in sedimentology (the study of sedimentary rocks formed by sand, silt, chalk and clay) and a course in magmatic petrology (the study of the mineralogical and chemical composition of rocks in deep Earth systems). In the sedimentology course, Ada again struggled with the interpretive type of reasoning and getting recognition from the teacher. In the second part of the academic year, the students began negotiating the subfields within the discipline differently; the interviewed students, including Ada, perceived sedimentology as less scientific, although in various degrees. The students considered sedimentology to be 'storytelling' and very different from igneous petrology and mineralogy, as these disciplines use mathematics and chemistry, hence other types of reasoning to support interpretations (Malm, 2014). The room for interpretation and creativity seemed larger in sedimentology and the students understood that interpretation can be discussed and several answers are possible. This clearly sets the type of reasoning apart from the experimental natural sciences (Cleland, 2001) and the students were negotiating this in various ways in relation to their science identity. For Ada, the negotiation was reinforced by not being recognised by the teacher:

> I found a proper definition in the book, and then I brought it the other day [to class] and the teacher was not entirely happy with me (Ada smiles). [The teacher] said: "Now I have to be careful with what I say, when I come over here" (4th interview, March).

Ada's approach to learning was contested by this teacher as the teacher hesitated to answer the (many) questions. Ada experienced that when the teacher did answer, the explanation was different from the definitions in the textbook. Ada therefore brings the textbook to class and asked directly how to understand the different explanations. In the situation, Ada experienced having gone too far in terms of questioning, it was not a legitimised practice in the learning situation to ask this type of question. In contrast to the previous course, Ada received no recognition for questioning and exploring, and was left with the impression of being 'too much'. After this, Ada focused more on the textbook than learning in class. At this time, Ada also chose to focus on the course in magmatic petrology.

As summer approached the students spent more time in the field. A seven-day trip to the island of Bornholm (with many sedimentological localities) was the final and the longest field trip of the year. On this field trip the students applied the interpretive type of reasoning many times in different situations, guided by the teachers. To illustrate the challenges of learning the interpretive type of reasoning we show the difference of presenting sedimentology as science in the making (as experienced in the field, see Fig. 7.1) and as a ready made science (as presented in the classroom, see Fig. 7.2).

Being in the field provides access to and experience with conducting observations and learning that the observations can motivate different interpretations.

Fig. 7.1 Stevns Klint at Højerup in Denmark. (*Note.* Fieldwork with students observing the face of the cliff and trying to make sense of what they observe, illustrating science in the making (Latour, 1987; Madsen et al., 2021). Photos taken by Rie Hjørnegaard Malm, published in Madsen et al. (2021))

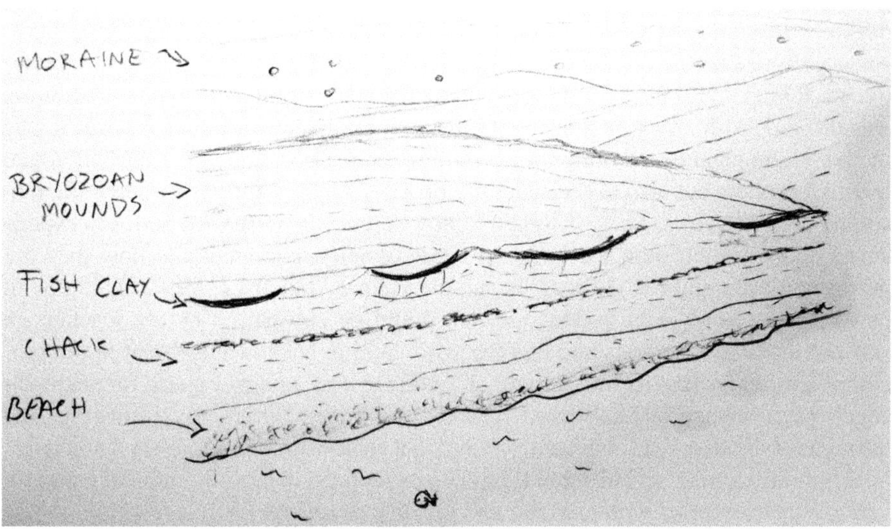

Fig. 7.2 Sketch of Stevns Klint at Højerup, matching the area in the photograph in Fig. 7.1. (*Note.* The sketch only displays the most prominent features of the cliff. Hence, there is an interpretation and a selection process behind what is included in the sketch. The reasoning process behind how the sketch is constructed (the transposition from the observed, illustrated with the photograph above) is often hidden for students in class and lecture-halls and shown as ready-made science (Latour, 1987). Sketch made by Rie Hjørnegaard Malm)

Hence, they have first-hand experience of the interpretive reasoning process. Ada says about the field trip:

> Sedimentology, I found out, is a discipline that works when you are out in the field, it does not work to sit and stare into some boring dry textbook, the discipline is simply not for that, (…) But when you're out and look at the outcrops, then it's pretty exciting actually, so I'm starting to like it a little better (5th interview, June).

In the above quote on fieldwork, Ada is on the way to developing the embodied type of thinking required for working in the field (Raab & Frodeman, 2002). The fieldwork provides access to learning about where to look, how to look, and what to infer from observing. For Ada, the experience in the field started re-negotiations of the discipline and sense of belonging. Ada realised why the teachers were not able to precisely answer the questions in class – because the same observations can prompt a series of possible interpretations, and to make solid inferences, one constantly needs to consider alternative interpretations.

At the end of the first year, Ada felt more comfortable in studying geology and reported being interested in focusing on mineralogy or petrology. Until this point, Ada's negotiations had indicated that deselecting sedimentology was in line with interpreting sedimentology as a 'less scientific' discipline. However, during the first year and especially during the fieldwork encounter, Ada learned to see how interpretations are done and thereby accessed some of the tacit knowledge that was not made visible in the beginning of the first year, and this expanded their understanding of geology as a discipline.

The analysis of Ada's story shows how identity work and negotiations of disciplinary practice are central in making sense of scientific content in relation to the different reasoning types in geology. Ada initially identified geology as a 'hard' science discipline (i.e., deductive and based on testing hypotheses). The interpretative reasoning styles in geology were new to Ada and learning this type of reasoning was difficult with no explicit access to the tacit knowledge it builds on. Ada's way of challenging and questioning the scientific practice was central to their individual learning process but also to their identity process. Understanding the scientific reasoning processes in geology is linked to how science is performed and Ada experienced and explored both the different types of science content and how they are performed differently in the various sub-disciplinary practises. Ada struggled with performing a knowledge production that could be recognised by the teachers in specific learning situations and thereby with gaining acknowledgement in the process. Expressions like feeling 'too much' indicate how a narrow space for acknowledged performance is established in the disciplinary culture. By following Ada throughout the first year, it became visible that establishing disciplinary knowledge in different learning settings and types of recognition constantly influence and, in various ways, interact with learning and identity processes.

7.4.2 Portrait 2: Tacit and Embodied Practices with Instruments

> Two students are standing very close to the wall of granitic gneiss with basaltic intrusions. The wall is part of a quarry that has been exploited for many years for road construction material. Trucks are driving by from the larger quarry below and other trucks are coming to the quarry where they are loaded with sorted rocks and then leave again. We have been directed to the wall where no quarry work is occurring today. We are all wearing helmets and luminescent vests.
>
> Along the wall the students begin describing and understanding the geological structures by measuring the spatial orientation of the different observable structures in front of them. I go towards two students engaged in the measurements. They are sharing one compass, one student holds the compass and moves their body to align with the wall, the other student asks questions about the practice of aligning one's body and holding one's hands in position.
>
> They are discussing the process of measuring to assure each other that they are placing themselves in the right spot and holding the compass at the right angle to do the measurement. Then they discuss what they are trying to measure [the spatial distribution of the different layers now completely eroded]. The two students are trying out the procedure of measuring with the compass in a number of places while they move along the wall to find out how the structures are spatially oriented. It seems like they have a common understanding of how they are measuring.
>
> A group of students gather around the teacher further down the wall of granitic gneiss in the quarry. By moving their body and being explicit about the position of their hands, the teacher uses the compass to show how to conduct the measurement, and the students, either in pairs or alone, are trying out the procedures. Students raise questions and the teacher answers by showing and doing the measurement over again with body and hands. The two students measuring along the wall reach the group of students and the teacher. They seek reassurance of their practices by asking the teacher a number of questions. The teacher reaches out for the compass and shows them how.
>
> After a while, we leave the granitic gneiss and walk towards the bus; the next stop is further north. Here, surrounded by forest we will find old weathered down volcanic cores and search for small pieces of mantle xenolith, the treasure of the day to bring home! (Fieldnotes, first author).

In the above, the students were learning to conduct a 'strike and dip' measurement of a geological structure in order to understand the spatial orientation of the observed structures. This consists of multiple tasks of both bodily doing things and using mental abstractions, as well as managing to combine the two. It includes both explicit and tacit knowledge of how to measure, what and how to interpret, as well as being able to formulate and articulate a geological structure. As Malm (2020) shows in a study of third-year BSc geology students, after weeks of practicing hitting rocks in a specific area, they are able to know which type of rock they are hitting with the hammer just by the bodily experience and sound created by the hitting. Learning this embodied way of using a hammer is a tacit type of geological knowledge. The measurement of strike and dip with a compass is another instance of embodied tacit geological knowledge.

Tacit knowledge implies two parts of knowledge; one part gives access to the other part – Polanyi (1966) formulates this as "We know the first term only by relying on our awareness of it for attending to the second" (p. 10). Hence, we are often not able to explicate the first part as it becomes a practice we perform with a purpose of understanding something else. It is only when our understanding of the phenomenon is flawed that our attention is directed to the first part of the tacit knowledge. Measuring strike and dip, that is, the spatial orientation of a geological feature with a compass, is a good example of this. When measuring several features in the same area, the geologist is able to tell how the features relate to each other. This information can for example be used to evaluate the timing of events creating the features, that is, constructing a timeline of relative age. This is part of constructing the geological story of the area – taking a small piece of information and relating it to the whole. Hence, the actual measurement of the strike and dip is in the terms of Polanyi the first term that we only know for attending to the second, namely the spatial orientation of the layers. Conducting the measurement includes choosing a place to measure that represents the true orientation of the feature, and this can be difficult in places where there are no clear indicators. Hence, you might not measure the true orientation and thereby gain data that do not make sense in the geological story. As students are repeatedly exposed to making these choices and evaluating them, over time they will learn 'where to measure', 'what makes sense' and 'seeing' in a geological setting. They might be able to practically do the measurement on many types of surfaces and only start to evaluate how they did it when something is not right or does not add up in the data. The practice becomes embodied. In this sense, they will have an embodied tacit understanding of doing the measurement and only notice it when something does not 'make sense'.

Let us return to the observations of the students in the field. A geologist must be able to describe and communicate the geological structure that they observe in the field. This involves a process of sketching, measuring and classifying. In the above observation of students and their teacher in the field, the focus is on practicing the ability to measure strike and dip as part of the process of describing the geological structure of the wall of granitic gneiss with basaltic intrusions. The actual measurement is the means to understand the geological structure. For the experienced geologist, in this case the teacher, the knowledge of how to measure the strike and dip is an embodied tacit knowledge used to describe the geological structure. For the teacher to teach the students to make strike and dip measurements the teacher shows the practice again and talks to the students about what it means. This is what Polanyi describes as "An ostensive definition" (Polanyi, 1966, pp. 5–6). This means to point something out for the student, a combination of a "Naming and pointing" (Polanyi, 1966, p. 5). In the case with the strike and dip measurement, we can describe how to make the physical measurement (how to use the compass), but how to translate that into a geological structure and be able to describe the geological structure is tacit. In the learning situation, we must trust the students to be able to see what "we have not been able to communicate" (Polanyi, 1966, p. 6). Similarly, in the work by Goodwin (2000) where he analyses the colour practises (determining the colour of the dirt they are excavating) of archaeologists he talks about the setting's *opaqueness* to outsiders. Only by becoming a member of the cultural community does this opaqueness dissolve.

This portrait of BSc students in the field demonstrates how the process of learning 'what makes sense' and how developing embodied tacit knowledge in the field – which is an important competence within geology – are strongly linked to access to the disciplinary culture held by teachers. In Portrait 3 we turn our attention to an MSc student collecting data in the field to show the ways in which the dimensions of performance and recognition play a central role in understanding the student's challenges of embodying a science identity.

7.4.3 Portrait 3: Being a Geologist

During their first days in the field, the student Ehm, and the two supervisors observe several outcrops together. The scientific content of the MSc thesis is discussed as new discoveries are made and new limitations found. The work environment is enthusiastic and the small group constantly work out new hypotheses to fit the observations. The supervisors and the student discuss how the different hypotheses can be approached and what the focus of the fieldwork can be.

At the different localities, all of them are making observations, taking notes and samples at the same time. They discuss what they see and how it can be interpreted. The group work long days in the field and use the evenings to look at maps, discuss the data and new ideas.

Then the student works alone for a few days. The group have agreed upon a preliminary plan for the fieldwork, and the type of observations and interpretations the student needs for their thesis. The student visits both new localities alone and returns to some previously visited localities. The student systematically collects data and makes notes of the questions that arise during the work. In these days Ehm often says "I have no idea what I am doing" and "It is so difficult to see what they [the supervisors] saw" but continues and seems somewhat comforted that the supervisors can help.

When Ehm and the supervisors meet in the field again, they continue to explore, discuss ideas and develop the hypothesis. The supervisors negotiate the responsibility with the student saying "So now, you show us what we need to see". The student is thus expected to take charge and to ask questions, however, although the supervisors express this shift in roles, they often get caught up in their own explorations, leaving the student with more and more questions unexplained. In the beginning, Ehm asks many questions, tries to explain the difficulties and participate in discussions. But as the days pass Ehm becomes quieter, and spends a lot of time checking their notebook. The research question is still open and the first half of the fieldwork has not provided a clearer idea of what to focus on.

This pattern continues, Ehm continuously tries to follow the supervisors' ideas and hypotheses, both in the field and in the late evening discussions over the map. The supervisors are enthusiastic, clearly enjoy these discussions and thrive in the field. The supervisors are keen on collaborating with Ehm and genuinely try to help and support. One of the supervisors tells me that this student does a great job, it is a difficult task but the student seems "Tough, and knows their way around the field". The supervisors clearly trust the student and see that the student performs well in the environment.

But insecurity seems to build up in the student. At this stage, Ehm has a few days alone, and these days in the field are completely different from the previous. The student is ruminating over the data collection, visits several sites to check the data, and at one point states: "Why did I set out to do this in the first place, when it is so unlikely that I will succeed? I remember when I was a child, I took a test, and it showed that I was not good at spatial thinking. Then why am I standing here and working with this? This is not who I am".

The student and I leave the field utterly exhausted (Fieldnotes, second author).

The geological fieldwork and collecting data from the field formed the backbone of Ehm's thesis work and part of the research group's further explorations of the area. Our analysis explores how the disciplinary culture is reflected in the working culture of the group and how both influence Ehm's individual experiences in the field. In the above, we first see how the group work with multiple hypotheses (Cleland, 2001) and the supervisors see connections at the localities based on their embodied experience (Raab & Frodeman, 2002). The group does an open exploration of the area, which aligns with a field-based research project in geology.

We see that Ehm is concerned about collecting 'the right data' and that being under a time constraint is stressful. The open research process and the lines of thinking are invisible and to some extent not available to the student. The supervisors rely on Ehm's previous training and experience and the fact that they recognise Ehm's performance in the field: Wearing the right clothes, keeping a notebook at hand, handling a compass correctly, asking sensible questions, being interested, taking lots of samples and working hard. This performance of fieldwork practices is recognised by the supervisors as valid within the disciplinary culture. The dimensions competence, performance and recognition (Carlone & Johnson, 2007), and particularly their intersections become visible here. The student performs as being competent, is recognised as being competent, but does not recognise themselves as being competent. In addition, the supervisors include the student in their work as an apparent equal, they often say that the student's observations are interesting and use them to advance their common ideas.

The student is being included in the community of practice and recognised as a legitimate participant (Lave & Wenger, 1991). However, when Ehm experiences difficulties with practicing geoscience, for example, keeping track of the hypothesis and seeing what the supervisors see in the field, Ehm engages in identity work by starting to question their own competence, which quickly leads to questioning both past, present and future ideas of becoming a geologist. This relates to feeling a sense of belonging (why did I set out to do this…), in relation to the disciplinary culture and embodied and tacit knowledge production (when the student starts questioning the observation and reasoning process).

> A year after being in the field, Ehm explains that many months were spent in the laboratory doing analysis, failing and trying, again and again. The results kept being 'wrong' and Ehm ruminates over the data collection in the field and feels convinced that mistakes were made during the fieldwork. The months pass by and in the end Ehm needs to show some of the analysis to the supervisors. When finally doing so, the supervisors react with great surprise; the analysis shows a new, unexpected pattern. Over the last month of the student's thesis process, the supervisors become more and more convinced that the data provides evidence for a new interpretation of the area. Ehm has not made a mistake, the fieldwork and following analysis in the laboratory has provided the research group with a new and exciting scientific story to tell. A story that later leads to a new publication about the area.
>
> At the time of the interview, Ehm acknowledges this as an exciting development and talks about the positive response during the exam, however, it is clear that Ehm still struggles with the experiences from the field (Fieldnotes, second author).

The insecurities about how to observe and 'seeing' what the supervisors see, combined with difficulties in understanding the complicated geological story, build up during the fieldwork. This feeling of being insecure about the fieldwork and data collection stays with the student in the laboratory, influencing the learning and research process, and continues to affect the student throughout the writing process and is still present after graduation. The student finds it difficult to acknowledge their own contribution to the results and continuously refers to 'making mistakes', not being prepared enough for the fieldwork and not being competent enough to 'see'. This ongoing negotiation of not belonging persists within the student and evidently the experiences in the field and reflections on the fieldwork has had a major impact on the way the student cannot identify as competent and as a legitimate participant.

By using science identity, we see how negotiations of competence in the field involve identity work and negotiations of becoming a geologist. This portrait shows how students have to 'be' and be able to 'see' in a specific way, to be recognised in the field by others. The specific performance of being competent in the field means that the students need to navigate the disciplinary culture in order to belong. As this portrait illustrates, this can amplify students' feeling of not 'fitting in' as they struggle to find ways of recognising *themselves* as competent aspiring geologists. Ehm performs in line with the disciplinary practice and culture, gets recognition from the supervisors and clearly has competence to produce valuable research data. However, Ehm still struggles with recognising their own achievements and feeling they belong. Why does this fieldwork become so problematic? Our analysis is that the student connects being competent in the field with being a 'good' geologist and having the experience of having failed in the field has consequences for the student's self-image as a future geologist. In this sense, the student experiences the disciplinary culture as excluding. The strong emphasis on the fieldwork practice and associated idea of how to be as a field geologist hinders the student in recognising their own accomplishments beyond the fieldwork and thereby hinders them in trusting the data collected in the field as they are analysed in the laboratory. Thus, the student has no strategies for separating competence in the field and being competent in the following laboratory, analysis and writing processes. The insecurities arising from the fieldwork made it difficult to negotiate a place within geology, in spite of having performed well, having received recognition and made an important contribution to research. The student does not work as a geologist today.

7.5 Discussion and Concluding Reflections

> It was not that any of us commented on the shoes, it seemed obvious that they were not aligning to the code of conduct and, apparently, so strongly that the student felt a need to express this as a first thing when entering the group. As more students arrived also more students with sneakers arrived. In the end, a mix of sneakers and hiking boots went on the field trip. However, not only the shoes were noticed, also types of trousers, jackets and hammers were noticed, I'm sure, although not verbally commented on (Fieldnotes, first author).

Returning to the wearing of sneakers, we wish to draw attention to the disciplinary aspect of recognition and competence in our understanding of how science identity is negotiated and performed in specific learning situations. In the three portraits given, we show how the three dimensions of science identity – competence, performance and recognition (Carlone & Johnson, 2007) – intersect and are negotiated in specific geological fieldwork learning situations. All three portraits illustrate how the students' learning experiences and identity work intersect with disciplinary knowledge and culture. In Portrait 1, the student is navigating teaching practices, disciplinary culture and knowledge production in relation to the interpretive nature of geology and includes ideas of science, in Portrait 2 the student is negotiating the tacit embodied practice of how to conduct strike and dip, and in Portrait 3 the student engages in identity work in relation to how to 'see' as a geologist in the field. Hence the tacit and embodied knowledge that are part of the disciplinary culture play into the students' identity work related to competence, performance and recognition within the discipline.

The portraits also illustrate how the embodied and implied knowledge are present across different teaching and learning practices throughout a study programme from first year of BSc to MSc level. In Portrait 1, the first-year student is simultaneously navigating how to understand the scientific approaches in geology, the different interactions with teachers and their patterns of recognition as well as finding a way to belong in the discipline. In Portrait 2, we see how the teacher holds embodied tacit knowledge, which per definition is difficult to teach BSc students orally and the teacher therefore constantly shares the practice of handling the instrument when conducting the measurement. The same mechanism is in play in Portrait 3, where the supervisors work side by side with the MSc student and by their doings show how they develop their understanding of the area and the hypothesis simultaneously. Based on these analyses, we demonstrate how including studies of tacit and embodied knowledge in concrete learning situations can unfold students' identity work when establishing their disciplinary knowledge in geology. Hence, we advocate including disciplinary culture in our understanding of competence, performance and recognition when we talk about science identity.

Science identity within a discipline takes many forms, both visible and tacit, and for the students it requires identity work in relation to performing in specific ways in order to be recognised as competent within the discipline. However, we have also shown how a disciplinary culture can become part of exclusion practices in specific learning situations by producing a narrow space for students to perform within, gain recognition from and feel competent in. In order to allow for wider participation in geology, we advocate a disciplinary culture that allows for an inclusion of different types of belonging and various ways of becoming a geologist.

Acknowledgements We wish to thank all the students and teachers who without any hesitations shared their practice and allowed us to participate in and study their fieldwork practices. We also wish to thank our colleague within science education Anders Johannson for very valuable and insightful comments and suggestions during the writing process. Finally, we thank two anonymous reviewers for their constructive comments that helped improve the chapter significantly. All usual disclaimers apply.

References

Adriansen, H. K. (2012). Timeline interviews: A tool for conducting life history research. *Qualitative Studies, 3*(1), 40–55. https://doi.org/10.7146/qs.v3i1.6272

Adriansen, H. K., & Madsen, L. M. (2009). Implications of doing insider interviews: Studying geography and geographers. *Norwegian Journal of Geography, 93*(3), 145–153.

Avraamidou, L. (2020). Science identity as a landscape of becoming: Rethinking recognition and emotions through an intersectionality lens. *Cultural Studies of Science Education, 15*(2), 323–345. https://doi.org/10.1007/s11422-019-09954-7

Boyle, A., Maguire, S., Martin, A., Milson, C., Nash, R., Rawlinson, S., Turner, A., Wurthmann, S. L., & Conchie, S. (2007). Fieldwork is good: The student perception and the affective domain. *Journal of Geography in Higher Education, 31*(2), 299–317. https://doi.org/10.1080/03098260601063628

Braun, V., & Clarke, V. (2006). Using thematic analysis in psychology. *Qualitative Research in Psychology, 3*(2), 77–101.

Carlone, H. B., & Johnson, A. (2007). Understanding the science experiences of successful women of color: Science identity as an analytic lens. *Journal of Research in Science Teaching, 44*(8), 1187–1218. https://doi.org/10.1002/tea.20237

Chamberlin, T. C. (1890). The method of multiple working hypotheses. *Science, 15*(366), 92–96.

Cleland, C. E. (2001). Historical science, experimental science, and the scientific method. *Geology, 29*(11), 987–990. https://doi.org/10.1130/0091-7613(2001)029%3C0987:HSESAT%3E2.0.CO;2

Danielsson, A. T. (2012). Exploring woman university physics students 'doing gender' and 'doing physics'. *Gender and Education, 24*(1), 25–39. https://doi.org/10.1080/09540253.2011.565040

Dodick, J., Argamon, S., & Chase, P. (2009). Understanding scientific methodology in the historical and experimental sciences via language analysis. *Science and Education, 18*(8), 985–1004. https://doi.org/10.1007/s11191-008-9146-6

Edwards, B. (2002). Deep insider research. *Qualitative Research Journal, 2*(1), 71–84.

Emerson, R. M., Fretz, R. I., & Shaw, L. L. (2011). *Writing ethnographic fieldnotes*. University of Chicago Press.

Enyedy, N., Goldberg, J., & Welsh, K. M. (2006). Complex dilemmas of identity and practice. *Science Education, 90*(1), 68–93. https://doi.org/10.1002/sce.20096

Feig, A. D. (2010). Technology, accuracy and scientific thought in field camp: An ethnographic study. *Journal of Geoscience Education, 58*(4), 241–251. https://doi.org/10.5408/1.3534863

Frodeman, R. (1995). Geological reasoning: Geology as an interpretive and historical science. *Geological Society of America Bulletin, 107*(8), 960–968. https://doi.org/10.1130/0016-7606(1995)107%3C0960:GRGAAI%3E2.3.CO;2

Gonsalves, A. J. (2014). "Physics and the girly girl - there is a contradiction somewhere": Doctoral students' positioning around discourses of gender and competence in physics. *Cultural Studies of Science Education, 9*(2), 503–521. https://doi.org/10.1007/s11422-012-9447-6

Gonsalves, A. J. (2020). Operationalizing intersectionality to investigate the role of recognition in the landscape of becoming. *Cultural Studies of Science Education, 15*(2), 347–357. https://doi.org/10/ggh6jx

Goodwin, C. (2000). Action and embodiment within situated human interaction. *Journal of Pragmatics, 32*(10), 1489–1522. https://doi.org/10.1016/S0378-2166(99)00096-X

Hastrup, K. (2004). *Kultur: Det fleksible fællesskab*. Aarhus University Press.

Hazari, Z., Sonnert, G., Sadler, P. M., & Shanahan, M. C. (2010). Connecting high school physics experiences, outcome expectations, physics identity, and physics career choice: A gender study. *Journal of Research in Science Teaching, 47*(8), 978–1003. https://doi.org/10.1002/tea.20363

Holmegaard, H. T., Madsen, L. M., & Ulriksen, L. (2014). A journey of negotiation and belonging: Understanding students' transitions to science and engineering in higher education. *Cultural Studies of Science Education, 9*(3), 755–786. https://doi.org/10.1007/s11422-013-9542-3

Johansson, A., Andersson, S., Salminen-Karlsson, M., & Elmgren, M. (2018). "Shut up and calculate": The available discursive positions in quantum physics courses. *Cultural Studies of Science Education, 13*(1), 205–226. https://doi.org/10.1007/s11422-016-9742-8

Kind, P. E. R., & Osborne, J. (2017). Styles of scientific reasoning: A cultural rationale for science education? *Science Education, 101*(1), 8–31. https://doi.org/10.1002/sce.21251

Latour, B. (1987). *Science in action: How to follow scientists and engineers through society*. Harvard University Press.

Lave, J., & Wenger, E. (1991). *Situated learning: Legitimate peripheral participation*. Cambridge University Press.

Lee, Y. J. (2012). Identity-based research in science education. In *Second international handbook of science education* (pp. 35–45). Springer.

Madsen, L. M., & Malm, R. H. (2011, September 5–9). *Learning in the field: Addressing the sequence of learning activities for geology students* [Abstract]. 9th ESERA International Conference, Lyon, France.

Madsen, L. M., & Malm, R. H. (2017). Geovidenskab skaber omverdensbevidste studenter. *Geografisk Orientering, 42*(2), 42–48.

Madsen, L. M., Evans, R., & Malm, R. (2021). Using a wicked problem for inquiry-based fieldwork in high-school geology: Addressing climate change and mass extinction events. In J. Dillon, M. Achiam, & M. Glackin (Eds.), *The role of out-of-school science education in addressing wicked problems* (pp. 167–188). Springer.

Malm, R. H. (2014). *Becoming a geologist, identity negotiations among first year geology students* (Publication no. 35). Master's thesis, University of Copenhagen.

Malm, R. H. (2020). *What is fieldwork for? Exploring roles of fieldwork in higher education earth science* (Publication No. 2349). Doctoral dissertation, University of Oslo. https://www.duo.uio.no/handle/10852/82828

Malm, R. H., & Madsen, L. M. (2014). *Geovidenskab på vej, en undersøgelse af Geovidenskab A på stx og htx*. University of Copenhagen.

Malm, R. H., & Madsen, L. M. (2015a). Brugen af geotoper i gymnasiefaget Geovidenskab A. In O. Fjær, P. J. Sætre, & R. Hokseggen (Red.). *Geografi på timeplanen: Rapport fra Geografikonferansen 2014* (pp. 57–69). Geografisk institutt, NTNU og Norsk Geografisk Selskap.

Malm, R. H., & Madsen, L. M. (2015b). *Geovidenskab – en undersøgelse af de første studenter*. University of Copenhagen.

Malm, R. H., Vinther, N., Ravn, M. W., & Madsen, L. M. (2015). Geodetektiv: Redskab til problemorienteret undervisning, abstrakt til Big Bang: Naturfag for fremtiden Konference 2015, Roskilde.

Malm, R. H., Madsen, L. M., & Lundmark, A. M. (2020). Students' negotiations of belonging in geoscience: Experiences of faculty-student interactions when entering university. *Journal of Geography in Higher Education, 44*(4), 532–549. https://doi.org/10.1080/03098265.2020.1771683

Neal, R., & Gordon, J. (2001). Fieldwork among friends. *Resources for Feminist Research, 28*(3–4), 99–113.

Nielsen, K. H., Harbsmeier, M., & Ries, C. (2012). *Scientists and scholars in the field: Studies in the history of fieldwork and expeditions*. Aarhus University Press.

Núñez, A. M., Rivera, J., & Hallmark, T. (2020). Applying an intersectionality lens to expand equity in the geosciences. *Journal of Geoscience Education, 68*(2), 97–114. https://doi.org/10.1080/10899995.2019.1675131

Polanyi, M. (1966). *The tacit dimension* (2009th ed.). The University of Chicago Press.

Polkinghorne, D. E. (1988). *Narrative knowing and the human sciences*. University of New York Press.

Posselt, J. R., & Nuñez, A. (2021). Learning in the wild: Fieldwork, gender, and the social construction of disciplinary culture. *The Journal of Higher Education, 93*(2), 163–194. https://doi.org/10.1080/00221546.2021.1971505

Pozzer, L. L., & Jackson, P. A. (2015). Conceptualizing identity in science education research: Theoretical and methodological issues. In C. Milne, K. Tobin, & D. DeGennaro (Eds.), *Sociocultural studies and implications for science education* (Vol. 12, pp. 213–230). Springer.

Raab, T., & Frodeman, R. (2002). What is it like to be a geologist? A phenomenology of geology and its epistemological implications. *Philosophy & Geography, 5*(1), 69–81. https://doi.org/10.1002/sce.20096

Rose, G. (1997). Situating knowledges: Positionality, reflexivities and other tactics. *Progress in Human Geography, 21*(3), 305–320. https://doi.org/10.1191/030913297673302122

Streule, M. J., & Craig, L. E. (2016). Social learning theories – An important design consideration for geoscience fieldwork. *Journal of Geoscience Education, 64*(2), 101–107. https://doi.org/10.5408/15-119.1

Varelas, M. (Ed.). (2012). *Identity construction and science education research. Learning, teaching, and being in multiple contexts* (Vol. 35). Springer.

Walford, G. (2009). The practice of writing ethnographic fieldnotes. *Ethnography and Education, 4*(2), 117–130. https://doi.org/10.1080/17457820902972713

Watson, R. A. (1969). Explanation and prediction in geology. *The Journal of Geology, 77*(4), 488–494.

Webster, L., & Mertova, P. (2007). *Using narrative inquiry as a research method. An introduction to using critical event narrative analysis in research on learning and teaching*. Routledge.

Wenger, E. (1998). *Communities of practice. Learning, meaning, and identity*. Cambridge University Press.

Chapter 8
Identity Perspectives in Research on University Physics Education: What Is the Problem Represented to Be?

Anders Johansson and Johanna Larsson

8.1 Introduction

Recent years have seen an increased focus on identity in physics, but what do we talk about when we talk about identity? In this chapter, we argue that critical, theoretical attention needs to be paid to the conceptualizations of *physics identities* that are used in research. We focus on physics in higher education settings, the physics that is taught to students in physics majors, bachelor's programmes or physics teacher programmes, but also to students in engineering, chemistry and life science. We discuss and highlight what physics education can learn from using identity perspectives, what problems are expected to be solved by bringing identity to the table, and how some under-used characterizations of the problem of identity in physics, using feminist epistemologies, provide new insights.

In physics education, identity perspectives are to a large extent used with the explicit goal of making physics studies better for students. For this interdisciplinary research to have impact, it needs to speak to both other education researchers and the physics community. In this complex situation, there is a need to discuss how the conceptualizations of identity may imply problematizations that run contrary to the intentions of the researchers. Building on a critical review of the literature and an analysis of empirical examples from our own work, we want to highlight how identity perspectives in physics education implies various problematizations, and how a careful consideration of our understanding of these problematizations are needed

---

A. Johansson (✉)
Chalmers University of Technology, Gothenburg, Sweden
e-mail: anders.l.johansson@chalmers.se

J. Larsson
Chalmers University of Technology, Gothenburg, Sweden

Uppsala University, Uppsala, Sweden

H. T. Holmegaard, L. Archer (eds.), *Science Identities*, Contributions from Science Education Research 12, https://doi.org/10.1007/978-3-031-17642-5_8

for identity perspectives to be an effective agent for change in physics education research.

The chapter is structured in four sections. The first gives a background of the field of research in higher education physics, identity research, and the approach that we take for analysing the literature. The second section presents and discusses the four problematizations that we find implied by the different ways of conceptualizing and using identity in the papers we review. The third section further illustrates the four problematization using an analysis of empirical examples from our own research projects, and is followed by a concluding discussion.

8.1.1 Studying Identities in University Physics Education

In science education, increased attention has been paid to identity issues over the last twenty-years, not least evidenced by the diverse contributions to this volume. This goes hand in hand with a growing concern for the sociocultural aspects of education (Lemke, 2001) where identity is seen for example as a way of connecting the social environment to issues such as students' attitudes and choices (Holmegaard, 2015; Tytler, 2014). Identity is also utilized to understand the large problems with equal representation and participation in science (Brickhouse, 2001), where the underrepresentation of women and minority groups is both a historical and present-day problem (Organisation for Economic Co-operation and Development (OECD, 2017a); United Nations Educational, Scientific and Cultural Organization (UNESCO, 2015). Here one influential tradition focuses on the possibilities for students' identification or disidentification with science in school (Archer et al., 2012; Brickhouse & Potter, 2001; Carlone, 2003; Godec, 2018). Other studies have focused on identity negotiations of university students and faculty (Carlone & Johnson, 2007; Ong, 2005; Tonso, 2006).

The underrepresentation of women in physics education has been a concern for many years. Even today, men make up 70–80% of the physics students in many Western countries (American Physical Society, 2018; OECD, 2017b; Universitetskanslersämbetet, 2016), even if the situation varies between cultural contexts (Hasse & Sinding, 2012; Moshfeghyeganeh & Hazari, 2021). The problem is even greater when considering the participation of ethnic or racial minorities, where for example black women are severely underrepresented in physics in the US (Coble et al., 2013; Ong et al., 2018). Apart from participation, many studies have documented gender disparities in results on standardized tests. Reviews of the quite extensive research highlighting gender in physics education have noted a lack of critical perspectives on gender, power and the culture of physics (Danielsson, 2010; Johansson, 2018a; Traxler et al., 2016). Here, a nuanced use of identity concepts is proposed as one way forward. However, along with the growing number of publications in physics education focusing on identity comes the need to analyze and evaluate how identity is used and to what effect. Hence, we want to highlight how different approaches, conceptualizations and methodologies in identity research in physics imply different representations of the "problem" identity are set out to solve.

8.1.2 What Is the Problem of Physics Identity Represented to Be?

In this chapter, we approach the question of identity perspectives in physics education with tools inspired by poststructuralist policy analysis (Bacchi, 2009; Bacchi & Goodwin, 2016) in order to initiate a discussion about the aims and consequences of our uses of identity. This is similar to what Vidor et al. (2020) has done for research on physics and gender and Beddoes (2011) for problematizations of underrepresentation in engineering education. In the framework of poststructuralist policy analysis, the central question for analysis is: "What is the problem represented to be?" (Bacchi, 2009) By phrasing the question in this somewhat awkward way, Bacchi highlights an understanding of "problems" not as given, but rather as inseparable from the process of "problematization", the social construction of some circumstances (and not others) as a problem. Constructions of problems, problematizations, are inherently political and contribute to a specific understanding of the world. Importantly, the point is not to ask how specific policy makers or researchers conceive of a problem, but rather which broader discourses shape a certain understanding (Bacchi, 2015). By taking a step back to question obvious 'problems', a 'what's the problem represented to be' approach urges us to ask how a certain problematization has come to be, which assumptions are inherent in the societal discourses it relates to, what effects this way of representing the problem has, and what questions are silenced or impossible to ask when we conceive of the area in this way.

One example of why this type of broader analysis is important is given by considering how the lack of diversity in STEM education is conceptualized by researchers as well as policymakers. Many governmental organizations highlight the need for a more extensive scientific workforce, where previously underrepresented groups are seen as an "untapped resource" (European Commission, 2004). Representing the problem in this way, as an issue of missed opportunities for expanding the scientific workforce, can be argued to stem from, and relate mainly to, discourses about national competitiveness (Lövheim, 2016; Lucena, 2005). Consequently, researchers have argued that focusing on getting previously underrepresented groups through STEM education (the "STEM pipeline") does not address issues of social justice sufficiently (Archer et al., 2017; Tajmel, 2019). The concerns of disadvantaged groups are not automatically addressed by more of their members getting STEM degrees (Cumings et al., 2014). Hence, while problematizing the composition of the STEM workforce puts diversity on the agenda, it also limits the ways in which social justice is considered and addressed.

Inspired by the "What's the problem represented to be" approach, we discuss the use of identity perspectives in physics education by analysing how different conceptualizations and uses of identity in research papers imply different ways of representing the problem this research is intended to solve. Our analysis constitutes a critical review (Grant & Booth, 2009), based on our knowledge of work in the field and complemented by readings of additional papers found through a literature search. In problematizing the problematizations of research on identity in physics education, we are writing from a specific position as academics within Swedish majority culture, fostered in a Scandinavian critical tradition and the academic fields of Swedish Physics Education Research and Gender Studies. We aim to develop a loving critique, where the role of the critic, following Foucault (1981/2000), is not to pass judgement but to analyse the assumptions of conventional ways of thinking and acting in order to explore how things could be different. In this we do not claim an objective outside position (Haraway, 1988), but are speaking from a position situated within the research discourses that we discuss, even though our perspective is a specific one within these traditions.

The papers included in our critical assessment of the field were in part found using a full-text search on the keywords 'physics', 'identity', and 'students' in Scopus that yielded 182 papers (as of March, 2020). Fifty-two of these discussed social aspects of education to some extent, and of these, thirty-four used identity as an analytical construct in higher education physics and were included in the review. We do not aim for a systematic or exhaustive review, and do not present a categorization of papers. Rather, we want to highlight how various 'problematizations' are implied in the conceptualizations and uses of identity in research, sometimes in multiple ways in the same text.

In studying problematizations with a primary focus on the conceptualization of the identity construct in published research articles, we were guided by three concrete questions:

1. How is the use of identity perspectives motivated and what explicit problems do these motivations point out?
2. What theoretical perspectives are used and what implicit problematizations do these imply?
3. How are the findings interpreted and discussed in relation to possibilities for change?

This last question represents one way of studying the 'effects' (Bacchi, 2009) of problematizations. However, our focus is on the 'discursive' (theoretical) effects, which means that we have not analysed how the research has been used for effecting change in practice, beyond what's stated in the papers. Building on the specific conceptualizations in single papers, our overarching analysis discusses the assumptions, connections and effects of the problematizations implied by broader conceptualizations of identity.

8.2 Four Problematizations

In this section we present and discuss four problematizations that we see as implied in the use of identity in research on physics education. These characterizations are the outcome of analysing the public discourse about research as represented in research papers. As such, the problematizations reflect how identity is conceptualized and used in the discussed papers, but do not describe the motivation or reasoning of the involved researchers. Our characterization thus presents idealized, even simplified, cases, and most researchers use several of these in different contexts (even in the same texts) for achieving the many goals of research in higher education physics. Our overall aim is thus to provide a background for wider discussion of these topics.

8.2.1 Students Fail to Develop Physics Identities

Research has traditionally represented the problem of physics education as one of understanding. Researchers have asked how, and why students (do not) understand physics content. Another focus has been on problem solving strategies, and here a classical take on the problem is to support students in learning to think like physicists (i.e., 'experts', Van Heuvelen, 1991). Inspired by the sociocultural tradition in education, in recent years some researchers have pointed out that we also need to attend to the social aspects of education to "expand the literature on physics expertise beyond the cognitive realm" (Rodriguez et al., 2015, p. 12). A popular way of doing this is by using the framework of situated learning, which points to how learning needs to be considered a question of gaining legitimacy in communities of practice, that is, a question of participation and of constructions of identity (Lave & Wenger, 1991; Wenger, 1998). Here, the issue is not only to 'think' like a physicist, but also to become one.

A number of studies focus on how students develop an (expert-like) physics identity. For example, Irving and Sayre (2015) discuss 'becoming a physicist' for upper-division physics students as acquiring a physics identity and being recognized by the community of physicists. It is common to use communities of practice and related theoretical frameworks to analyse students' development of physics identities in specific contexts and courses. Some examples are interdisciplinary experiences (Rodriguez et al., 2015; Sawtelle & Turpen, 2016), advanced physics courses (Irving & Sayre, 2014, 2015), or the experiences of teaching or learning assistants (Close et al., 2016; Gretton et al., 2017). One motivation for the need to research identity that is present in several of these studies is the problem with attrition in physics; students who develop a physics identity are more likely to stay in the field (Rodriguez et al., 2015). These studies commonly identify the complex negotiations going on as students navigate membership in different communities, for example moving beyond expectations of performing in classroom communities (doing 'schooling') towards more meaningful central participation in communities of professional physicists and physics instructors (Close et al., 2016). The representation of the problem that an identity perspective is set out to solve, as it is presented and discussed in these papers, can be summarized in a simplified way as: *Students fail to develop physics identities*. Students may well learn physics content, but that is not enough, they need to identify as part of the physics community to be able to persist and thrive in physics.

In reading these papers, we note that the role and character of the physics community in shaping students' identities is discussed very little. Shanahan (2009) suggests that this may be a general weakness of the communities of practice framework, where a focus on individuals' participation "tends to take the norms as given and attends to how individuals navigate those norms" (p. 57). This may represent a foregrounding of 'agency' without a critical assessment of the 'structures' that influence students' agency. One example of a critical assessment of structure can however be found in a relatively extensive case study by Irving and Sayre (2016). Here, the authors note that the tendency of universities to try to "Create more physicists in the mold of those that they already have (professors, the majority of who conduct research)" and "the absence of a community of practice of physics teachers within the degree track" (Irving & Sayre, 2016 p. 1195) creates severe problems for one of the students in crafting an identity as a physics teacher.

These works represent a socially informed view of the processes of physics education. However, if the question about physics education and identity is primarily represented as being about students' "Development of an appropriate subject-specific identity" (Irving & Sayre, 2015, p. 1) other important questions, like the role of the culture of physics, may be passed over in silence. Another issue that may be treated with silence is that the possibilities for developing a physics identity or to navigate the norms of the discipline is not the same for all students. In the problematizations discussed in the following sections, these issues are brought to light more explicitly.

8.2.2 Underrepresented Students Do Not Develop Enough of a Physics Identity

Research in physics education has employed identity perspectives to attend to the glaring issues of underrepresentation of women and other groups in physics. In particular, a careful consideration of identity has been put forward as a way of overcoming the 'binary gender deficit model' of studying gender in physics only by comparing men and women (Traxler et al., 2016). In line with this, and parallel to the approaches discussed in the previous section that are concerned with the whole student group, many studies pay particular attention to the identity development of underrepresented students. This can be seen as a more 'sociopolitical' (and not only 'sociocultural', Gutiérrez, 2013) framing of the questions of identity, where the differing opportunities of students are highlighted.

Quantitative work in this field asks what factors help students (and women in particular) to develop a physics identity (Hazari et al., 2010, 2013). Based on the notions of science identity introduced by Carlone and Johnson (2007), Hazari et al. (2010) developed a model for 'physics identity' used in statistical analysis of questionnaire data that has since been adopted in many studies and also used in other STEM fields (see, for example, Hazari et al., 2017; Kalender et al., 2019a, b; Lock et al., 2013; Verdín et al., 2019; Verdín & Godwin, 2017; Wang & Hazari, 2018; Seyranian et al., 2018). This research is generally motivated by documented problems of underrepresentation and sometimes underperformance in physics. Large-scale quantitative studies provide solid evidence for the critical state of participation and belonging for diverse groups in physics, evidence that is very useful in convincing and calling physicists and physics departments into action. On the other hand, quantitatively comparing students' identities means that identity is operationalized as something that students possess more or less of, where more physics identity correlates with more success in physics (Kalender et al., 2019a; Seyranian et al., 2018). This means that the complex interplay between structure and agency risks being lost in the reporting of such results; the lower measures of physics identity of some students may be read as a deficit of those students. Hence, one problematization implied by the operationalization and measurement of student identity is: Underrepresented students do not develop enough of a physics identity. In particular, this problematization is implied if simultaneous attention is not paid to the role of the culture of physics. When student identification is measured, and the physics environment is treated as a number of factors that can affect identification, the perception of the problem is directed towards students and the amount of identity they possess. This is especially important given the preference for quantitative results among the physics community. Accordingly, there is a risk that presenting differences in physics identity without at the same time examining and critiquing hegemonic structures may limit how physicists and physics educators n conceptualize the responsibility for change.

This problematization does not fully characterize all quantitative work mentioned here. It is however important to consider the consequences of this possible interpretation for which actions are taken. Based on quantitative results showing that explicit discussions of underrepresentation have a statistically significant effect on the measure of physics identity for women (Hazari et al., 2010, 2013), Lock and Hazari (2016) analysed two high school classroom discussions about underrepresentation. They found that the discussion challenged the student's assumptions about the world of science, "Subsequently, the norms in students' figured worlds may change or become less rigid allowing for a new openness to physics identity development amongst female students" (Lock & Hazari, 2016, p. 1). In this example, a positive change was seen in how students figure science. A similar focus can be seen in a study, also focusing on the high school level, that investigated an intervention for physics-interested young women participating in the Physics Olympiad competition, "In order to support their physics engagement and their physics identity development" (Wulff et al., 2018, p. 2). While these approaches highlight important areas for encouraging physics participation, there is a risk that a focus on the identity development or refigured worlds of these students emphasizes this as the only thing that needs to change, absolving science itself.

One way of understanding the silences inherent in this problematization is to consider the broad dimensions of whether identity is seen as something possessed by individuals or rather something that is negotiated in interaction (Pozzer & Jackson, 2015). A focus on the identity development of students may promote a view of identity as a 'personal characteristic' possessed by individuals. This view can make it difficult to simultaneously see more structural, social, and political issues.

While it is common to highlight how physics can be hostile to women along with cultural ideas about 'who fits' in physics, a too strong focus on the development of physics identity among underrepresented students (including women) risks implying (particularly for the uninformed reader) that the solution is to change these students. We may thus implicitly construct the problem as a problem of 'deficits' among these students (Traxler et al., 2016). Silences around the role the physics community plays in reproducing unequal participation in physics risk reproducing the gender problem in physics as a 'women's problem', where the majority culture is left unproblematized. We also risk missing the perspectives of students who resist certain physics identities and do not want to be a part of the physics community as it is presently formed.

8.2.3 Normative Physics Identities Impede Equal Participation

Some socio-politically oriented research on identity in higher education physics uses approaches with an explicit focus both on students' identity development and on the systemic barriers hindering equal participation. This research is often inspired by feminist and intersectional analysis frameworks to investigate power in the culture of physics, which often includes a 'negotiation' rather than 'possession' view of identity (Pozzer & Jackson, 2015). This involves explicitly aiming to avoid the previously discussed risk of framing diversity issues in terms of 'deficit frameworks', for example by adopting what Hazari and Potvin (2005) refer to as the 'culture bias viewpoint' on underrepresentation of women in physics.

The representation of the problem that identity analysis can help us solve here can be characterized as: Normative physics identities impede equal participation. Here, students' identities are contextualized and analysed in relation to a critical view of what is considered normal or desirable in the physics community, highlighting conflicts, power, and resistance.

This framing of equity issues in physics is not new. It was the background for Ong's (2005) longitudinal in-depth study of the identity negotiations of women of colour in physics education. Ong (2005) attends to the complexity in processes of gendering and racialization as well as their intersection, and focuses explicitly on both the agency of the informants to negotiate limiting structures, and how these structures can ultimately be changed. This study highlights the importance of studying identities with a view of the intersection of different power structures (Crenshaw, 1989). Aiming towards understanding the problems of participating in the physics community with a sensibility about the interaction of physics culture, gender and class constructions, Danielsson (2009, 2014) combines the communities of practice framework with discourse analysis and gender theory. This makes it possible to ask for example how the 'doing' of physics in combination with a 'doing' of working-class masculinity "May threaten the idealised portrayal of physics as objective, rational, and disembodied" (Danielsson, 2014, p. 492). Danielsson (2014) asks whether certain identity performances associated with minoritized students can be reconciled with expected or normative physics identity. A similar analysis is done by Gonsalves (2014), who examines 'recognizable' forms of being a physicist in the context of doctoral education. Gonsalves (2014) finds that as competence in experimental physics seems to bear strong masculine connotations, the women physicists always seem to stand out as 'other'. At the same time, stereotypical femininity is constructed in opposition to the purported 'neutrality' of physics, and this means that a recognizable woman physicist position relies on difference both to 'other women' and 'ordinary physicists'.

Hyater-Adams et al. (2018, 2019) draw on earlier notions of physics and science identity (Carlone & Johnson, 2007; Hazari et al., 2010) as well as critical race studies to construct a framework to study race and physics identity. They claim that "Understanding the ways in which Black folks identify as physicists can provide useful information about the facets of the physics discipline that perpetuate systems of oppression" (Hyater-Adams et al., 2019, p. 1). The authors analyse the physics identification of black physicists and students, and recommend that we provide positive recognition for students from marginalized populations. This speaks to the problematization of underrepresented students' troubles with developing a physics identity, as outlined in the previous section. At the same time, the authors also claim that "We must make a significant and intentional effort to break down the harmful ideologies inherent in physics culture", like the idea that physics is "Not influenced by the systems of oppression and marginalization inherent in society" (Hyater-Adams et al., 2019, p. 13). Framing the problem in this way problematizes the culture of physics and thus speaks to the silence in the previous problematization. This illustrates how the problematizations we have sketched here do not necessarily exclude each other.

In all these examples, a central question is how physics culture is biased to find particular groups of students lacking, and how this can be changed. This can be compared to the reinterpretations of who can be an 'expert' or hold a 'science identity' that have been documented by researchers in school settings, where young disadvantaged students take on identities such as a 'community science expert' (Barton et al., 2013; Barton & Tan, 2010). A starting point for reconceiving physics is to rethink the role of arrogance and implicit masculinity that discourage many women (Bremer & Hughes, 2017). Hyater-Adams et al. (2019) also point out that welcoming "Alternative modes of participation" (p. 13) would allow more people to feel at home in physics. This involves thinking about how different physics subjects are valued. Hazari and Potvin (2005) highlight the problem that "Traditional teaching of physics perpetuates elitist elements and does not expose students to all the ways in which physics can be pursued in the world" (p. 17). Some examples are the minimization of subfields such as biophysics, geophysics, and atmospheric physics in introductory teaching. Could more applied physics, or working in a more practical 'embodied' way (as related by the students with working class background in Danielsson, 2014) be brought into the canon of 'celebrated' physics, where theoretical and 'disembodied' fields are accorded higher status (Martin, 2017)? Similar concerns are raised in relation to the result that physics identity is strongly predicted by intrinsic fulfilment career motivations, more available to those "Who come from backgrounds with the luxury of affording knowledge-based motivations" (Hazari et al., 2010, p. 994). Changing the norms of what is seen as required for physics, what motivations are seen as legitimate, what subfields are seen as better, and what identity performances are expected are some of the important changes needed. In these calls for broadening the culture, the problem concerns how we can get more (diverse) people into physics. This is similar to the second problematization; the difference is the explicit emphasis that physics has to change to allow this. However, an additional question that is left out in these approaches concerns the relationship between physics identities and the knowledge produced in physics.

8.2.4 Normative Physics Identities Have Consequences for What Physics Knowledge Is Produced, and the Role of Physics in Society

The problematization highlights the connections between epistemology, identity and social justice. Research in physics education has traditionally focused on students' acquisition of the correct physics knowledge, but also on their understanding of physics in terms of epistemology. Taking an identity perspective means shifting the problematizations to sociocultural or sociopolitical concerns, recognizing that the social environment is inseparable from learning, in particular when it comes to the social justice aspects of education. Both these perspectives are concerned with possible ways of *doing* physics. However, they largely keep silent on the question of what physics *does,* in terms of how physics knowledge operates in the world. We argue that here lies an additional possibility for thinking critically about issues of social justice in a broader perspective than the participation of a more diverse group of people in physics. A possibility to analyse the production and reproduction of physics in line with feminist epistemologies. The question is: Can we analyse identity, knowledge, and social justice together?

As one goal of physics education is to educate knowledge workers with high status and influence in modern societies, it is important to ask how physics education affects the knowledge produced, conveyed, and used by physicists. We draw inspiration from the now 35-year old characterization of feminist approaches to analysing science proposed by Harding (1986). Harding describes five research programmes, from 'equity studies', which ask why there are so few women in science, to elaborations of feminist epistemologies, which puts a critical eye to the very root of knowledge production. Can we adopt a perspective of feminist epistemologies, and articulate it together with identities in physics? Apart from a broadening of physics culture that is welcoming to a more diverse group of people, this would also involve thinking about what this broadening entails for the operation of physics in society.

A central point made by feminist science studies is that science is always linked with social concerns, in motivations, processes and consequences, no matter how 'pure' a field is purported to be (Harding, 1991). It is important to ask whose problems the education of more physicists or the development of new physics will solve. The problem of identity in physics can thus be represented as: Normative physics identities have consequences for what physics knowledge is produced, and the role of physics in society. Attending to this problem, some theoretical arguments for how physics could be reconceived in a feminist way have been proposed (see Bug, 2003; Harding, 1991; Rolin, 2008; Whitten, 2012), but we have hitherto seen very few studies that explicitly and based on empirical data discuss questions of identity and epistemology in physics education.

One study that looks at physics education with this lens is Hasse's (2015) exploration of "The material co-construction of hard science fiction and physics". In this paper, Hasse draws from observations and interviews in an undergraduate physics programme to discuss how the 'future imaginaries' encoded in common ways of performing as a 'good' physics student point to certain ideas about the goals of physics. Hasse argues that certain science fiction-fantasies often motivate physics students as well as physicists, and that these fantasies often are traditionally masculine. In the physics education context she studied, ideas about colonizing Mars served as a motivation for the selected course content. Hasse points out that "If the selected facts are embedded in problems of how to build houses on Mars rather than how to create a less polluted planet Earth it is likely that these choices also guide our scientific concerns" (Hasse, 2015, p. 934). Hasse suggests that we ask "Whose visions of tomorrow's worlds' and 'whose science fiction imaginaries' are actually promoted in Western science education" (Hasse, 2015, p. 936). In this way, concerns about the identities of physics students points to bigger questions about the role of physics in creating a better society.

Another epistemological take on the issue of identities and inequalities in physics is given by Prescod-Weinstein (2020), who discusses who is seen as an objective 'observer' in physics. Prescod-Weinstein points out that racial and gender privilege confer epistemic authority, to the point where black women are seen as less competent observers of their own experiences of racism in physics than their white male colleagues. This represents a kind of epistemological double standard. High levels of evidence are expected for accounts of racism or sexism (which are often dismissed as anecdotal), while the developments of string theory by (overwhelmingly male and white) theoretical physicists are accepted in physics even though the prospect of corroborating experimental evidence for string theory currently seems distant. Perhaps the accusation of racism represents a bigger threat to the epistemology of physics than string theory, since it threatens the notion of physics as a "Culture of no culture" (Traweek, 1988, p. 162), an objective field where the race or gender of its practitioners should not matter. In addition to the question of how to support 'black physics identity' explored by Hyater-Adams et al. (2019), the account given by Prescod-Weinstein implies that it is not necessarily sufficient to build positive identities for previously underrepresented groups in physics. We also need to counter the inherent racism and sexism in conceptions of who a knower is. That is, we need a reform of the implicit biases in the epistemology of physics.

The arguments of Hasse (2015) and Prescod-Weinstein (2020) show how social epistemology perspectives are fruitful when reasoning about social justice in physics. Questions of equity concern both a widened recruitment of underrepresented students, and whose knowledge and priorities are allowed to point out the future direction of physics. Hence, the multiple problematizations and research strategies outlined in the last four sections all contribute to a better understanding and better tools for creating a more just physics education. Of concern however, is where and how we can invoke the different problematizations. The two papers we have taken as examples in this section are published in *Cultural Studies of Science Education* and *Signs*, two journals which are not the go-to venues for most researchers in physics education, but that can be expected to be welcoming for this kind of analysis. In what way is it possible to have these conversations with mainstream physics education researchers and physicists?

8.3 Reimagining Physics Together with Students

In this section, we provide three empirical examples to illustrate the ways of using and conceptualizing identity in physics education we have discussed in the previous sections. We discuss the stories of the trainee physics teacher Emily and the physics master's student Eugenia, as well as the experiences of five students of undergraduate quantum physics. With these examples, we illustrate how the four ways of representing the problem of identity outlined above yield complementary yet intertwined interpretations of the social justice issues at stake in physics. Note that the research was originally conducted in Swedish; quotations have been translated and all names used herein are pseudonyms.

Emily is a trainee physics teacher in her sixth semester, studying to become an upper secondary school teacher. She was interviewed as part of a project investigating what is involved in being recognized as a legitimate physics teacher-to-be in a Swedish physics teacher programme (Larsson, 2021). At this programme, trainee teachers study most physics courses together with the physics bachelor's programme (physics majors, usually with an elective in mathematics). Emily wanted to become a physicist but could not get into the bachelor's programme. She chose to study physics teaching with the explicit plan to later switch to physics. At the point of the interview, she had however decided to stay in physics teaching. Emily's choice between teaching and other ways of doing physics depended on both her social belonging in the student group and her possibilities of recognizing herself as smart enough to aim for a research career (Larsson & Danielsson, 2022). These negotiations are situated at a physics department where trainee teachers were generally perceived as less competent and ambitious compared to students on a more research-oriented track on the physics bachelor's programme (Larsson et al., 2021). When asked about her dreams about future work, Emily replies:

> Oh, I don't know, I mean, I think I would find it fun to work with research as well. Uh, it's just, hmm. So, number one: That I still think I'm a little too stupid, which uh, which I don't think is really true logically, and I think that if I had heard someone else [say so about themselves] then I would just have said 'no way'. But I still think like that.

That it is important to be intelligent enough to fit in physics, and that women are often deemed as not possessing the right kind of intellect, is a common theme in literature on identity in physics (Francis et al., 2017; Gonsalves & Seiler, 2012; Johansson, 2020).

Throughout the interview, Emily expressed that she was aware of these gendered norms, and in the quote, we see her taking on a reflective stance towards her own experience of not feeling smart enough. Emily's negotiations are also heavily burdened with the status differences between physics and physics teaching. She describes how her choosing teaching puts her in the "Stupid gang", which means choosing to become a less prestigious physics person (Larsson & Airey, 2021; Larsson & Danielsson, 2022). Emily also reasoned about how she wants to use physics to make a difference for other people, and becoming a teacher is one such choice. At the same time, she appreciates the "Exciting stuff in modern physics" and the possibility of "Really getting to understand stuff".

Interpreting Emily's story using the first or second problematization, which focus students' identity development, we may perceive her situation as a student or woman in need of support for developing her physics identity. What can be done to bolster Emily's appreciation that she does have what it takes to become any kind of physicist? Arguing from a 'culture bias viewpoint' (Hazari & Potvin, 2005) and the third problematization, would instead suggest a need to change some of the cultural perceptions of who a physicist should be. Emily's choices are complicated by the strong cultural ideals of expected brilliance (Leslie et al., 2015; Storage et al., 2016) coupled with the perception of these traits as masculine.

The high prestige of research compared to a teaching career also affects the recognition of Emily as a fully appreciated member of the physics community. A further elaboration of this picture comes from the perceived status differences between research fields in physics. In general, the more theoretical fields in physics have higher status, being supposedly purer and more fundamental (Becher & Trowler, 2001; Martin, 2017; Traweek, 1988; Whitten, 1996). Among physics bachelor students it is also common to state the motivation for studying physics as wanting to understand how the world works in some sense (Bøe & Henriksen, 2013; Levrini et al., 2017). Even when student goals are directed towards more practical topics, these are often connected to some kind of frontier thinking, like the dreams of colonizing Mars documented to be common in the physics program studied by Hasse (2015).

In a study based on interviews with international master's students in physics at two Scandinavian universities, an assumed hierarchy between fields could be shown to influence students' negotiations of identities, capabilities, passions and choices (Johansson, 2020). One example is Eugenia, who was conscious of how her interest in solving problems in material physics stood in contrast to more philosophical approaches. Eugenia was interested in creating materials for energy production, aiming at goals like mitigating climate and energy crises. These goals can definitely be considered grand, but are still more 'down-to-earth' than exploring Mars or understanding the universe. Solving societal problems is a more common motivation for studying science among women, but in general less common for physics students than in other STEM disciplines (Eccles, 2007; Sax et al., 2016; Schreiner, 2006). The motivations for physics studies of Eugenia or Emily are thus not the most common or normative among physics students, who are often more 'physics for physics sake' oriented (Levrini et al., 2017).

In light of the fourth problematization, suggesting a feminist reimagination of physics, we see how these stories contain a seed for a reconceived appreciation of the role of physics in society. Trainee teachers like Emily are preparing to teach a new generation of students what the subject of physics is all about. They can empower students to learn physics on their own terms, and make it useful in students' own contexts (Barton & Tan, 2010). Countering the prestige asymmetry in physics (Martin, 2017) and allowing for other motivations or 'modes of participation' (Hyater-Adams et al., 2019) means to value Emily's and Eugenia's motivations and choices as at least as legitimate as others. A reevaluation of what is seen as prestigious or proper ways of doing physics, may also hold the potential for redrawing epistemological lines. A vision of physics aligned with feminist epistemologies (Harding, 1991; Whitten, 2012) is achieved when social, cultural or political matters of concern are seen as a part of physics proper.

Some parts of physics have a particular role to play in relation to the dreams of many physics' students. Among these, quantum mechanics is a subject that is often entered with high expectations. These expectations often do not match the practical realities of courses, and this mismatch has been shown to be unsettling for students (Johansson, 2018b). The special role of quantum mechanics is a result of its extensive pop-cultural aura, its connection to the greatest 'heroes' of twentieth century physics, its sometimes apparently enigmatic results, and its importance for modern technology. Despite this wide appeal of the subject, the teaching of undergraduate quantum mechanics around the world often follows traditional patterns of physics teaching, with lectures, textbooks and tutorials emphasizing the mathematical formalism and students' competence in solving textbook problems (Greca & Freire, 2014). This was to a large extent the case in quantum mechanics classes at two Swedish universities observed in the research project reported in Johansson et al. (2018). In one of these undergraduate classes, negotiations between idealized pictures of quantum mechanics versus the reality of the course played out in a group interview with five male, white students with Swedish background:

Gabriel: In almost every documentary about physics they talk about quantum physics, but they can't explain all the cool phenomena that they talk about because then you say "No, quantum physics" and like "It's too advanced for ordinary people". So there's been expectations and hopes like "Oh, it will be [interesting] learning that some time". Those were the expectations for quantum physics.

Elliot: Then you were also prepared that, at least I was that it would be very difficult because I guess that's what you've heard before. That it is one of the most difficult courses you take, so you were prepared for that, and I think it has been, but also, a terribly interesting course, to kind of finally get to delve a bit deeper into physics.

AJ: But in what way is it difficult?

Ian: I mean it feels very, very abstract in some way. That you know that particles have spin and such, but you don't really know how you should really apply this with bra- and ket-states and all that, if you try to think about it more in reality, kind of. Like, you sit and calculate on it, but what its meaning in real life is, that's what I think can be quite abstract. [...]

Bob: It can seem very cool and magnificent when you talk about it in a popular science way, but when you are really there and are doing quantum physics and sort of calculate your raising-, lowering- operators, then it's not that, what should I say, glorious. So, there is a certain difference. I still think it is fun to do it, but you have a completely different idea about what quantum physics is now, when you've been doing it yourself, than you had before.

Using the first problematization, we can understand the last student to speak, Bob, as a majority student who is working on developing a successful physics identity. He has reconciled his previous ideas with a, perhaps, more realistic picture of what doing physics involves: "I still think it is fun to do it, but you have a completely different idea about what quantum physics is now". He is thus moving closer to being a part of the physics community.

Understanding this quote in terms of the third problematization, the culture of physics and its effect on equal participation, we can analyse what lies in Gabriel's reflection that quantum physics is "Too advanced for ordinary people" but that he still wants to learn it. Once again, this reflects the "Widely taken-for-granted notion that physics is difficult and only for the brainy" (Archer et al., 2017, p. 93). Who is excluded by this notion of physics in general and quantum physics in particular as a 'too advanced' subject? And what does it mean to become a physicist as opposed to being one of the 'ordinary people'? The students' discussion also helps us focus the practice of physics in quantum mechanics courses by asking which different modes of doing physics are enabled by the course, and who is included or excluded by this. The course is theoretical, 'abstract', and difficult to relate to everyday phenomena (even though quantum mechanics is the basis for much modern technology). Theoretical ways of doing physics are prioritized, without connecting the content to the various concerns of the students present. At the same time, as Bob reflects, the "cool" or "glorious" phenomena played up by popular scientific accounts of quantum mechanics does not really materialize in the course. Rather, the course allows for very little philosophical exploration in favour of a 'shut up and calculate' approach to quantum mechanics (Johansson et al., 2018). For whom are courses like this designed? Meteorology students, future researchers in physics, trainee teachers, and engineers could all benefit from approaches to quantum mechanics more clearly aligned with their future work.

If we consider this model of teaching quantum mechanics as a reflection of contemporary attitudes in physics, we can ask what wider epistemological consequences this structuring of 'legitimate' ways of doing physics has. That is, asking questions drawing on social epistemology, the fourth problematization. A look at the history of quantum mechanics teaching suggests part of an answer. Kaiser (2002, 2014) describes how the teaching of quantum mechanics in the US changed during the Cold War arms race (in which physics was very much enlisted). From initially being concerned with foundational and interpretational issues, the teaching of quantum mechanics turned to a more 'shut up and calculate' approach (Kaiser, 2014), with a focus on applicability in the atomic domain (Kaiser, 2002). Kaiser (2002, 2014) argues that this aligned with training physicists for employment in military research. Allowing some extrapolation, the experiences of our Swedish quantum physics students can be understood as the continuation of this practice of imagining physics as part of a solution for defending the nation (Lucena, 2005). Rather than seeing these students' dissatisfaction with the 'shut up and calculate' approach as an obstacle to their formation of physics identities, it can perhaps be understood as an argument for reimagining physics. We need not draw the consequences as far as claiming that standard quantum mechanics education creates uncritical 'quantum mechanics', ready for creating weapons of mass-destruction. However, asking whose dreams and aspirations for physics are taken seriously, valued, and eventually financed in research is an important tool for realizing the full critical potential of physics identity research.

8.4 Conclusion

We want to return to the general question of what research in physics education can learn from using identity perspectives, and what some directions for further development of this research can entail. In this chapter, we have discussed the public discourse around identity in research papers in order to highlight the representation of the 'problems' that identity perspectives are set out to solve.

In contrast to the common focus on the cognitive learning of students, physics identity research insists that learning needs to be understood from a social perspective. Identity studies have emphasized that everyone's relations to physics are not the same; the fabric of physics student experiences is convoluted by unique configurations of identities. In-depth investigations of gendered, classed, and racialized identities in physics have illuminated both how underrepresented students struggle with identifying and being identified as physics people, and how the specific culture of physics is a part of the problem which needs to be challenged if we are committed to a socially just education in physics.

The four problematizations and the various approaches employed by the reviewed papers are important in providing a rich and complex image of physics education that challenges the historical legacy and on-going inequalities in physics. We want to emphasize the importance of continuously questioning how the 'problems' empirical research projects set out to solve are represented, in published papers as well as other discourse. This means a research reflexivity on the level of political implications of research. Discussing 'radical' re-imaginations of physics may be more or less strategic given the audience, even though the variety of approaches included in the mainstream physics education journals has increased in recent years. We suggest that important strategic questions are made possible by keeping the four problematizations in mind. For example, will the culture of physics necessarily change if more diverse generations of physics students are recruited, or do these students risk being 'assimilated' into reproducing the norms of the field as it is presently formed?

We believe an important, if strategically complicated, opportunity lies in simultaneously examining identity and epistemology in physics. Given the wide-ranging societal influence of the physics discipline, knowledge construction in physics needs to be considered in relation to the 'identity production' going on at the same time. The examples of analysing physics education with the help of feminist epistemologies that we have discussed here can serve as a starting point for further developing these thoughts. Research on physics education committed to a more just and equitable physics stands to gain from asking questions about 'whose knowledge' and 'whose interests' are given precedence in physics.

Acknowledgements We want to thank our many colleagues who have commented on drafts of the text. A special thanks to Klara Goedecke as well as the engaged, critical, and keen-eyed reviewers for engaging in the process of creating this chapter with us.

References

American Physical Society. (2018). *Percentage of women in physics*. American Physical Society. https://www.aps.org/programs/education/statistics/womenphysics.cfm

Archer, L., DeWitt, J., Osborne, J., Dillon, J., Willis, B., & Wong, B. (2012). "Balancing acts": Elementary school girls' negotiations of femininity, achievement, and science. *Science Education, 96*(6), 967–989. https://doi.org/10/f4cdk9

Archer, L., Moote, J., Francis, B., DeWitt, J., & Yeomans, L. (2017). The "exceptional" physics girl: A sociological analysis of multimethod data from young women aged 10–16 to explore gendered patterns of post-16 participation. *American Educational Research Journal, 54*(1), 88–126. https://doi.org/10/gctkdj

Bacchi, C. (2009). *Analysing policy: What's the problem represented to be?* Pearson.

Bacchi, C. (2015). The turn to problematization: Political implications of contrasting interpretive and poststructural adaptations. *Open Journal of Political Science, 5*(1), 1–12. https://doi.org/10/gfx844

Bacchi, C., & Goodwin, S. (2016). Poststructural policy analysis. *Palgrave Macmillan*. https://doi.org/10.1057/978-1-137-52546-8

Barton, A. C., & Tan, E. (2010). We be burnin'! Agency, identity, and science learning. *The Journal of the Learning Sciences, 19*(2), 187–229. https://doi.org/10/db56x6

Barton, A. C., Kang, H., Tan, E., O'Neill, T. B., Bautista-Guerra, J., & Brecklin, C. (2013). Crafting a future in science: Tracing middle school girls' identity work over time and space. *American Educational Research Journal, 50*(1), 37–75. https://doi.org/10/gctkps

Becher, T., & Trowler, P. (2001). *Academic tribes and territories: Intellectual enquiry and the culture of disciplines* (2nd ed.). Open University Press.

Beddoes, K. (2011). Engineering education discourses on underrepresentation: Why problematization matters. *International Journal of Engineering Education, 27*(5), 1117–1129. https://www.ijee.ie/contents/c270511.html

Bøe, M. V., & Henriksen, E. K. (2013). Love it or leave it: Norwegian students' motivations and expectations for postcompulsory physics. *Science Education, 97*(4), 550–573. https://doi.org/10/f42pp9

Bremer, M., & Hughes, R. M. (2017). How novices perceive the culture of physics. *Journal of Women and Minorities in Science and Engineering, 23*(2), 169–192. https://doi.org/10/ggn4ww

Brickhouse, N. W. (2001). Embodying science: A feminist perspective on learning. *Journal of Research in Science Teaching, 38*(3), 282–295. https://doi.org/10/c85ktx

Brickhouse, N. W., & Potter, J. T. (2001). Young women's scientific identity formation in an urban context. *Journal of Research in Science Teaching, 38*(8), 965–980. https://doi.org/10/bx4t4x

Bug, A. (2003). Has feminism changed physics? *Signs, 28*(3), 881–899. https://doi.org/10/d5ggb3

Carlone, H. B. (2003). (Re)producing good science students: Girls' participation in high school physics. *Journal of Women and Minorities in Science and Engineering, 9*(1), 17–34. https://doi.org/10.1615/jwomenminorscieneng.v9.i1.20

Carlone, H. B., & Johnson, A. (2007). Understanding the science experiences of successful women of color: Science identity as an analytic lens. *Journal of Research in Science Teaching, 44*(8), 1187–1218. https://doi.org/10/cpprr8

Close, E. W., Conn, J., & Close, H. G. (2016). Becoming physics people: Development of integrated physics identity through the learning assistant experience. *Physical Review Physics Education Research, 12*(1), 010109. https://doi.org/10/gctkh7

Coble, K., Cunningham, B., Freeland, E., Hodapp, T., Hodari, A. K., Ivie, R., Martínez-Miranda, L. J., Ong, M., Petty, S., Seestrom, Seidel, S., Simmons, E., Throennessen, M., Urry, M., & White, H. (2013). Many steps ahead, a few steps back: U.S. women in physics. *AIP Conference Proceedings, 1517*, 162–163. https://doi.org/10/gg2vkz

Crenshaw, K. (1989). Demarginalizing the intersection of race and sex: A black feminist critique of antidiscrimination doctrine, feminist theory and antiracist politics. *University of Chicago Legal Forum, 1989*(1), 139–167.

Cumings, K. M., Welton, A. D., & Grogan, M. (2014). "Truth or consequences": A feminist critical policy analysis of the STEM crisis. *International Journal of Qualitative Studies in Education, 27*(9), 1155–1182. https://doi.org/10/gft5bs

Danielsson, A. (2009). *Doing physics – Doing gender: An exploration of physics students' identity constitution in the context of laboratory work.* Doctoral dissertation. Uppsala University. https://urn.kb.se/resolve?urn=urn:nbn:se:uu:diva-98907

Danielsson, A. (2010). Gender in physics education research: A review and a look forward. In M. Blomqvist & E. Ehnsmyr (Eds.), *Never mind the gap! Gendering science in transgressive encounters*. Uppsala University. https://urn.kb.se/resolve?urn=urn:nbn:se:uu:diva-127361

Danielsson, A. (2014). In the physics class: University physics students' enactment of class and gender in the context of laboratory work. *Cultural Studies of Science Education, 9*(2), 477–494. https://doi.org/10/f22r4p

Eccles, J. S. (2007). Where are all the women? Gender differences in participation in physical science and engineering. In S. J. Ceci & W. M. Williams (Eds.), *Why aren't more women in science? Top researchers debate the evidence* (pp. 199–210). American Psychological Association. https://doi.org/10/cp6w4z

European Commission. (2004). *Europe needs more scientists: Report by the high-level group on increasing human resources for science and technology in Europe*. Office for Official Publications of the European Communities.

Foucault, M. (2000). So is it important to think? In J. D. Faubion (Ed.), *Essential works of Foucault, 1954–1984* (Vol. 3., pp. 454–458). The New Press. (Original work published 1981).

Francis, B., Archer, L., Moote, J., DeWitt, J., Mac Leod, E., & Yeomans, L. (2017). The construction of physics as a quintessentially masculine subject: Young people's perceptions of gender issues in access to physics. *Sex Roles, 76*(3–4), 156–174. https://doi.org/10/f9nnjt

Godec, S. (2018). Sciencey girls: Discourses supporting working-class girls to identify with science. *Education Sciences, 8*(1), 19. https://doi.org/10/ggpk98

Gonsalves, A. J. (2014). "Physics and the girly girl - there is a contradiction somewhere": Doctoral students' positioning around discourses of gender and competence in physics. *Cultural Studies of Science Education, 9*(2), 503–521. https://doi.org/10/gctkz2

Gonsalves, A. J., & Seiler, G. (2012). Recognizing "smart super-physicists": Gendering competence in doctoral physics. In M. Varelas (Ed.), *Identity construction and science education research* (pp. 157–172). Sense Publishers. https://doi.org/10/cs4t

Grant, M. J., & Booth, A. (2009). A typology of reviews: An analysis of 14 review types and associated methodologies. *Health Information & Libraries Journal, 26*(2), 91–108. https://doi.org/10/ftbpbr

Greca, I. M., & Freire, O. (2014). Teaching introductory quantum physics and chemistry: Caveats from the history of science and science teaching to the training of modern chemists. *Chemistry Education Research and Practice, 15*(3), 286–296. https://doi.org/10/gctknp

Gretton, A. L., Bridges, T., & Fraser, J. M. (2017). Transforming physics educator identities: TAs help TAs become teaching professionals. *American Journal of Physics, 85*(5), 381–391. https://doi.org/10.1119/1.4978035

Gutiérrez, R. (2013). The sociopolitical turn in mathematics education. *Journal for Research in Mathematics Education, 44*(1), 37–68. https://doi.org/10/f4zm26

Haraway, D. (1988). Situated knowledges: The science question in feminism and the privilege of partial perspective. *Feminist Studies, 14*(3), 575–599. https://doi.org/10/bvtwq4

Harding, S. (1986). *The science question in feminism*. Cornell University Press.

Harding, S. (1991). *Whose science? Whose knowledge?* Open University Press.

Hasse, C. (2015). The material co-construction of hard science fiction and physics. *Cultural Studies of Science Education, 10*(4), 921–940. https://doi.org/10/gctks3

Hasse, C., & Sinding, A. B. (2012). The cultural context of science education. In D. Jorde & J. Dillon (Eds.), *Science education research and practice in Europe* (pp. 237–252). Sense Publishers. https://doi.org/10/cs4v

Hazari, Z., & Potvin, G. (2005). Views on female under-representation in physics: Retraining women or reinventing physics? *Electronic Journal of Science Education, 10*(1).

Hazari, Z., Sonnert, G., Sadler, P. M., & Shanahan, M. C. (2010). Connecting high school physics experiences, outcome expectations, physics identity, and physics career choice: A gender study. *Journal of Research in Science Teaching, 47*(8), 978–1003. https://doi.org/10/fwt79h

Hazari, Z., Potvin, G., Lock, R. M., Lung, F., Sonnert, G., & Sadler, P. M. (2013). Factors that affect the physical science career interest of female students: Testing five common hypotheses. *Physical Review Special Topics - Physics Education Research, 9*(2), 020115. https://doi.org/10/gctkvg

Hazari, Z., Brewe, E., Goertzen, R. M., & Hodapp, T. (2017). The importance of high school physics teachers for female students' physics identity and persistence. *The Physics Teacher, 55*(2), 96–99. https://doi.org/10/gc5tr7

Holmegaard, H. T. (2015). Performing a choice-narrative: A qualitative study of the patterns in STEM students' higher education choices. *International Journal of Science Education, 37*(9), 1454–1477. https://doi.org/10/gctkn7

Hyater-Adams, S., Fracchiolla, C., Finkelstein, N., & Hinko, K. (2018). Critical look at physics identity: An operationalized framework for examining race and physics identity. *Physical Review Physics Education Research, 14*(1), 010132. https://doi.org/10/gdkxg7

Hyater-Adams, S., Fracchiolla, C., Williams, T., Finkelstein, N., & Hinko, K. (2019). Deconstructing black physics identity: Linking individual and social constructs using the critical physics identity framework. *Physical Review Physics Education Research, 15*(2), 020115. https://doi.org/10/ggh6jv

Irving, P. W., & Sayre, E. C. (2014). Conditions for building a community of practice in an advanced physics laboratory. *Physical Review Special Topics - Physics Education Research, 10*(1), 010109. https://doi.org/10/gctkrb

Irving, P. W., & Sayre, E. C. (2015). Becoming a physicist: The roles of research, mindsets, and milestones in upper-division student perceptions. *Physical Review Special Topics - Physics Education Research, 11*(2), 020120. https://doi.org/10/gctknm

Irving, P. W., & Sayre, E. C. (2016). Identity statuses in upper-division physics students. *Cultural Studies of Science Education, 11*(4), 1155–1200. https://doi.org/10/gctkfg

Johansson, A. (2018a). *The formation of successful physics students: Discourse and identity perspectives on university physics*. Doctoral dissertation. Uppsala University.. https://urn.kb.se/resolve?urn=urn:nbn:se:uu:diva-357341.

Johansson, A. (2018b). Undergraduate quantum mechanics: Lost opportunities for engaging motivated students? *European Journal of Physics, 39*(2), 025705. https://doi.org/10/gctj9n

Johansson, A. (2020). Negotiating intelligence, nerdiness, and status in physics master's studies. *Research in Science Education, 50*(6), 2419–2440. https://doi.org/10/gfjxxr

Johansson, A., Andersson, S., Salminen-Karlsson, M., & Elmgren, M. (2018). 'Shut up and calculate': The available discursive positions in quantum physics courses. *Cultural Studies of Science Education, 13*(1), 205–226. https://doi.org/10/f3rq2b

Kaiser, D. (2002). Cold War requisitions, scientific manpower, and the production of American physicists after World War II. *Historical Studies in the Physical and Biological Sciences, 33*(1), 131–159. https://doi.org/10/c9qzd6

Kaiser, D. (2014). History: Shut up and calculate! *Nature, 505*(7482), 153–155. https://doi.org/10/gctkpn

Kalender, Z. Y., Marshman, E., Schunn, C. D., Nokes-Malach, T. J., & Singh, C. (2019a). Why female science, technology, engineering, and mathematics majors do not identify with physics: They do not think others see them that way. *Physical Review Physics Education Research, 15*(2), 020148. https://doi.org/10/ggh6jq

Kalender, Z. Y., Marshman, E., Schunn, C. D., Nokes-Malach, T. J., & Singh, C. (2019b). Gendered patterns in the construction of physics identity from motivational factors. *Physical Review Physics Education Research, 15*(2), 020119. https://doi.org/10/ggh6jt

Larsson, J. (2021). *Trainee teacher identities in the discourses of physics teacher education: Going against the flow of university physics*. Doctoral dissertation. Uppsala University.. https://urn.kb.se/resolve?urn=urn:nbn:se:uu:diva-440006.

Larsson, J., & Airey, J. (2021). On the periphery of university physics: Trainee physics teachers' experiences of learning undergraduate physics. *European Journal of Physics, 42*(5), 055702. https://doi.org/10/gmkbt6

Larsson, J., & Danielsson, A. (2022). The unexceptional physics girl: Trainee teachers crafting constructive positions of learning physics. (Unpublished manuscript).

Larsson, J., Airey, J., & Lundqvist, E. (2021). Swimming against the tide: Five assumptions about physics teacher education sustained by the culture of physics departments. *Journal of Science Teacher Education, 32*(8), 934–951. https://doi.org/10/gmkj85

Lave, J., & Wenger, E. (1991). *Situated learning: Legitimate peripheral participation*. Cambridge University Press.

Lemke, J. L. (2001). Articulating communities: Sociocultural perspectives on science education. *Journal of Research in Science Teaching, 38*(3), 296–316. https://doi.org/10/bhsggh

Leslie, S.-J., Cimpian, A., Meyer, M., & Freeland, E. (2015). Expectations of brilliance underlie gender distributions across academic disciplines. *Science, 347*(6219), 262–265. https://doi.org/10/f6ttq3

Levrini, O., Ambrosis, A. D., Hemmer, S., Laherto, A., Malgieri, M., Pantano, O., & Tasquier, G. (2017). Understanding first-year students' curiosity and interest about physics—Lessons learned from the HOPE project. *European Journal of Physics, 38*(2), 025701. https://doi.org/10/gctkb9

Lock, R. M., & Hazari, Z. (2016). Discussing underrepresentation as a means to facilitating female students' physics identity development. *Physical Review Physics Education Research, 12*(2), 020101. https://doi.org/10/gctkc8

Lock, R. M., Hazari, Z., & Potvin, G. (2013, January). Physics career intentions: The effect of physics identity, math identity, and gender. In *AIP Conference Proceedings* (Vol. 1513, no. 1, pp. 262–265). American Institute of Physics.

Lövheim, D. (2016). *Naturvetarna, ingenjörerna och valfrihetens samhälle: Rekrytering till teknik och naturvetenskap under svensk efterkrigstid* [Scientists, engineers, and the society of free choice: Recruitment to technology and science in post-war Sweden]. Nordic Academic Press (Kriterium). https://doi.org/10.21525/kriterium.7.

Lucena, J. C. (2005). *Defending the nation: U.S. policymaking to create scientists and engineers from Sputnik to the 'war against terrorism'*. University of America Press.

Martin, J. D. (2017). Prestige asymmetry in American physics: Aspirations, applications, and the purloined letter effect. *Science in Context, 30*(4), 475–506. https://doi.org/10/gf9wbz

Moshfeghyeganeh, S., & Hazari, Z. (2021). Effect of culture on women physicists' career choice: A comparison of Muslim majority countries and the West. *Physical Review Physics Education Research, 17*(1), 010114. https://doi.org/10/gjfxhw

Ong, M. (2005). Body projects of young women of color in physics: Intersections of gender, race, and science. *Social Problems, 52*(4), 593–617. https://doi.org/10/bpb6tz

Ong, M., Smith, J. M., & Ko, L. T. (2018). Counterspaces for women of color in STEM higher education: Marginal and central spaces for persistence and success. *Journal of Research in Science Teaching, 55*(2), 206–245. https://doi.org/10/gctkbh

Organisation for Economic Co-operation and Development. (2017a). *Education at a glance 2017: OECD indicators*. OECD Publishing. https://doi.org/10/gctkbh

Organisation for Economic Co-operation and Development. (2017b). *Distribution of graduates and entrants by field: Share of tertiary graduates by field and gender, Education at a glance*. OECD Publishing. http://stats.oecd.org//Index.aspx?QueryId=79587.

Pozzer, L. L., & Jackson, P. A. (2015). Conceptualizing identity in science education research: Theoretical and methodological issues. In C. Milne, K. Tobin, & D. DeGennaro (Eds.), *Sociocultural studies and implications for science education* (pp. 213–230). Springer. https://doi.org/10/cs4w

Prescod-Weinstein, C. (2020). Making black women scientists under white empiricism: The racialization of epistemology in physics. *Signs: Journal of Women in Culture and Society, 45*(2), 421–447. https://doi.org/10/ggnnzs

Rodriguez, I., Goertzen, R. M., Brewe, E., & Kramer, L. H. (2015). Developing a physics expert identity in a biophysics research group. *Physical Review Special Topics - Physics Education Research, 11*(1), 010116. https://doi.org/10/gctkgg.

Rolin, K. (2008). Gender and physics: Feminist philosophy and science education. *Science & Education, 17*(10), 1111–1125. https://doi.org/10/d6rvvf

Sawtelle, V., & Turpen, C. (2016). Leveraging a relationship with biology to expand a relationship with physics. *Physical Review Physics Education Research, 12*(1), 010136. https://doi.org/10/gc5s5w

Sax, L. J., Lehman, K. J., Barthelemy, R. S., & Lim, G. (2016). Women in physics: A comparison to science, technology, engineering, and math education over four decades. *Physical Review Physics Education Research, 12*(2), 020108. https://doi.org/10/gctq4v

Schreiner, C. (2006). *Exploring a ROSE-garden: Norwegian youth's orientations towards science – Seen as signs of late modern identities*. Doctoral dissertation. University of Oslo.. https://urn.nb.no/URN:NBN:no-12326.

Seyranian, V., Madva, A., Duong, N., Abramzon, N., Tibbetts, Y., & Harackiewicz, J. M. (2018). The longitudinal effects of STEM identity and gender on flourishing and achievement in college physics. *International Journal of STEM Education, 5*(1), 1–14. https://doi.org/10/gfvrn7

Shanahan, M. C. (2009). Identity in science learning: Exploring the attention given to agency and structure in studies of identity. *Studies in Science Education, 45*(1), 43–64. https://doi.org/10/dq3fw8
Storage, D., Horne, Z., Cimpian, A., & Leslie, S.-J. (2016). The frequency of "brilliant" and "genius" in teaching evaluations predicts the representation of women and African Americans across fields. *PLoS One, 11*(3), e0150194. https://doi.org/10/f8whhw
Tajmel, T. (2019). Diversity, human rights and physics education: Theoretical perspectives and critical awareness. In M. Pietrocola (Ed.), *Upgrading physics education to meet the needs of society* (pp. 239–252). Springer. https://doi.org/10.1007/978-3-319-96163-7_16
Tonso, K. L. (2006). Student engineers and engineer identity: Campus engineer identities as figured world. *Cultural Studies of Science Education, 1*(2), 273–307. https://doi.org/10/bs8dt3
Traweek, S. (1988). *Beamtimes and lifetimes: The world of high energy physicists*. Harvard University Press.
Traxler, A. L., Cid, X. C., Blue, J., & Barthelemy, R. (2016). Enriching gender in physics education research: A binary past and a complex future. *Physical Review Physics Education Research, 12*(2), 020114. https://doi.org/10/gctkdz
Tytler, R. (2014). Attitudes, identity, and aspirations toward science. In N. G. Lederman & S. K. Abell (Eds.), *Handbook of research on science education* (Vol. 2, pp. 82–103). Routledge.
United Nations Educational, Scientific and Cultural Organization. (2015). *UNESCO science report: Towards 2030*. UNESCO Publishing.
Universitetskanslersämbetet. (2016). *Kvinnor och män i högskolan* [Women and men in universities] (No. 16). https://www.uka.se/publikationer%2D%2Dbeslut/publikationer%2D%2Dbeslut/rapporter/rapporter/2016-10-14-kvinnor-och-man-i-hogskolan.html
Van Heuvelen, A. (1991). Learning to think like a physicist: A review of research-based instructional strategies. *American Journal of Physics, 59*(10), 891–897. https://doi.org/10/bnc6n5
Verdín, D., & Godwin, A. (2017, June). Physics identity promotes alternative careers for first-generation college students in engineering. In *ASEE Annual Conference and Exposition, Conference Proceedings* (Vol. 2017). American Society for Engineering Education.
Verdín, D., Godwin, A., Sonnert, G., & Sadler, P. M. (2019). Understanding how first-generation college students' out-of-school experiences, physics and STEM identities relate to engineering possible selves and certainty of career path. In *2018 IEEE Frontiers in Education Conference (FIE)*. Institute of Electrical and Electronics Engineers. https://doi.org/10/ggn4wk
Vidor, C. D. B., Danielsson, A., Rezende, F., & Ostermann, F. (2020). What are the problem representations and assumptions about gender underlying research on gender in physics and physics education? A systematic literature review. *Revista Brasileira de Pesquisa Em Educação Em Ciências*, 1133–1168. https://doi.org/10/gjjsjs
Wang, J., & Hazari, Z. (2018). Promoting high school students' physics identity through explicit and implicit recognition. *Physical Review Physics Education Research, 14*(2) https://doi.org/10/gd8dqn
Wenger, E. (1998). *Communities of practice: Learning, meaning, and identity*. Cambridge University Press.
Whitten, B. L. (1996). What physics is fundamental physics? Feminist implications of physicists' debate over the superconducting supercollider. *NWSA Journal, 8*(2), 1–16.
Whitten, B. L. (2012). (Baby) steps toward feminist physics. *Journal of Women and Minorities in Science and Engineering, 18*(2). https://doi.org/10/gctj9q
Wulff, P., Hazari, Z., Petersen, S., & Neumann, K. (2018). Engaging young women in physics: An intervention to support young women's physics identity development. *Physical Review Physics Education Research, 14*(2), 020113. https://doi.org/10/gfgxrr

Part IV
Science Teachers' Identities and Practices

Chapter 9
Exploring the Connections Between Student-Teacher-Administration Science Identities in Urban Settings

Rachel Askew and Katie Wade-Jaimes

It's a pretty typical middle school science class: students sit at small tables with notebooks and textbooks. Wearing uniforms, they quietly watch a teacher deliver a PowerPoint, writing in their notebooks when she tells them to. The room has cabinets and sinks in the back, but there are no pieces of science equipment out. There are some pre-made posters on the front wall with cartoons of the scientific method and various science vocabulary, and a set of rules. On the back wall, there is a "data wall" with graphs from students' most recent district-level benchmark exam. Class is disrupted a few times because the door has to remain locked and closed (school policy) so students have to knock and wait to get in. The class is almost eerily quiet – no time for group discussions or student questions, no labs or any movement around the room. Most students work diligently on their notes, but a few have their heads resting on their desk and are not engaged in the class. One group of three in the back of the room is particularly dedicated to finishing all of their work and getting the right answers, raising their hands to answer questions and seeking the teacher's approval. Most of the students in the class aren't interested in science careers, but one of the students, Bobbie, says she is a science person because she always gets As and wants to be an archaeologist. She has loved science since she was in elementary school, and tells the following story to explain her initial interest in science:

> … everybody was just laughing, and this one boy he dropped something and the glass broke. It was in like a little cylinder. And for some reason the glass just started dissolving. So, everybody was like, whoa! But the good thing is that, since I was a good student, and I was one of the helpers, and I told everybody to put their jackets on before.

R. Askew (✉)
Freed Hardeman-University, Henderson, TN, USA
e-mail: raskew@fhu.edu

K. Wade-Jaimes
University of Nevada, Reno, NV, USA

H. T. Holmegaard, L. Archer (eds.), *Science Identities*, Contributions from Science Education Research 12, https://doi.org/10.1007/978-3-031-17642-5_9

This experience in an urban classroom is not dissimilar to classrooms seen throughout literature on science learning in urban areas (i.e., Calabrese Barton, 2001; Carlone et al., 2014; Wade-Jaimes & Schwartz, 2019). In this study, we define urban as a school located in a large metropolitan area. Specifically situated in the United States, we draw from previous work on the characteristics of urban schools presented by Welsh and Swain (2020) in an analysis of the definitions used in research to describe urban schools. They note "six categories are typically used to define urban education:

(a) population/location/ geography,
(b) enrollment,
(c) demographic composition of students,
(d) resources in schools,
(e) disparities and educational inequality, and.
(f) social and economic context (p. 94).

Acknowledging the realities and assets of urban schools situated our exploration on how science is perceived and performed inside these contexts. For example, in the vignette above, students reflect on what it means to be a good science student, they focus on following directions and putting on the proper attire (i.e, lab coat) before participating. In this context, it's easy to see why many students don't identify with science - it's not very exciting or relevant to their lives. Bobbie's description of her experience with science in elementary school (in the U.S., ages 5–10) is active and exciting, possibly even a little dangerous, with Bobbie having a specific role in the action by telling students to wear their jackets (lab coats). This is in sharp contrast to what occurs in her middle school classroom (ages 10–13), where isn't much evidence of *doing* science, and as a result students are rewarded (usually through grades) for generic "good student" characteristics, such as finishing work quickly, not talking, and getting high test scores. Previous work has shown how this limits student identity to that of a Good Student (Archer et al., 2012; Carlone et al., 2014; Wade-Jaimes & Schwartz, 2019), instead of a scientist. In this paper, we view classroom actions as based on specific choices the teacher made, including choices that may seem passive, such as to conform with school culture or adopt existing curriculum. Although it's clear in the opening vignette that the teacher is making decisions that impact the classroom and, ultimately, the students, such as her choice of how to deliver content, this chapter argues that the teachers and students are both part of a larger context and subject to macro level discourses that constrain both teacher and student science identity. *What does it mean to be a good science student? Who decides who is and is not recognized as a scientist in the classroom?* An understanding of how science teachers negotiate these discourses, and position themselves as science teachers, will enable us to understand their choices and support teachers in disrupting the oppressive discourses limiting both themselves and their students. This chapter builds on previously used frameworks for identity to consider the interplay between teacher and student identities within the same meso-level context of the figured world of urban school science.

9.1 Theoretical Framings

9.1.1 Science Identity

We conceptualize science identity as how one sees or senses themselves within the discourses of science. While the experiences constituting one's science identity are not limited to the walls of a classroom, work on science identity in education commonly takes place inside school walls. Work in science identity often cites Gee (2000) for a reference point of conceptualizing identity as being a certain 'kind of person', involving both performance and recognition (Avraamidou, 2016; Carlone & Johnson, 2007; Wade-Jaimes & Schwartz, 2019). Some other commonalities of science identity work include the idea that science identities are constantly changing and occurring in relation to aspects outside of themselves, such as the environment, language, classroom, and/or community.

9.1.2 Science Teacher Identity

Similar to our conceptualization of science identity, we discuss teacher identity as the ways in which one sees her or himself relating to their position as a teacher - or their views of themselves as a certain kind of teacher. Combining an identity in science as well as an identity of a teacher, the merging of various subjectivities takes place when describing one's sense of self as a science teacher. This adds to the subject position of teacher as well as the discourses that accompany science. Avraamidou (2016) defined teacher identity as having three aspects: "(a) teacher identity is socially constructed and constituted; (b) teacher identity is dynamic and fluid, constantly forming and reforming; (c) teacher identity is complex and multifaceted, consisting of various sub-identities that are interrelated" (p. 154). These aspects of identity lead to an understanding of the importance of it within classrooms, for both teachers and, we argue, for students as well. How teachers see themselves and their beliefs about teaching lead to opportunities or alienation for students (Avraamidou, 2016; Emdin, 2010). This definition suggests that science teacher identities are continuously being reconstituted due to contexts, situations, and experiences, for example those who teach elementary or secondary, and those who teach in rural, suburban, or urban schools.

9.1.3 Urban Science Teacher Identity

In this chapter, we are using the descriptor "urban" to refer to teaching in a large metropolitan area. In the United States, urban areas, and schools in these areas, often have larger numbers of people of color than suburban or rural areas, and schools are often under-resourced compared to their suburban (but not rural) counterparts (Milner, 2012). Much research has been reported on that identifies the challenges of teaching in urban settings (e.g., Coffey & Farinde-Wu, 2016; Kokka, 2016). There has also been a call for research in urban education specifically in science education, which Fraser-Abder (2010) describe as an essential need in education research. Science teachers in particular have been the focus of much research around the challenges in urban education settings. For example, Ingersoll (2001) identified that math and science teachers have higher job dissatisfaction than other subject areas and Rinke (2009) explored the high rates of attrition for urban science teachers. A significant focus of the research on urban science education has focused on preservice (e.g., Freedman & Appleman, 2008; Rivera Maulucci, 2011; Tobin et al., 1999, 2001), new science teachers (e.g., Davis et al., 2006), and alternative certification programs (e.g., Proweller & Mitchener, 2004). This research largely focuses on preparing teachers for urban settings. While many studies have focused on preservice urban education, there is little literature that examines the experiences and identity development of in-service urban science teachers, particularly those who are no longer preservice or new teachers. The few exceptions (e.g., Calabrese Barton & Berchini, 2013; Fraser-Abder, 2010) generally focused on a small number of teachers. As Freedman and Appleman (2008) note, "Issues of power are particularly relevant for programs that most often prepare white, middle-class teachers from small towns or suburbs to teach in communities of color in high poverty urban schools" (p. 112). They continue to point out that research shows teachers generally have little experience in places different from their own and that this contributes to deficit views of urban schools.

9.2 Multi-level Discourse Framework

This research illustrates a particular need to consider the larger societal forces that impact both student and teacher identity development in science classrooms. This entails an examination of multiple levels of discourse (Fig. 9.1). This chapter builds on the multi-level discourse framework presented in Wade-Jaimes and Schwartz (2019), including both macro and micro levels of discourse, which are connected through a meso level, and also considers the connections between teacher and student identity in context.

At the Macro-level, we acknowledge that societal-level discourses and ideologies impact identity work by controlling what individuals can know, say, or do (Foucault, 1976), creating subjects that can or cannot be recognized as certain "types" of people, or identities (Butler, 1994). Importantly, these discourses work through both inclusion, defining who can be seen as a science person for example, as well as exclusion of individuals from identities based on characteristics such as race or gender. This framework considers how discourses of science intersect with those of race and education to define narrow types of science identities and science teaching identities.

Fig. 9.1 Multi-level framework for understanding teacher and student identity development

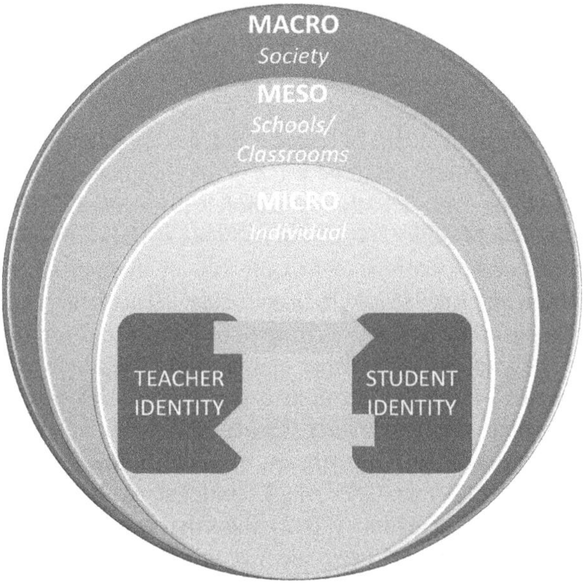

In Western society, science is viewed as elite, specialized, and objective (Aikenhead, 1996; Lemke, 1990). Western modern science is based on white, middle class, masculine ideals and discourses that frame it as authoritative, powerful, difficult (Lemke, 1990) and emotion and value-free (Aikenhead, 1996). Lemke (1990) also describes the discourse of school science as the view that science is a powerful and specialized way of talking about the world, which he refers to as the "mystique" of science. This mystique includes myths like science is much more complex and difficult than other subjects, science is in conflict with common sense, and scientists are superhuman elites. Lemke (1990) argues that these myths are perpetuated through the unwritten rules of science taught in science classrooms. Scientific knowledge is often presented as if there were one right way to know the world, where people disappear as actors. Science then becomes a description of the way the world is as opposed to the views of humans trying to make sense of the world. This discourse is perpetuated in science classrooms through the use of abstract, decontextualized, technical language. Emotion, values, and humor are all left out of scientific language, which is presented as serious and dignified. Another discourse that Lemke (1990) identifies is the authoritative and difficult nature of science, conveyed through objective, special truths that are only available to select experts. Students are taught not to trust their commonsense ability to figure things out and instead rely on knowledge conveyed through textbooks and other authorities. Like Aikenhead (1996), Lemke (1990) argues that the discourse of school science is defined by white, male, middle class, native English speakers who are committed to the values of middle-class culture: emotional control, orderliness, rationalism, achievement, punctuality, and social hierarchy. Discourses of school science are also influenced by discourses of education, in particular neoliberal discourses of accountability, individuality, and student responsibility (Bazzul, 2012; Tobin, 2011), leading to schools and science classrooms where the emphasis is on test scores, standards, and individual student work instead of inquiry, understanding, and collaboration.

These discourses of science and school science intersect with discourses of education and race in unique ways in urban schools and classrooms. Research in urban education has identified the increasing criminalization of schools, including the presence of police, or school resource officers, in schools (Theriot, 2009) and high levels of suspensions and expulsions of students of color, leading to the "school to prison pipeline" (Wald & Losen, 2003). This leads to a discourse of urban education that positions students as in need of increased discipline and punishment, compared to their suburban or white counterparts. Research has also shown that deficit discourses of urban education and urban schools are commonplace in US society, including among teachers (Jacobs, 2015; Milner, 2012). Indeed, the word "urban" is often used as a euphemism or code to describe under-resourced public schools that serve Brown and Black students (Calabrese Barton, 2001; Milner, 2012; Mutegi, 2013). Milner (2012) notes that at one school, in referring to how the teachers define urban, "They seemed to classify the school as urban because of their perceived shortcomings of students and parents in the school" (p. 558).

These macro level discourses intersect in schools and classrooms to create figured worlds, or cultural models that define and support certain types of people (Gee, 2014; Holland et al., 1998). Figured worlds define what counts as significant, who is recognized as a certain type of person, and, by exchange, who is invisible or unrecognizable. Previous research has shown how figured worlds of school science influenced by the macro level discourses described above create positions for students as science people that are narrowly defined and difficult to access (Wade-Jaimes & Schwartz, 2019). For example, the privileged science student position identified by Wade-Jaimes and Schwartz (2019) was a student who worked quickly and quietly (influenced by deficit discourses of students in urban schools being too loud or unruly) and did well on tests (influenced by discourses of accountability).

Within figured worlds, individuals act and interact with each other through micro-scale interactions. In the figured world of school science, these interactions occur between teachers and students and provide the performance and recognition that form the basis of identity development. We can see how these multiple levels of discourse impact student identities when we consider the commonly identified position of the "good science student" (e.g., Carlone et al., 2014). Based on macro-level discourses of students in urban schools, and particularly Brown and Black students, being violent and in need of control intersecting with discourses of accountability privileging standardized testing over engagement in science, figured worlds of school science become places where student behavior is equated with learning and science is relegated to textbooks and facts to memorize. This creates the celebrated position of the "good science student" who is able to work quickly and quietly without getting into trouble and makes good test scores.

We argue that to understand science or science teaching identity development, it is important to consider how both teachers and students are constrained by macro level discourses through the positionalities created in the figured world of urban school science. It is also important to consider how the performance and recognition of identities by teachers and students form a cyclical relationship: teachers' performance of science teaching identities in figured worlds of school science includes how they recognize students as students and/or science people; students' performance of science identities in figured worlds (a meso-level discourse) of school science includes recognition for teachers. This chapter explores that relationship, how it is impacted by macro-level discourses, and how it impacts teachers' science teaching identity development.

9.3 Intersections of Identities

9.3.1 Relationship Between Teacher and Student Identity

Teachers and students are in classrooms together - and how teachers view science influences their teaching and subsequently students' ideas surrounding science and their science identities (Bolshakova et al., 2011; Hazari et al., 2015). Likewise, the ways in which science is presented and who is recognized as a science person within classrooms influences how students see themselves in science (Carlone et al., 2014; Wade-Jaimes & Schwartz, 2019). In a study with elementary teachers participating in a professional learning community, Kane and Varelas (2016) noted the shifts in science teacher identity among participants and the corresponding shift in how they positioned their students within science.

"(...) they (re)constructed their roles as teachers vis-à-vis their students' roles as learners and as scientists. Thus, student identity–in the sense of roles that students were seen by their teachers as taking on or should be taking on in science class–was an integral part of the construction of teacher-of-science identities" (p. 190).

The authors note a specific relationship between the teachers' identities as "teacher-of-science" and students' positions in science classrooms. Further exploring connections between teacher-student-science identities can show us more about how both teachers and students take up science identities, and how the different ways in which they do so inform each other.

Within science classrooms, students and teachers are interacting with various ideas and conceptualizations of what it means to do science. How teachers identify within science influences how they teach science who they recognize as scientists (Bergman & Morphew, 2015). Likewise, students' experiences in science classrooms constitute their views of themselves within science. According to Brown et al. (2005), "Science, like all content areas in education, involves the balancing of interpersonal relationships among student, teacher, and social group. These relationships are constructed through the discourse processes in classrooms (...)" (p. 99). Supporting this quote from Brown et al. (2005), in a longitudinal study, found the influential relationship between teachers and students in science to be of high impact on students choosing science careers. Exploring the impact of an after-school science program for girls from low socioeconomic, single-parent households pursuing STEM careers, data from program records, surveys, and interviews were analyzed.

Interestingly, the area with the highest perceived impact was not anything specific to content but rather in the relationships built with teachers and mentors. In another longitudinal study, Archer et al. (2012) explored elementary girls' science identities, consisting of over 9000 surveys and 170 interviews. Citing the international phenomenon of the underrepresentation of women in science, this study was conducted in England and the authors state that by age 10/11, most participants did not identify with science. In this study, they explored more in-depth the girls who *did* identify with science, probing for further explanation and possible barriers. Of the girls who did identify with science, most were from middle- or upper-class families, tying again to the exclusivist stereotype surrounding science. They conclude by noting that girls are not less capable of pursuing science careers, but they are led to different paths due to "social, cultural, and structural factors" (p. 968)..

These "social, cultural, and structural factors" relate to the various conceptualizations of science identity and the importance of noting how such factors interact within science classrooms. If students do not see themselves in science, alienation occurs (Emdin, 2010). Emdin referred to alienation as "the feeling an individual experiences when he or she is excluded from the discourse in school or in other social settings" (p. 8). Brown et al. (2005) and Emdin (2010) presented the classroom as a site of possible alienation when suggesting the language and presentation of content impact how students view themselves in relation to science. This alienation removes students from the content of the science setting. Science discourse (e.g., Brown et al., 2005; Lemke, 1990) can itself alienate students. Students' cultural identity can often be pushed away or marginalized when in a science classroom based on the discourse (Elmesky, 2005). In each of these studies, the focus on relationships between students and teachers/mentors is imperative to understanding students' science identities. To fully explore students' science identities, we must acknowledge the connection and importance of the teacher-student relationship.

9.3.2 Relationship Between Teacher Identity and Administration

When discussing school science, it is important to acknowledge the contextual factors that are also at play (i.e., curriculum, standards, tests, evaluations, etc.). In the current public school system in the United States, test scores and accountability measures influence the content and often pedagogical choices within classrooms. These contextual pieces impact the content taught and systems such as teacher evaluations put specific pressure on teachers to gain certain scores and perform in certain ways. Just as teachers are responsible for teaching students, the current context places administration personnel (including principals, assistant principals, and content coaches) as responsible for evaluating teachers. In the current system, administrators are responsible for teacher evaluations, which tie directly to teachers' job status (for example, poor evaluation scores can lead to loss of teaching position) and often bonus pay. While administrators are following local rules and directives, no teachers in this study mentioned district personnel or standard writers. However, administrators' views and evaluations came up often. But, how do those in administrative roles conceptualize science? And how might this influence the ways in which teachers take up science in their classrooms?

Teachers are evaluated based on observations and students' performance on standardized tests. Despite pushes for science education reform (to be more inquiry-based) and the implementation of new standards (NGSS), science teaching is still often limited by the expectations from the testing culture (Aydeniz & Southerland, 2012). Discussing the impact of standardized testing on science education, Aydeniz and Southerland (2012) state that "While teachers are being scrutinized to improve students' test scores, there is not a system through which we hold science teachers accountable for promoting the goals of science education reform in the classroom or to use instructional and assessment practices". The authors are acknowledging the tension between efforts towards reform-based practices happening without training of evaluators (i.e, principals, coaches) and traditional standardized tests. As administrators are looked to for schoolwide scores, their knowledge of goals for science education reform play a role in holding teachers accountable for teaching science in ways that promote science inquiry rather than memorization. Currently, the standardized tests given to students do not include assessing the goal of science education reform, rather they continue to focus on memorization of facts.

In general, research shows that teachers' relationships, and perceived relationships, with administration impact their teacher identity, agency, and resilience (Richmond, 2016; Saka et al., 2013). However, there is little research surrounding administrators' ideas about science teaching and specifically, how that impacts science teachers' professional identities.

9.3.3 Methodology

To explore how teachers engaged with macro-level discourses and how they positioned themselves within the figured world of urban school science through performance and recognition of identity, we used Gee's (2014) approach to discourse analysis, as described below.

9.4 Context and Data Collection

This study took place in a large city in the southeast United States. Most of the participants are k-12 science teachers in the public school system in this city; however, some participants are from independent schools or neighboring school districts.

Teachers who self-identified as teaching in an "urban area" were invited to participate in the study.

Data reported here was collected through open-ended questionnaires. In the questionnaire, participants provided basic information about themselves and their schools (i.e., years of teaching experience, size of school) as well as providing more detailed answers to questions such as, "What makes a good science teacher?" This chapter presents an overview of discourses engaged in across questions and then focuses on data specifically from questions focused on recognition and recognizing, including:

(a) How would you describe an excellent science teacher?
(b) How do you think your administration would describe an excellent science teacher?
(c) How would you describe an excellent science student?
(d) How do you think your administration would describe an excellent science student?

Sixty-four participants completed these questionnaires. Questionnaires were completed by hand by the participants and then transcribed and entered into NVivo by the authors.

9.4.1 Data Analysis

In this work, we used Gee's (2014) method of discourse analysis, in particular the tools of Identities and Figured Worlds. Based on these tools, the following questions guided data analysis: "What identity or identities is this piece of language being used to enact? What identity or identities is this piece of language attributing to others, and how does this help the speaker or writer enact their own identity?" (Gee, 2014, p. 34) and "What figured worlds are relevant here? Are there differences here between the figured worlds that are affecting exposed beliefs and those that are affecting actual actions and practices?… What sorts of texts, media, experiences, interactions, and/or institutions could have given rise to these figured worlds?" (Gee, 2014, p. 115). Data collected through the questionnaires was first used to identify the discourses that teachers engaged in when describing and explaining their teaching experiences. This data was coded using discourses that have previously been identified in the literature and in urban schools (Wade-Jaimes & Schwartz, 2019) and generally fell into categories of science, education, and race. Codes included, for example, the idea that science is exclusive, the importance of accountability and responsibility in education, goals of individuality in education, and views that urban schools are violent and lack community resources. The questions focused on recognition and recognizing were then examined more closely to explore how teachers recognized excellent science teachers and students, how they perceived their administration recognizing excellent science teachers and students, and what conflicts existed between the two views. In this way we sought to further explore the complex interplay between science teacher identity development and science student identity development.

9.5 Findings

9.5.1 *Engagement in Discourses*

The participants in this study negotiated two equally prominent, but conflicting discourses: science as authentic understanding of the natural world, and deficit views of their students, schools, and communities. They were also influenced by discourses of accountability in education, which, although they were critical of, impacted their views of science teaching.

9.5.2 *Urban Education: Deficit Discourses*

Many teachers appeared to be negotiating real, logistical challenges of teaching in an urban setting with deficit views of the students and community. Across all questions, 39 teachers expressed some type of deficit perspective, for example about students' lack of motivation or desire to learn, families not valuing education, or students not caring about school. When asked simply to describe their schools, 19 of the teachers expressed deficit views, either explicitly or using coded language. For example, one teacher answered the question, "What's unique about teaching in an urban school?" by saying, "Lower desire to learn. Lack desire to be challenged." Another teacher described her school as, "Students are below grade level and have behavioral problems." Other teachers used coded language (e.g., Kailin, 1999), describing their schools as simply, "Title 1″ or their students as on "free and reduced-price lunch," both of which indicate high poverty levels in the school and are used to invoke images of urban schools and their students as resource scarce. Another 36 teachers described their schools in terms of what is missing, for example saying "Lack of time (focus is ALL on test scores, math, and reading), Lack of resources (see parentheses above), Lack of parental involvement." Only six of the teachers expressed asset-based views of their schools. One teacher explained that a unique aspect of teaching in an urban school is, "The knowledge that students do have is unique to them and is not always something I would have thought of," and another teacher said, "Students really value and respect their teachers who push them hard."

The deficit views expressed by the teachers in this study largely focused on students, however, they also expressed deficit views of families (seven references), previous teachers (five references), and the school as a whole (four references). One teacher expressed a deficit view of families by saying, "But in my neck of the woods, the majority of the homes expect for the students to have somewhere to go

Mondays to Fridays, 7:00 am to 2:30 pm. From these homes, where the level of educational achievement is high school level or below, it is difficult to expect more and desire more." Participants occasionally referred to previous teachers and schools as a challenge to overcome, saying, "It doesn't seem like a lot of teachers willingly choose to work in an urban setting. Urban students deserve great teachers," and "I believe that students in urban schools need a good teacher/education. That they should be allowed to experience everything people in better schools experience." A few teachers directly made the connection between these deficit views and their desire to work in an urban school, for example focusing on "giving back" (seven teachers) and "making a difference" (two teachers). Several teachers also mentioned the sacrifices associated with teaching, such as working long hours or spending their own money.

While 20 teachers indicated that the reason they chose to teach in an urban school district was because they themselves attended an urban school, many of them in the same district they now teach, over half of those teachers (11) still expressed deficit views of their students and schools. For example, one teacher said, "I chose to work in an urban school district to fill a much-needed occupancy. If someone does not step in, no one will show those students how it is supposed to be done. A caring, effective teacher is needed to fill this void. I grew up and went to school in an urban district, so this was also a way to give back to the community in which I was raised."

9.5.3 Science: Authenticity

Overall, the teachers in this study held authentic, meaningful ideas of what science is and what they believed science teaching should look like. The most coded phrase across all questionnaires was "real world" (96 references from 53 teachers), with teachers stressing the importance of connecting science and science teaching to the real world over and over. One teacher explained, "Rewarding moments for me is seeing the light bulb go off within the students who understands the content through everyday living - Aug 21st solar total eclipse was amazing." Teachers talked about "bringing science to life" and "engaging students with everyday phenomena", with 58 teachers mentioning doing labs or engaging in the science and engineering practices in some way 147 times across the questions. Teachers also mentioned traits such as curiosity and persistence in science (80 references by 53 teachers) and the need to take risk (six teachers) and learn from failure in science (14 teachers). Some teachers indicated a deep understanding of the goals of the new standards, saying they felt their job was to, "let students BE scientists and DO science." Teachers also directly confronted damaging discourses of exceptionalism in science, saying that one challenge they faced was in undoing stereotypes students had been taught about science be extra hard or scary.

In some cases, teachers indicated that although they held strong, authentic views of science, those views conflicted with deficit discourses around urban education, indicating that although science was about understanding the "real world," that world did not include urban areas and students. These participants mentioned that students were not exposed to science outside of school, and indicated that a challenge of teaching in an urban environment was the lack of science present in the area. "The students almost start off as a blank canvas. What they get is what the teacher teaches. I think it is unique to help students build upon limited knowledge, helping them realize that science is a part of everything done on earth." Another teacher explained the reason she taught in urban schools as, "It happened by chance. I have stuck with it in hopes of helping students who do not always have the opportunity to explore. Everyday things happen that these students have no idea about and with science explorations they have the opportunity to gain knowledge and information about the world around them." Finally, a teacher explained that teaching in an urban area was unique because, "Their exposure is vastly different. Holes and gaps in learning or experiences makes it a challenge." It is also evident that the deficit discourses engaged in by teachers intersected with elite discourses of science, i.e., the students were specifically viewed as lacking science knowledge and experience because they were at urban schools.

9.5.4 Education: Accountability

Operating parallel to the teachers' authentic views of science were educational discourses of accountability. Thirty-six of the participants mentioned either standards or standardized tests (or both) a total of 84 times across the questionnaires, with four teachers mentioning them five or more times in their answers. However, 39 of the instances were included in questions asking participants to reflect on what they perceived their administration would feel is important and another five references were critiquing the importance of standardized testing and explaining the challenges created by a focus on testing. This indicates that although teachers acknowledged the influence of discourses of accountability on their teaching, these discourses created conflict for them.

9.6 Recognizing/Recognition

As recognition is a key piece of science identity (see Carlone & Johnson, 2007), we explore how recognition was discussed for the teacher-student and teacher-administration relationships. We focused specifically on questions around recognition (of science teacher identity) and recognizing (of science student identity) to understand how teachers' negotiation of the macro-level discourse identified above impacted their science teaching identity and recognition of students' science identity.

9.6.1 Recognition for Science Teachers

Teachers were asked to describe an excellent science teacher, whether they saw themselves as an excellent science teacher, and to explain how they thought their administration would define an excellent science teacher. These questions were meant to elicit the ways in which science teachers positioned themselves and how they were positioned by significant others (their administration). Unsurprisingly, many teachers did not identify themselves as excellent science teachers because they viewed ongoing learning and growth as necessary pieces of teaching; we felt answers stating that they were not excellent science teachers *actually* indicated that the teachers probably were excellent teachers. What was surprising about the answers to these questions was the conflict between how teachers defined excellent science teaching and their perceptions of their administration's view of excellent science teaching. While the teachers described an excellent science teacher as having many science-specific characteristics, such as knowing science specific content (15), being able to connect the content to the real world (eight), and engaging students in labs and other hands-on learning activities (seven), the description they perceived their administration having of excellent science teaching revolved primarily around standardized testing (24 teachers). Teachers also indicated the state evaluation rubrics, which are standardized for all teachers and have no specific science-related measures, would be important to administration. Strong discrepancies in how teachers defined excellent science teaching compared to how they perceived their administration would perceive it were seen in 38 of the responses, with only four teachers indicating they thought they would be in complete agreement with their administration. A few teachers stated that they didn't know how their administration would define an excellent science teacher or that they would not be able to identify an excellent science teacher, for example saying, "Most principals aren't aware of the importance of science."

Another source of recognition for teachers was seen in answering the prompt, "Describe a rewarding experience you have had as a science teacher." For this prompt, the most common response (15) had to do with student appreciation, either in the moment or from students who had finished the class and came back to share their appreciation with the teacher. For example, one teacher said a rewarding experience was, "when a student returns to show gratitude or tell me how well they are doing now because of my class." Teachers also described the reward of students having an "Ah-ha!" or "lightbulb" moment (10), mastering content (eight), and making connections to the real world (seven). All of these responses indicate some type of recognition for the teacher through the student – either explicitly in the form of appreciation, or implicitly as students learn.

In a few cases, when discussing recognition from students, the influence of discourses of accountability was present. For example, one teacher said, "A very bright student reported to me the next year that he made a 100 on [standardized test] and wanted to thank me for teaching him the previous year." Another teacher said she felt like a good science teacher because, "To some extent yes because my students come back to tell me how they are doing in science in middle school, having learnt science the elementary grades. My students do well in state exams."

There was also evidence that deficit views of students influenced teachers' perceived recognition from students. One teacher described a rewarding moment as, "Any time a student comes to you after they leave your class to say thank you. I've had students (former) come to me 10-years after I taught them to say thank you. One student told me, 'I was doing wrong and messing up but you wouldn't give up on me. Thank you!'" In this acknowledgement, the teacher is noting the student remembering themselves as "doing wrong" and "messing up" but that the recognition for the teacher was that they would not give up on a student who needed such support.

Another teacher explained, "It is unique to teach at an urban school because of making difference in a child's life. You/I may be all they have to look for and that students will look back in the future, never forget it." In this quotation, the teacher was acknowledging how much difference teachers can make in urban schools – as opposed to rural or suburban areas. The common idea that teachers at urban schools are the only people in the students' lives that care for them and/or will not give up on them perpetuates a deficit mindset through saviorism (Matias, 2016).

9.6.2 Recognition by Science Teachers

Participants were also asked to describe an excellent science student and describe how their administration would define an excellent science student. Teachers again demonstrated authentic meaningful discourses around science in their descriptions, with 35 responses containing some kind of reference to being curious, creative, persistent, or understanding how and why things happen. The number one characteristic that teachers identified as indicating an excellent science student was asking questions/curiosity (29 teachers). By contrast, only 11 participants thought their administrators would include the same characteristics of science in their descriptions of excellent science students. Fifteen teachers indicated test scores would be important to their administration with seven teachers indicating that the *only* metric their administration would use to determine an excellent science student would be test scores. By contrast, only one teacher mentioned test scores in their description of an excellent science student, indicating that, at least in theory, they value more authentic characteristics of science. This dissonance between how teachers perceive administration would define an excellent science student by test scores, and the perceptions of teachers will be explored further in the discussion section. Many of the teachers also perceived that administration would value students behaving well (seven) and working hard (11).

Although the descriptions by teachers of excellent science students indicated authentic views of science, some teachers also highlighted student responsibility, indicating that excellent students already exist and are not created in the classroom. While one teacher responded to the question, "I refuse to answer, any student can be an excellent science student!" other teachers highlighted the responsibility of the student to challenge themselves ("An excellent science student has a love of learning and challenges themself") (eight) or do extra research and learning outside of school ("Investigates and researches beyond the school setting") (eight).

9.7 Discussion

In the all too typical classroom described in the vignette, the way in which Bobbie saw herself as a good science student was related to her adherence to the specific lab rules – not to the processes of science. In this story from an urban classroom, we see the discourses of urban education and accountability reflected in what it means to be a good science student: a quiet classroom, PowerPoint presentations, using science words, and a data wall, which all interacted to show students, like Bobbie, what counted as "science" in that room. Aspects of classroom decor, lab rules to follow, expectations of student behavior and presentation of science content contextualize what it means to do science within classroom walls. We know that the student-teacher-science identity interaction is at play within these walls but, based on questionnaires with teachers, we call in to question the relationship between administration-teacher-science teaching identities as well.

9.7.1 Impact of Macro-level Discourses: Teachers' Negotiation of Discourses of Science, Education, and Urban Education

Research shows the impact of teacher-student relationships on students' science identities (Brown et al., 2005; Fadigan & Hammrich, 2005). Due to this, we wanted to find out more about the ways in which urban science teachers were thinking of and presenting science to their students. In this study, we found that within the science classroom, teachers' negotiations of discourses of science, education, and urban education were often deficit based. Teachers did often disrupt the discourses of science as special, elite, or difficult and presented it more as authentic and real world. However, the teachers did not necessarily include urban in their conceptualization of the real world. For example, on teacher noted, "My students don't have experiences of common things like the Mississippi River." Although the teacher is highlighting science in the real world, they are dismissing the experiences of the students because they are not "common." Noting this, teachers were not disrupting deficit discourses surrounding urban schools. Rather, "urban" was used as a code word to trigger a deficit mentality. Missing from these discussions were asset views of students, social justice language, abilities to discuss systemic challenges instead of local, logistic challenges (i.e., lack of resources in their school vs. overall injustices in funding education). For example, one teacher gave African American as a nationality of her students, indicating that while she was acknowledging the race of her students, she did not have the appropriate language, or was not comfortable using the appropriate language (i.e., African American is not a nationality, it is a race and/or ethnicity). With this in mind, we question how it is that we are supporting urban science teachers' identities specifically within urban settings. In what ways are pre-service education programs preparing teachers to even *talk* about urban schools and their students? And more so – how are we continuing this conversation with in-service teachers? As previously mentioned, teachers participating in these discourses ranged in years of experience in urban schools from novice to veteran. Opportunities to reframe and support teachers' views of urban students and schools are needed to challenge the deficit discourses.

In describing excellent science teachers, students' discourses of accountability were found. Teachers did not accept these discourses, pushing back on test scores being the only qualifier for excellent science teaching and learning, however they did not discuss attempting to disrupt these discourses but felt as if they had to work within them. This tension shines a light onto the relationship between administrators and teachers. As evaluators, administrators observe teachers' lessons and provide scores and feedback on their teaching. Previous research shows that in the current testing culture, there is no recognition for teaching science practices - only test scores (Aydeniz & Southerland, 2012). Urban schools are impacted disproportionately by this use of evaluation as they are subject to being labeled "failing" (Sirrakos Jr & Emdin, 2017). In this study, teachers described an excellent science teacher as very specific to science - knowing the content, exploring, doing labs, etc. However, they thought administrators would have a much more generic view, revolving around test scores/evaluations. Their descriptions of how administrators would describe excellent science teachers and students, revolving around test scores, could be describing any class. While this data shows discrepancies between how administrators and teachers would recognize excellent science teachers, it is interesting to note that it shows teachers value science recognition from students - specifically the "A-ha!" moments and student appreciation. These differences in recognition bring up questions with the previously discussed framework (Fig. 9.1).

As shown below in Fig. 9.2, recognition is a piece of science teacher identity. How teachers are, or are not recognized, influences the ways in which they see themselves. In this study, we note the discrepancies between teachers' definitions and their perceptions of their administration's definitions of excellent science teachers. This discrepancy could be a way to change the narrative of what it means to be a good science teacher – or it could be a source of frustration and reason to leave science teaching. Recognition from administrators may come in various forms, but evaluations cannot be ignored. Evaluations are one of few ways teachers receive feedback, even if they don't think they're a good measure, they're still powerful when there aren't other measures in place for teachers to receive recognition from administration. In exploring these data we ask – How does administration convey their standards/views of what an excellent science teacher is? Is it only through test scores? Are teachers left guessing?

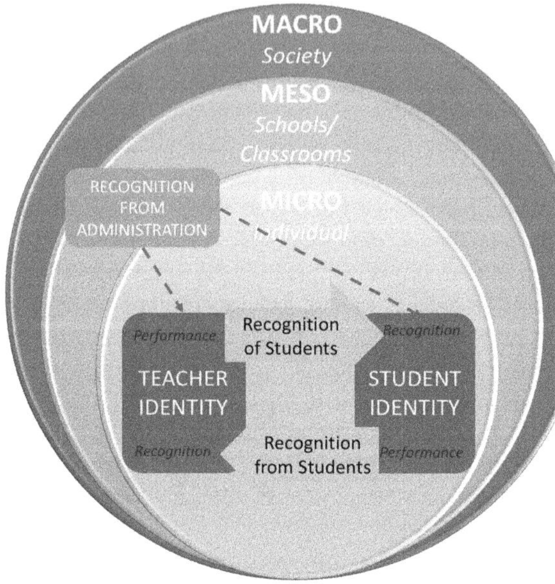

Fig. 9.2 Updated framework for understanding science teacher identity

Another change to this framework from the initial model is that recognition from students is important for teachers. In this work, we are asking specifically – how do macro level discourses impact who a teacher can see as a good student? When thinking of teacher recognition, we consider how teachers themselves are recognizing students within science. Simultaneously, we suggest that the ways in which administrators recognize science teachers and students may influence teachers' recognition of their students.

9.8 Implications for Research and Practice

This chapter has highlighted the need to consider aspects of context in understanding science teacher identity, and in particular we have highlighted the role of discourses of urban education and science in how teachers position themselves and their students, as well as how they perceive their positioning from administrators. Although teachers held authentic meaningful views of science, many engaged in deficit discourses often found in urban areas.

Teachers need to understand, and develop tools to disrupt, these discourses in their teacher preparation programs. Considering the unique context of urban schools, many of the teachers in this study held realistic views of the challenges of teaching in urban education that were not far from deficit views. For example, almost all teachers mentioned the lack of resources in their schools. However, teachers did not have the tools to critique this lack of resources as part of larger systems of oppression (no teachers mentioned the injustice of having fewer resources for their students) and, as a result, we worry they could come to embrace the dominant deficit discourses of urban education. Similarly, the known biases of standardized tests make it very difficult for students in urban areas to do well on them consistently without a strong "teach to the test" mentality. Knowing this, when teachers perceive that their administration values test scores above other measures of teaching or learning, there is a danger that teachers will never be able to meet these standards. What is the impact on the teachers' identity if they cannot gain any recognition from their administration? We worry this lack of recognition could result in teachers embracing deficit discourses to explain students' test scores or abandoning authentic science discourses in favor of test preparation.

We have also highlighted the role of recognition in science teacher identity, drawing off of work by Avraamidou that suggests recognition is often underanalyzed and theorized in science identity work (e.g., Avraamidou, 2020, 2021). Recognition is acknowledged in science identity research as an important piece, however, the attention to explicit and implicit recognition needs to be further explored, and as we suggest, within various relationships in schools (i.e., student-teacher, teacher-administration). Although many teachers found meaningful recognition from students, few described administrators who held similar views of science teaching. Recognition from students is of course important, but is unlikely to reflect the deep knowledge of science teaching pedagogy that would support growth as a teacher. Administrators need a better understanding of not just what excellent science teachers say and do, but also how to provide recognition to their science teachers that goes beyond test scores and generic evaluations. Even teachers who do well in terms of standardized tests and teaching rubrics need to know the work that they value is also valued in their schools.

References

Aikenhead, G. S. (1996). Science education: Border crossing into the subculture of science. *Studies in Science Education, 27*, 1–52. https://doi.org/10.1080/03057269608560077

Archer, L., DeWitt, J., Osborne, J., Dillon, J., Willis, B., & Wong, B. (2012). "Balancing acts": Elementary school girls' negotiations of femininity, achievement, and science. *Science Education, 96*, 967–989. https://doi.org/10.1002/sce.21031

Avraamidou, L. (2016). *Studying science teacher identity: Theoretical, methodological and empirical explorations*. Sense Publishers.

Avraamidou, L. (2020). Science identity as a landscape of becoming: Rethinking recognition and emotions through an intersectionality lens. *Cultural Studies of Science Education, 15*(2), 323–345.

Avraamidou, L. (2021). Identities in/out of physics and the politics of recognition. *Journal of Research in Science Teaching*. https://doi.org/10.1002/tea.21721

Aydeniz, M., & Southerland, S. A. (2012). A national survey of middle and high school science teachers' responses to standardized testing: Is science being devalued in schools? *Journal of Science Teacher Education, 23*(3), 233–257. https://doi.org/10.1007/s10972-012-9266-3

Bazzul, J. (2012). Neoliberal ideology, global capitalism, and science education: Engaging the question of subjectivity. *Cultural Studies of Science Education, 7*, 1001–1020. https://doi.org/10.1007/s11422-012-9413-3

Bergman, D. J., & Morphew, J. (2015). Effects of a science content course on elementary preservice teachers' self-efficacy of teaching science. *Journal of College Science Teaching, 44*, 73–81. https://www.jstor.org/stable/43631942

Bolshakova, V. L., Johnson, C. C., & Czerniak, C. M. (2011). "It depends on what science teacher you got": Urban science self-efficacy from teacher and student voices. *Cultural Studies of Science Education, 6*(4), 961. https://doi.org/10.1007/s11422-011-9346-2

Brown, B. A., Reveles, J. M., & Kelly, G. J. (2005). Scientific literacy and discursive identity: A theoretical framework for understanding science learning. *Science Education, 89*, 779–802.

Butler, J. (1994). *Contingent foundations: Feminism and the question of "postmodernism"*. The postmodern turn: New perspectives on social theory (pp. 153–170).

Calabrese Barton, A. (2001). Science education in urban settings: Seeking new ways of praxis through critical ethnography. *Journal of Research in Science Teaching, 38*, 899–917.

Calabrese Barton, A., & Berchini, C. (2013). Becoming an insider: Teaching science in urban settings. *Theory Into Practice, 52*, 21–27. https://doi.org/10.1002/tea.1038

Carlone, H. B., & Johnson, A. (2007). Understanding the science experiences of successful women of color: Science identity as an analytic lens. *Journal of Research in Science Teaching, 44*, 1187–1218. https://doi.org/10.1002/tea.20237

Carlone, H. B., Scott, C. M., & Lowder, C. (2014). Becoming (less) scientific: A longitudinal study of students' identity work from elementary to middle school science. *Journal of Research in Science Teaching, 51*, 836–869.

Coffey, H., & Farinde-Wu, A. (2016). Navigating the journey to culturally responsive teaching: Lessons from the success and struggles of one first-year, Black female teacher of Black students in an urban school. *Teaching and Teacher Education, 60*, 24–33. https://doi.org/0.1016/j.tate.2016.07.021

Davis, E. A., Petish, D., & Smithey, J. (2006). Challenges new science teachers face. *Review of Educational Research, 76*, 607–651. https://doi.org/10.1002/tea.20237

Elmesky, R. (2005). "I am science and the world is mine": Embodied practices as resources for empowerment. *School Science and Mathematics, 105*(7), 335–342. https://doi.org/10.1002/tea.20237

Emdin, C. (2010). Affiliation and alienation: hip_hop, rap, and urban science education. *Journal of Curriculum Studies, 42*, 1–25. https://doi.org/10.1080/00220270903161118

Fadigan, K. A., & Hammrich, P. L. (2005). Informal science education for girls: Careers in science and effective program elements. *Science Education Review, 43*, 83–90.

Foucault, M. (1976). *The archaeology of knowledge*. Harper & Row.

Fraser-Abder, P. (2010). Reflections on success and retention in urban science education: Voices of five African-American science teachers who stayed. *School Science & Mathematics, 110*, 238–246. https://doi.org/10.1111/j.1949-8594.2010.00031.x

Freedman, S. W., & Appleman, D. (2008). "What else would I be doing?": Teacher identity and teacher retention in urban schools. *Teacher Education Quarterly, 35*(3), 109–126. https://www.jstor.org/stable/23478984

Gee, J. P. (2000). Identity as an analytic lens for research in education. *Review of Research in Education, 25*, 99–125.

Gee, J. P. (2014). *An introduction to discourse analysis, theory and method*. Routledge. https://doi.org/10.4324/9780203005675

Hazari, Z., Cass, C., & Beattie, C. (2015). Obscuring power structures in the physics classroom: Linking teacher positioning, student engagement, and physics identity development. *Journal of Research in Science Teaching, 52*(6), 735–762. https://doi.org/10.1002/tea.21214

Holland, D., Lachicotte, W., Skinner, D., & Cain, C. (1998). *Identity and agency in cultural worlds*. Harvard University Press.

Ingersoll, R. M. (2001). Teacher turnover and teacher shortages: An organizational analysis. *American Educational Research Journal, 38*(3), 499–534.

Jacobs, K. B. (2015). "I want to see real urban schools": Teacher learners' discourse and discussion of urban- based field experiences. *Perspectives on Urban Education, 12*(1), 18–37.

Kailin, J. (1999). How White teachers perceive the problem of racism in their schools: A case study in "liberal" Lakeview. *Teachers College Record, 100*(4), 724–750.

Kane, J. M., & Varelas, M. (2016). Elementary school teachers constructing teacher-of-science identities. In *Studying science teacher identity* (pp. 177–195). Sense Publishers. https://doi.org/10.1007/978-94-6300-528-9_9

Kokka, K. (2016). Urban teacher longevity: What keeps teachers of color in one under-resourced urban school? *Teaching and Teacher Education, 59*, 169–179. https://doi.org/10.1016/j.tate.2016.05.014

Lemke, J. L. (1990). *Talking science: Language, learning, and values*. Ablex Publishing Corporation.

Matias, C. E. (2016). On the "flip" side: Unveiling the dangerous minds of white teacher candidates. In *Feeling White* (pp. 9–24). Brill Sense. https://doi.org/10.1007/978-94-6300-450-3_2

Milner, H. (2012). But what is urban education? *Urban Education, 47*, 556–561. https://doi.org/10.1177/0042085912447516

Mutegi, J. W. (2013). "Life's first need is for us to be realistic" and other reasons for examining the sociocultural construction of race in the science performance of African American students. *Journal of Research in Science Teaching, 50*, 82–103. https://doi.org/10.1002/tea.21065

Proweller, A., & Mitchener, C. P. (2004). Building teacher identity with urban youth: Voices of beginning middle school science teachers in an alternative certification program. *Journal of Research in Science Teaching, 41*, 1044–1062. https://doi.org/10.1002/tea.20036

Richmond, G. (2016). Making sense of the interplay of identity, agency, and context in the development of beginning science teachers in high-poverty schools. In *Studying science teacher identity* (pp. 219–235). Sense Publishers. https://doi.org/10.1007/978-94-6300-528-9_11

Rinke, C. R. (2009). Finding their way on: Career decision-making processes of urban science teachers. *Science Education, 93*(6), 1096–1121. https://doi.org/10.1002/sce.20339

Rivera Maulucci, M. S. (2011). Language experience narratives and the role of autobiographical reasoning in becoming an urban science teacher. *Cultural Studies of Science Education, 6*, 413–434. https://doi.org/10.1007/s11422-011-9323-9

Saka, Y., Southerland, S. A., Kittleson, J., & Hutner, T. (2013). Understanding the induction of a science teacher: The interaction of identity and context. *Research in Science Education, 43*(3), 1221–1244. https://doi.org/10.1007/s11165-012-9310-5

Sirrakos, G., Jr., & Emdin, C. (Eds.). (2017). *Between the world and the urban classroom*. Springer.

Theriot, M. T. (2009). School resource officers and the criminalization of student behavior. *Journal of Criminal Justice, 37*, 280–287. https://doi.org/10.1016/j.jcrimjus.2009.04.008

Tobin, K. (2011). Global reproduction and transformation of science education. *Cultural Studies of Science Education, 6*, 127–142. https://doi.org/10.1007/s11422-010-9293-3

Tobin, K., Seiler, G., & Smith, M. W. (1999). Educating science teachers for the sociocultural diversity of urban schools. *Research in Science Education, 29*(1), 69–88. https://doi.org/10.1007/bf02461181

Tobin, K., Roth, W. M., & Zimmermann, A. (2001). Learning to teach science in urban schools. *Journal of Research in Science Teaching: The Official Journal of the National Association for Research in Science Teaching, 38*(8), 941–964.

Wade-Jaimes, K., & Schwartz, R. (2019). "I don't think it's science:" African American girls and the figured world of school science. *Journal of Research in Science Teaching*. https://doi.org/10.1002/tea.21521

Wald, J., & Losen, D. J. (2003). Defining and redirecting a school-to-prison pipeline. *New Directions for Student Leadership, 2003*, 9–15.

Welsh, R. O., & Swain, W. A. (2020). (Re)defining urban education: A conceptual review and empirical exploration of the definition of urban education. *Educational Researcher, 49*(2), 90–100. https://doi.org/10.3102/0013189X20902822

Chapter 10
Science Teacher Identity Work in Colonized and Racialized Spaces

Gale Seiler and Hildah Kwamboka

Concerned with questions of success in and access to science, researchers have been drawn to the concept of identity, and we share that interest in how the concept of identity can help us understand people's experiences in relation to science. We are particularly interested in the identities and experiences of science teachers. Even more specifically, we are interested in the identities of science teachers from historically nondominant groups, because as Joseph (2010) reminds us, "It is for those at the margins that identities matter most" (p. 17).

Our examples come from two continents. However, we believe they are relevant to (although not identical to) many other contexts that share histories of colonization, oppression, and racism. Therefore, we use the term *nondominant* to refer to members of groups that are outside the historical and current normative forces structuring educational and other systems in particular locations, and therefore experience great power imbalances. In many contexts, as in the ones we illustrate, the history of colonialism brought domination and subordination of peoples based on a system of racial difference, thus, our use of *nondominant* conveys this positioning and its racialized aspects.

Science identity is constructed as one works to see oneself and be seen by others as one who knows and/or does science. Similarly, science teacher identity entails efforts to be and be seen as a science teacher, which in most cases involves being someone who not only knows science, and perhaps does science, but also teaches science effectively.

As will be shown below, there has been inadequate attention to race and the colonized and racialized nature of classrooms and schools in research on science teacher identity. Thus, as a science education community, we have much to learn about the

G. Seiler (✉)
Shepherdstown, WV, USA

H. Kwamboka
Michigan State University, East Lansing, MI, USA

H. T. Holmegaard, L. Archer (eds.), *Science Identities*, Contributions from Science Education Research 12, https://doi.org/10.1007/978-3-031-17642-5_10

impact of racialized spaces on the identity work of nondominant science teachers and the particular threats to identity work that they experience. Along with other theories, we will use Fanon to "articulate a social theory … that helps us to understand internalized oppression and the way it works on different bodies in schooling and education" (Dei, 2010, p. 3).

To begin, we will consider two questions: How has race been incorporated into science teacher identity research? What theoretical ideas and research approaches can help us move forward in this area? We will then use two examples to illustrate how we might bring a focus on schools as colonized and racialized spaces to the study of science teacher identity among nondominant science teachers.

10.1 Research on Science Teacher Identity

A reflective approach to studying science identity has been prevalent in the literature, and the focus has been on personal histories and how they shape and reveal science identities. This emphasis on personal stories is present in research on science teacher identity as well. However, we believe that this emphasis has masked the social forces that constrain science teacher identity work and has severely limited our understanding of the racialized nature of identity work by nondominant science teachers.

Most research on science teacher identity has utilized data sources such as teacher interviews and written or oral reflections, as was documented in a review of empirical studies specifically about science teacher identity (Avraamidou, 2014). Avraamidou stated that most of the studies lacked consideration of "the nature and characteristics of the contexts in which these studies took place and how these contexts may (or may not) have impacted upon the participants' identities" (p. 165). While sociocultural practice theory, an often-used theory of identity, draws attention to how we form identities during activities and interactions in the social world (Urrieta, 2007), a methodological reliance on interviews and reflections means that few studies have actually examined how science teacher identities are constructed in action. We recognize that teacher reflection through interviews, narrative writing, or journaling can be valuable, in that it can reveal the stories we tell about ourselves and may even aid the storyteller in making sense of their positioning. However, with these kinds of data sources we often miss the interplay between social and institutional forces and possibilities for individual actions. In other words, we miss the socially constructed nature of identity.

We suggest that, as a result of the focus on personal, reflective accounts of identity, the kinds of actions and interactions that might support or rupture the construction of a science teacher's identity in-practice are largely unknown and unexplored. Furthermore, for nondominant teachers, it is likely that the power of these ruptures is even greater and more significant. This echoes Fanon's (1967) reminder of the importance of the "inner relationships between consciousness and the social context" (p. 100) and the need to recognize how context is colonized and racialized.

Fanon also challenged us to ask ourselves: How are we complicit in Western hegemony? (Dei, 2010). It may be that our reluctance to engage in more substantive ways with race and racism in science identity research reveals some complicity by us, as science identity researchers.

In this chapter, we will document and explore the processes of identification and dis-identification with science and science teaching by members of racial groups that are outside the historical and current normative forces structuring educational systems in particular locations. Through this, we can begin to understand the fractures and (re)formation of identities that does/can/cannot occur when nondominant teachers teach science in today's educational systems. These ruptures occur even when nondominant teachers are serving nondominant students or teaching in predominantly nondominant communities or countries, because the systems are largely shaped and powered by White Eurocentric norms. For example, this might occur with African American teachers in schools in the United States, Indigenous teachers in Northern Canada, Kenyan teachers in Kenya, or Zulu teachers in South Africa. Lastly, these ruptures can constrain and shape identity work such that trauma results, that is, feeling overwhelmed and isolated, the effects of which may be both immediate and long-lasting.

10.1.1 Theoretical Framework

Identity entails the way we place ourselves or get placed by others into pre-existing categories (often called social identities) in a social setting. Thus, we do some of the work of identity construction ourselves, while some of it is done to/for us (Holland et al., 2001). On one scale, these categories are based on constructs such as race and gender, but also include categories of teacher, scientist, and parent. Each of these social categories provides particular resources (i.e., knowledge and ways of talking, being, acting, and thinking) with which we can construct/position ourselves thereby doing identity work. Thus, a person does identity work when they employ the knowledge and ways of talking, being, acting and thinking that are associated with that social category and community in an attempt to be recognized as a legitimate community member. A person can also try to use resources from other categories, groups or settings in novel ways to get others to accept them as legitimate community members. These attempts at legitimate participation and identification (identity work) may be recognized or rebuffed, in overt or subtle ways, and we will see that this can have immediate and long-term consequences for identity construction. At times, other people may construct/position us inside or outside a social category, whether we are attempting to position ourselves within it or not. Thus, identity work is always both individual and collective.

The outcomes of this identity work are carried forward over time and sometimes into other settings as successful resources are accumulated and used in on-going identity work, and messages are received about resources not to use. The idea of identity trajectories (Jackson & Seiler, 2013) helps us envision this temporal and spatial aspect of identity work, and the consequences of identity work being truncated or interrupted. By that, we mean the constraints and even trauma experienced when one tries to use their available resources to be recognized as a legitimate member of a community, but they are not, or when the norms and expectations inhibit one's attempts to even try to use their resources. Our research is based on the idea that the norms and expectations of particular communities and settings are held, taken up, and used locally, but that they emanate from the larger historical and social context, and they must be studied and theorized accordingly.

10.2 Figured Worlds and Cultural Models

To aid us in linking identity work on a micro scale (e.g., in classroom interactions) to the larger social context, we utilize the concepts of figured worlds and cultural models. Holland et al. (2001) described a figured world as "a socially and culturally constructed realm of interpretation in which particular characters and actors are recognized, significance is assigned to certain acts, and particular outcomes are valued over others" (p. 52). Thus, access to a particular figured world is regulated, as some people are recruited or allowed access while others are not. A figured world is socially organized and performed through a set of circulating storylines and other resources that figure or construct that world in a particular way. Because it is socially organized, it is the work of the participants that produces or reproduces the figured world. Thus, a figured world can be changed or novel figured worlds produced, which can open up possibilities for those marginalized or not recognized as legitimate participants. However, figured worlds tend to be durable and are more likely reproduced that transformed or created anew.

A figured world is built on cultural models, that is, the storylines and all that is taken to be typical or expected in a given context (e.g., a school or a science classroom). Cultural models are the "mental/emotional knowledge structures" (Holland et al., 2001, p. 297) associated with a particular figured world and on which individuals rely to interact and frame their experiences in a figured world. Thus, cultural models serve as resources that fuel or constrain identity construction and recognition by others as legitimate participants in a figured world. Cultural models that permeate schools usually mirror those in the wider world, for example, they may include ideas about teaching and learning, science and scientists, as well as ideas about racial hierarchies and stereotypes of White excellence and failure of Others.

Figured worlds vary in scope and location and can be thought of as nested or overlapping. We believe that this way of thinking about figured worlds and their constituent cultural models allows us to pay attention to larger historical and social contexts related to race, power, and privilege, and how they play out in localized figured worlds. For example, we can think of schools as figured worlds that are constructed in particular ways and shaped by particular stories, norms, and expectations. While any particular school or classroom may look, sound and feel somewhat different from another, most share a common set of knowledge structures, owing to histories of colonialism and now increased globalization. Science classrooms are a specific kind of classroom figured world, with stories and expectations shaped by other figured worlds, and cultural models from multiple sources figure those worlds and assign recognition and significance to people and actions in them. These cultural models include those that depict scientists as White males, all science knowledge as presented in Western textbooks, and proper science classroom interactions as based on White norms. Of course, within particular science classrooms there are local variations. For example, there will be differences in what resources students and teachers bring with them. As will be shown later, contestation arises when these do not align with the norms and expectations that hold power and figure each classroom to align with larger figured worlds.

As shown here, using the constructs of figured worlds and cultural models allows us to move between scales of social context, from classroom interactions to the larger figured worlds of science, education, and science education (Jackson & Seiler, 2013), with their gendered and racialized cultural models that exert power over science teaching and teachers. Urrieta (2007) reminded us that figured worlds are "intermeshed with trans-local systems of power and privilege" (p. 109) and that cultural resources within figured worlds are tied to powerful, trans-local institutions. While many recognize the role of power and privilege in figured worlds and the identity construction that occurs within them, there has been little specific attention to the racialized aspects of figured worlds or to the history of particular figured worlds as sites of colonization.

As explained below, the figured worlds of schools and science classrooms are rooted in colonial conquest and Whiteness. Therefore, they are heavy with forces and structures that constrain the use of racialized identity resources. We are interested in the consequences of those forces and the possibilities found in resistance to them.

10.3 Colonialism and Whiteness

In our attempt to heighten attention to the racialized forces that shape identity work among nondominant science teachers, we draw from writings on colonialism and Whiteness, and from the work of Frantz Fanon who wrote about links between colonialism, colonial racism, and identity.

Colonialism is the policy or practice of acquiring political control over another country, occupying it with settlers, and exploiting it economically. While colonialism took on (and continues to take on) different forms in different locations, the economic exploitation that led to the enslavement of 12.5 million people from African countries in North America, the Caribbean, and South America, and the economic and social control exerted by White minorities in African countries and other colonial contexts was a horror with a far-reaching toll. For this reason, we recognize commonalities between experiences of nondominant teachers across such locations. We also need to consider that throughout history and across continents, education has been used as a means of control and assimilation of nondominant people, and colonial interests continue to play a role in the shaping of school systems in many contexts even today (Guo, 2007). For example, Shizha (2010) recounted how—through structural adjustment packages, various conventions and declarations, funding conditions and other means—Western countries shaped post-independence science education in many African countries. Aikenhead (2008) noted the deleterious consequences of importing science curricula and teaching materials from Europe and North America as part of the global homogenization of education.

Given this context of historic and continuing colonization of nondominant bodies and minds, the structural forces that have shaped most institutions and systems (including education) are based on Whiteness, and this recognition is particularly crucial in considering the identity work of nondominant science teachers. By Whiteness we mean how White norms permeate society while appearing commonplace and value-neutral.

Frankenberg (1993) explained Whiteness as a social construction designed purposefully to maintain systemic advantage of Whites over other groups. More specifically, Picower (2009) referred to Whiteness as a tool for maintaining "Hegemonic stories and dominant ideologies of race, which in turn, uphold structures of White supremacy" (p. 204). Whiteness is a set of norms and perspectives embedded in an oppressive system. These norms and expectations are often taken up by all participants in that system—those who are White as well as those who are not, those who benefit greatly from the system and those who benefit less or not at all. Thus, the fundamental power of Whiteness lies in its insidiousness, invisibility, pervasiveness, and perceived neutrality. This is the controlling mechanism that maintains colonial relations as natural occurrences (Fanon, 1967), and we will see this illustrated in the science teachers represented below.

Recognizing this connection between colonial forms of education and Whiteness lets us see that Western science and Eurocentric forms of teaching continue the oppression of members of nondominant groups, students and teachers alike. Narratives and artifacts of Whiteness embedded in the education system are further examples of racially oppressive cultural models that structure the figured world of science education. In this way, the history of colonialism and the concept of Whiteness help us to understand the racialized nature of science classrooms today (Seiler, 2019).

10.4 Science Teachers in Colonized, Racialized Spaces

In our research, we recognize the need for analysis that crosses micro and macro scales to show how widespread racialized and gendered cultural models significantly constrain identification with science and science teaching for nondominant teachers. While our analysis is often at the micro level, the theoretical framework of figured worlds and cultural models allows us to broaden our understanding of teacher identity to include larger cultural models—for example, cultural models about race, teaching, and science. Although contexts may be geographically disconnected and distinct, they are undergirded by strikingly similar educational experiences of colonialism and Whiteness.

We now turn to two examples of nondominant science teachers working in educational systems rooted in a Eurocentric view of teaching, a Western view of science, and an ideology of Whiteness. In doing so, we focus on identity trajectories (Jackson & Seiler, 2013) and how cultural models are produced/accessed and can/cannot be used as resources for on-going identity work, in-the-moment and over time and place. We will draw from a co-author's experiences as a Kenyan teaching science in Kenya and as a science teacher educator in the United States, and the experiences of an African American science teacher in Baltimore. Holland et al. (2001) explained that "ones history-in-person is the sediment from past experiences upon which one improvises in response to the subject positions afforded one in the present" (p. 18). In both cases, we will see that these sediments from past experiences are both personal and collective, but they confront storylines, artifacts, and norms that are historic and can be powerful at interrupting identity work.

In the first case, the co-author uses memory work, which is a social constructionist and feminist research method (Crawford et al., 1992) that draws upon remembered moments perceived to have significantly impacted the trajectory of a person's life. Our use of memory work may seem to contradict our earlier comments on the hazards of over-reliance on reflective accounts of science teacher identity. However, to connect reflection with identity-in-action, we focus specifically on memories of teaching, that is, we pay attention to what Hildah said or did and threats to her identity work while teaching. This enables us to use memory work to focus on activity and the importance of power in identity work in particular figured worlds (Urrieta, 2007).

10.5 Hildah

Co-author Hildah Kwamboka was a Ph.D. student in science education at the time of this writing and a co-instructor in courses in the science teacher preparation program at a predominantly White mid-Western university. Hildah was born and grew up in Kenya. Before coming to the United States, she was a science and agriculture teacher for four years in Kenya, where national examinations are the marker of success or failure in school. In a previous publication (Seiler & Kwamboka, 2019), Hildah wrote about her experiences as a science teacher in Kenya. We include an excerpt from that here:

> One day I decided to do something different in the Agriculture course that I taught. In Agriculture, there was less strictness to adhere to science terminology and spelling, as long as students showed understanding of the concepts in their explanations, albeit in English. In one formative assessment, I asked my students to write the test both in English and their tribal language where possible. The task was to identify common farm weeds, the economic importance of those weeds, and their scientific names. To my surprise, my students knew a

> lot about weeds, including information that was missing from the textbook. They expressed firsthand knowledge of how to control weeds and the economic importance of weeds that I, as their instructor, was not familiar with. My students showed a great understanding and enthusiasm of the agricultural content and had a lot to say when they were not limited to English or scientific vocabularies and spelling. After grading, all the students scored at least 18 out of 20. These were the same students who were struggling in my Biology class, and whom I would rate under average on Biology tests. I sat in my office for a moment and asked myself, can I teach Biology like this, at least sometimes? As you can guess, the answer I arrived at was no. I said to myself, "Biology is a science and there is no way you can do science like this and I have to prepare my students for the tests. Agriculture is an art and science, but Biology is not." I did not think about it again.

Hildah recalled that, while teaching in Kenya, she conformed to the course syllabus, textbook, and national assessment standards and to the colonized figured world that these artifacts reproduced. As was expected, her teaching emphasized mastery of facts and scientific language, including proper use and spelling of scientific vocabulary. These were the only cultural models for science teaching that were recognized as legitimate in this figured world, and they were rooted in Western science, Eurocentric education, and a history of colonial oppression, illustrating the translocal systems of power and privilege (Urrieta, 2007) that often shape figured worlds. They did not offer Hildah other ways of doing the work of a science teacher. Rather, they represented what Fanon described as oppression of the mind, which is the basis of alienation and internalized oppression.

The power of language that Hildah alludes to was described by Fanon in his writings. In fact, he began his book, *Black Skin, White Mask* (1967) with a chapter called *The Negro and Language.* In it, he laid out his ideas on "colonialist subjugation" (p. 8) and the resulting self-division it creates. He wrote that to speak "means above all to assume a culture, to support the weight of a civilization" (p. 8). Thus, we see the colonial project carried out through the suppression and silencing of language, such as the tribal languages of Hildah and her students. Consequently, science teaching and learning becomes an alienated endeavor that was disconnected from the tribal ways of Hildah and her students.

While in her doctoral program, Hildah had opportunities to learn about concepts such as identity, cultural models, funds of knowledge (Moll et al., 1992), structure and agency (Sewell, 1992), and colonialism, and she came to ascribe different meanings to her teaching experiences in Kenya. She questioned why Black teachers in African countries continued to teach science in ways that emulate the colonial masters' education system and why textbooks lacked local examples and decentered cultural ways of thinking and talking. More recently Hildah wrote about new cultural models that became available to her and new possibilities for becoming (Holland et al., 2001).

> As a science teacher teaching about energy flow in ecosystems in Kenya, I was not able to recognize that just outside my classroom there was a full ecosystem in which various organisms interacted, or that my students from their diverse locations had knowledge about other ecosystems. Instead, I used the classic ecosystem example from the textbook, with animals and plants that my students might not be familiar with—because that was how I was expected to teach. As a young science teacher beginning to make my way, that was the only

path available to me. When I look back, I now see my students as resources for learning, and I see my own out-of-school knowledge and experiences as resources for teaching. But that way of thinking did not exist for me then, in the figured world of a Kenyan science classroom.

In the figured world of graduate studies, Hildah found space to construct herself in another way, as her knowledge and experiences were valued by significant others (e.g., professors and advisors) and she was encouraged to draw upon her own cultural resources.

Graduate school learning has supported me in recognizing the cultural forms of doing science in my life in Kenya, such as, predicting weather patterns and seasons by observing the sky, and the process of respiration in fermentation of porridge malt and traditional beer brewing. Taking courses that incorporated Indigenous knowledge influenced my recognition and appreciation of the science practices that my family engaged in. I read critical science scholars who positioned science as a culturally and socially laden subject. This gave me the opportunity to recognize the potential of incorporating ways of being and doing science that are generally outside of the Eurocentric view.

Although she was beginning to figure herself in new ways (Holland et al., 2001), Hildah still doubted whether out-of-school knowledges, such as those of her tribe in Kenya, could earn legitimacy in science education and position her as a competent science educator. Hildah realized that, despite the opportunities availed to her by some of her professors, how others positioned her in the field of science education mattered to her identity. In a science teaching methods course that she co-taught, she was encouraged to share her knowledge during the planning for the course and in the classes. However, the resultant identity work was not without hazards.

As a graduate teaching assistant and co-instructor of science teaching methods courses, I sometimes used my grandmother to show how science is embedded in the daily lives of people, even though it is not presented that way in Western science textbooks. However, I am often not confident that my cultural repertoire is as legitimate as those deeply ingrained Eurocentric science practices. I have seen other people cast doubt on such knowledge, since it is not documented in their reference materials as part of mainstream science. For example, I used my grandmother's practice of breaking apical dominance in vegetable plants to illustrate how she engaged in science practices even though she did not attend formal schooling and couldn't name the hormones or processes as they are labelled in science texts. One of the preservice teachers (a white male) in my class challenged this example indicating that science is about explaining HOW things happen. According to him, my grandmother did not understand the plant hormone interactions that lead to what we see in apical dominance, hence, her practices did not count as real science.

As I try to unpack and incorporate other ways of doing science in my teaching by using my passed-down knowledge, I anticipate that my students might not take me seriously. I fear that I will be seen as less "sciencey" if I suggest or use ideas that diverge from Western textbook ways of seeing and doing science. Hence, I am constantly deliberating what to say, when, and how much. Such self-policing stems from: 1) images of what is generally considered and recognized as real science teaching; 2) the perception that few people might accept this kind of teaching; 3) the reaction I receive from students. Frequently I have questioned my efforts to draw on local knowledge in my teaching because the culture of

> science does not welcome it. And even in my own mind, the question is there, because this is not the way I was taught to see or teach science until I got to graduate school.

Hildah experienced the dilemma of wanting to be seen as a competent science teacher and teacher educator, while also wanting to leverage her cultural resources and expand the notions of science prevalent in the figured world of science education. The figured world of the doctoral program allowed her to perform new understandings of herself in relation to science and science teaching, and that seemed to sustain her identity work among others who shared these ways of thinking and recognized her knowledge. However, when she was teaching the preservice teachers in courses in their certification program, she was the only member of a nondominant group; all the students and the other instructor were White. In this figured world, the cultural models and Eurocentric views of science and knowledge held fast and, at that time, derailed her recognition as a competent science teacher if she drew from her "passed-down knowledge" and experiences. The strong negative reaction of students illustrates the powerful repressive presence of colonialism that Fanon (1967) described. Dei (2010) wrote that decolonizing is "about bringing to the fore subjugated voices, histories, and experiences. It is about legitimizing practice and experience as the contextual basis of knowing" (p. 9). Hildah's experiences show the toll of trying to do this while teaching in a figured world based on White norms of teaching science.

More recently, Hildah wrote about other experiences related to her identity work as a science educator and her struggles in being recognized as such.

> What is troubling to me is how some colleagues, who are from the dominant group (i.e., White), tend to position me. While co-teaching a science teaching methods class with an instructor who is White, I was frequently asked to take the lead on multicultural and social justice science teaching-related assignments, planning, and classroom activities. On the other hand, I was not asked to take the lead when it came to teaching about core practices of science or science standards, or in tasks that demonstrated content area expertise. To combat this and to obtain recognition, I am forced to assert my expertise as a competent science educator—something that is physically and psychologically draining. Often, it does not feel worth the fight to reposition myself. Being pushed to the periphery of science and positioned as a non-scientist impacts my identity work both at the moment, as well as in future positioning in science. As the fear of being seen as less competent and less sciencey sets in, I am less likely to try to position myself as competent in science down the road.

Hildah shared that it was hard for her to write the above narrative, and she hesitated to include it, as it is still very "raw and emotional" for her to even think about. It shows the consequences of Hildah trying to construct her identity as a competent science educator, while having the identity of social justice educator thrust upon her by someone with more power in that figured world. The identity thrust upon Hildah drew on larger cultural models (e.g., racialized people should be good at teaching social justice) and her target identity ran counter to those models. This disidentification and lack of recognition for the identity she aspired to construct created physical and psychological pain for Hildah, as she described. Hildah's reflection also illustrates how sediments from past experiences (Holland et al., 2001) can negatively impact identity trajectories and future identity work.

In the wake of George Floyd's murder and protests against police brutality in the United States and beyond, Hildah again struggled with how to position herself as a science educator, while recognizing her experiences as a Black woman in the United States and a social justice advocate.

> Following the killing of George Floyd, I cannot stop thinking about how his and the deaths of other Black men at the hands of the police relate to social justice science teaching. I can see how the events in the video can be used to teach core ideas of science (e.g., body systems, homeostasis, cellular respiration), while at the same time engaging students in issues of racial injustice. However, based on my past experiences, I am not sure how it will be received by White educators and students, and I fear how I will be positioned by them. As much as I worry about this, I also feel guilty if I do not seize this opportunity to educate others about how to link science with social justice. Science has been at the center of the construction of race and racism; it is my belief that it should be at the center of dismantling racism and teaching about racial injustices. But I struggle with how to do that, yet also be positioned as competent in science.

Hildah understands that asserting the importance of social justice teaching in science might lead her to be positioned as a non-scientist by preservice teachers or colleagues. On the other hand, not sharing her knowledge and resources as it pertains to science and social justice creates a conflict between who she is as an educator and who she is as a wife of a Black man and mother of a Black son. In the figured world of science education, where Whiteness is pervasive, discussions of race and, in particular, racial injustice are seen as having no place. To construct and teach this kind of social justice science lesson positions Hildah against powerful forces of status and privilege.

To further explore both the fragility and resilience of science teacher identity work, we now draw from another case—Donna, an African American science teacher in Baltimore. Here we look more closely at her teaching practices and utilize analysis of videorecordings to look at identity work in the process of science teaching. This example provides evidence of radical possibilities when cultural models from nondominant experiences are used in science teaching. However, as with Hildah's memory work, it also reveals the racialized limits and constraints experienced by nondominant science teachers as they construct their identities around science teaching in colonized figured worlds of schooling.

10.6 Donna

Powerful cultural models related to science and science teaching are rooted in colonialism and Whiteness, and because they are so pervasive, we acquire them through participation in a many figured worlds. They may lead science teachers to act in certain ways and to enact certain practices and not others. They may also bring teachers from nondominant racial groups to question their own fit as scientists or teachers of science and to leave resources untapped while teaching. As we will see with Donna, she often felt discomfort with the idea of mixing who she was out-of-school with her identity as a science teacher and, similar to Hildah, she often hesitated to use those cultural resources in her teaching.

Donna was an African American student teacher in a predominantly African American high school in Baltimore. As previously described (Seiler, 2011), she was a returning student who had worked in the business world for a number of years. As such, she did not have a science background. So, when she decided to change careers and become a biology teacher, she had to take a number of required science courses before she entered the teacher education program to become certified to teach biology. Because Donna had not been a science major as an undergraduate, her own experiences with science were limited. She never had the opportunity to work in a lab, or collaborate with scientists, or participate in forms of science that are widely recognized as legitimate. Thus, her experiences with science were narrowly constructed as a specific set of practices and knowledges that she could not easily lay claim to.

Co-author Gale Seiler knew Donna as a very competent speaker, with a background in public relations. However, when Donna began her student teaching, she seemed nervous in the classroom; she didn't smile or laugh much while leading a lesson; she tended to face the chalk board; her transitions were awkward; and her explanations were hesitant. While this might be expected for many preservice teachers, it was a surprise for Donna, as she was such an accomplished public speaker in other contexts. In conversations with her, Donna shared how her limited experiences with science made it difficult for her to see herself as a scientist or part of a collective of people who teach science. In addition, she recalled the single African American science teacher she had when she was a student and how this also made her feel like an outsider.

Donna's classroom discourse illustrated these constraints. She frequently utilized knowledge level questions that elicited brief, one-word answers from students. This familiar pattern of initiation-response-evaluation (Lemke, 1990) has been identified as common in traditional science classrooms based on White norms. In conversation with me and her cooperating teacher, Donna recognized that this was different from her interaction patterns as a mother. She described how she elicited meaningful responses while helping her nine-year-old with homework: "All the time, I say, 'I don't know.' She says, 'But you go to college.' I say, 'I still don't know.' It makes her explain it to me." Donna's classroom use of the interaction patterns she used with her own children was stymied by cultural models that she held about her role as a science teacher.

> I wanted to know that the students understood what I was trying to teach them. I thought the only way to get an affirmation that I was effective was to hear the "right" answer. At the time, that meant I was a teacher. Getting the information to them as quickly as possible was the goal, because biology was a tested area and can be seen as having complicated aspects to it. At home, I understand my role as a mother is to develop as much as I can in my children. Every opportunity must teach more than the obvious lesson. It just didn't occur to me that this motherly practice could or should be applied in the classroom.

Being an African American who grew up and still resided in Baltimore, Donna shared much with her students, especially related to racial, cultural, educational, and community experiences, however, she initially did not draw upon these aspects of who she was. Cultural resources from her out-of-school figured worlds did not have status within the science classroom. Her everyday performances were constrained by Western notions of how science teaching should be done, which did not include aspects of who she was as an African American woman and mother.

The usual power dynamics between student teacher and mentor were likely present in Donna's case, but the fractures and contradictions Donna experienced, as an African American woman, were heightened by her cooperating teacher's maleness and Whiteness. However, Donna's cooperating teacher, a White male, was absent from class one day. On that day, as Donna did a review of previously taught material near the end of a class, she enacted a whole class, call-and-response activity. Call-and-response is an oral participation form common in West Africa and widely present in parts of the Americas impacted by the trans-Atlantic slave trade; it is commonly heard in settings such as African American churches. In this version of call-and-response, correct student answers were greeted with a loud "Sho'nuff" to which the largely African American students came alive, clamoring to produce an answer that would earn a shout-out. Donna described "Sho'nuff" as an "old school" response from "back in the day," and the call-and-response interaction pattern was well known to Donna and to the African American students in the biology class.

> We [Donna and the students] just thought of that there; I didn't plan it. What happened was what I had planned ended early, and there was time to review. I thought they'd be more open to cooperating if I made it something where they got to get up, and sit down, and move, and shout and, you know, do some stuff. And while I was standing there, we created the "Sho'nuff" shout out. They really got into that. They loved being affirmed with "Sho'nuff." Plus, I related it to a TV show, because they watch TV Land. The saying is from some 1970s show; I don't know if it was *Good Times* or *What's Happening*, one of those shows. So when I said "Sho'nuff," they knew it [the phrase] exactly, right from the old TV show.

In those minutes at the end of class, Donna resisted the identity of a traditional science teacher and constructed a new identity by combining non-science cultural models with science teaching probably for the first time. Whether Donna would have taken this step and done this if her cooperating teacher had not been absent on this day, or taken a similar step on another day, is impossible to say. However, that Donna made these first inroads into using her racialized cultural models on a day when her cooperating teacher (a White male) was not present provides insights into the contestation of cultural models that was beneath the surface of this science classroom and the power of the of White Eurocentric science class cultural models in truncating Donna's identity work. In addition, Donna's identity reconstruction was possible only in collaboration with her students, and their collective attunement with her actions was crucial to the success of the activity and reshaping the classroom figured world. After this positive recognition from students, Donna began to rely more on the cultural models that she shared with the students and a new kind of figured world began to form in the classroom. Donna's teaching visibly changed to include more body movement such as swaying and hand gestures, livelier voice patterns, and oral practices from African American culture (Boykin, 1986).

10.7 Story of Chainsaw Charlie and Carpenter Carl

When her cooperating teacher returned the next day, Donna told him about how she had led the review lesson and the positive response she had gotten from students, and he encouraged her to keep trying similar things. While we do not know what might have happened if he did not respond in this way, it is clear that his recognition of her identity bid was important to Donna's trajectory as a science teacher. Coming from a White male, who was also the person who would evaluate her student teaching, his endorsement of her identity work expanded the set of significant reifying stories (Sfard & Prusak, 2005) that existed for Donna about how to be a science teacher. As the following example illustrates, Donna began to figure herself differently as a science teacher, as she drew upon previously excluded cultural models.

While teaching about molecular genetics the following week, Donna presented herself as a storyteller. In doing this, Donna relied on her orality, a culturally acquired disposition of storytelling that she also used as a mother and which holds a central role in African American culture (Boykin, 1986). Building on the often-used reference to DNA as a ladder, Donna created a story of Chainsaw Charlie who cut people's ladders in half and Carpenter Carl who helped the victims by building new ladders from each half, and this evolved in a two-day lesson on DNA replication. Donna said that she had the basic story outlined in advance, but the classes didn't go particularly well until, hesitatingly, she made changes in how she used the story in the last class period of the day. In a conversation, she recalled it this way.

> I didn't do it as a storyteller so much in the earlier periods. By 7th period I just thought to myself, you know what, I'm just going to go and do it. And it turned out so good.
>
> Yeah. Instead of whatever I'd normally do, I tried to enter into it [the story]. This time I just came out and I said, "There was this guy (clapped hands); his name was Charlie." And I just went into the story and I was adding stuff and they were all into it. In the earlier periods I just threw the story out there, but didn't really make it a story, but by 7th period I elaborated and went on and it became a story at that point. That's when it worked. Then the next day I had to go on with the story and explain about the guy that's going to repair the ladders and what happened to them, because you can't have a story without an ending.

A videorecording showed that Donna opened the next day's lesson by reminding the students of the story that she had begun to tell them on the previous day. Donna presented herself as a storyteller when she used the first person ("I told you a story"), and her actions were aligned with that role, as her body swayed as she walked with her hands behind her back. In response to her query about the story from the previous day, the students' initial response of "Chain-saw Charlie" was mumbled, low and not in unison. Not content with the response, Donna verbally cued them by modeling "Chain-saw Charlie" in three distinct parts and she used explicit directions ("All together now"). She also signaled the rhythm with her hands as they repeated "Chain-saw Charlie" the second time. This time they were in unison, but still not loud. The third time she raised her voice slightly and evoked mimicry (she pronounced "Pit-i-ful" in three distinct parts), and the students responded by chanting "Chain-saw Charlie" and this time in loud, three-part unison and un-cued.

The recognition and positive responses that Donna received from the students as she engaged in these practices emanating from their shared, out-of-school cultural models were central to the positive outcome of the identity work that Donna was undertaking and the reconstruction of the classroom figured world. Together they ascribed new meanings to cultural norms outside of the traditional White norms of science classrooms, illustrating how figured worlds can be transformed through the work of the participants (Holland ct al., 2001). Who Donna was as a science teacher expanded to include aspects of who she was as an African American, Baltimorean, mother, and churchgoer. This identity work included physical movement as well as vocabulary, cadence, and patterns of talk, which illustrates the importance of embodied knowing in identity work, something that Fanon (1967) wrote about.

During her student teaching, Donna initially did not mix cultural models from the figured worlds of her home and community with the figured world of science teaching. The norms of Eurocentric schooling and Western science were powerful and weighed on Donna every day in the classroom. It took a bit of luck (i.e., the absence of her cooperating teacher one day and having a cooperating teacher who was open to including new norms) for Donna to feel as if she could draw upon non-dominant cultural models. Donna found time and space to construct herself in new ways and (at least temporarily) resist the Whiteness embedded in science classrooms. She came to claim a sense of ownership and responsibility for the knowledges that were hers and her students', which is a part of creating a decolonizing space. Dei (2010) reminded us that only "educators' pedagogic, instructional, and communicative practices truly lead to decolonizing education" (2010, p. 8).

10.8 What We Learned

Our original goal in this chapter was to call attention to the colonized and racialized nature of schools and science classrooms and the impact of those spaces on non-dominant science teacher identity work. Hall's (1990) concept of diaspora provides a way of making sense of the experiences of non-dominant teachers like Hildah and Donna in schools built on Whiteness and Western science. Diaspora speaks not only to physical dispersion or removal from ones homeland, but also to a sense of displacement, a sense of being taken away from what one knows and values.

Constructing ones identity under the interplay of history, culture, and power that permeates our schools is challenging for all teachers, but it creates particular fractures and conflicts for those who have experienced life in figured worlds often considered lesser than those of the educational system and science. Thus, becoming a science teacher is in many ways a diasporic experience for nondominant teachers and has elements of the traumatic character of the colonial experience that Hall and Fanon talk about.

When teaching in Kenya, Hildah conformed to a figured world shaped by Eurocentric teaching, Western science, and ideologies of Whiteness. She knew that local knowledge of plants, soil, or fermentation was not real science in the figured world of schooling there. Such is the repressive presence of colonialism. Thus, who she was as a tribe member, granddaughter, gardener, and community member was not valued in school, and she was expected to perform Eurocentric Whiteness in the science classroom, despite not being European or White. For Donna, stories with which she constructed herself included being African American, Christian, and a mother, all of which were rooted in her experiences in the African American community in Baltimore. Yet in the science classroom as a teacher for the first time, she experienced dissonance; her cultural models and stories about herself did not fit with the culture of science and science teaching. While engaged in teaching, she experienced contradictions, fractures, and feelings of misplaced parts as she worked to become a science teacher. To use Fanon's words, "He becomes whiter as he renounces his blackness" (1967, p. 9), and this is what Hildah and Donna were expected to do, but this can have traumatic consequences both for the teachers and their students.

As shown at the beginning of this chapter, there has been inadequate attention in identity research to how nondominant science teachers are positioned in colonized and racialized educational contexts, and consequently little focus on the threats to their identity work as they teach in those spaces. We have said that we believe that we must center race and the racialized nature of science classrooms when exploring the experiences and identity construction of nondominant science teachers. But what do we mean by centering race? Simply including nondominant teachers in our research is not enough. Simply asking them to reflect on their experiences with science in decontextualized ways is not enough. As researchers, we must assume responsibility for interrogating schools and classrooms as racialized and colonized spaces, and we must do that in ways that reveal how Whiteness is embedded in them. Figured worlds and cultural models, though widely used to study science identity, have not been used in ways that interrogate the role of racism and colonialism in science identity work of teachers in schools. We hope that we have illustrated how figured worlds and cultural models can be combined with other theories and ways of thinking about race and racism to reveal historic and current ideologies of Whiteness that permeate our educational systems and too often truncate the identity work of teachers from nondominant groups. Urrieta noted that figured worlds "happen as social processes and in historical time" (2007, p. 109), thus, they are well suited for exploration of colonialism and racism in schools.

The spiritual and symbolic violence that Hildah and Donna experienced is what Fanon (1967) referred to as dehumanization. The alternative was described by Dei as "a process of healing and becoming healthy and whole again" (p. 9) and is part of decolonization. Fanon also noted that decolonizing is a subversive act, and this returns us to the question of our complicity in or subversion of the Western hegemony and the role of our research decisions in that complicity. Will we be subversive or complicit in the face of science classrooms that remain colonized and racialized? Will we recognize the significance of teachers showing up Black (or nondominant) in science classrooms in a White supremacist world? Dei wrote, "decolonization, that is a struggle for an anticolonial space, entails educators and learners working together to address questions of power, history, knowledge, identity, and representation" (p. 8). We suggest that this should include researchers as well, particularly those interested in science identity and science teacher identity.

References

Aikenhead, G. (2008, February). *Importation of science programs from Euro-American countries into Asian countries and regions: A recipe for colonization?* [Conference presentation]. Conference of Asian Science Education, Kaohsiung, Taiwan.

Avraamidou, L. (2014). Studying science teacher identity: Current insights and future research directions. *Studies in Science Education, 50*(2), 145–179. https://doi.org/10.1080/03057267.2014.937171

Boykin, A. W. (1986). The triple quandary and the schooling of Afro-American Children. In U. Neisser (Ed.), *The school achievement of minority children: New perspectives* (pp. 57–92). Lawrence Erlbaum Associates.

Crawford, J., Kippax, S., Onyx, J., Gault, U., & Benton, P. (1992). *Emotion and gender: Constructing meaning from memory*. Sage.

Dei, G. J. S. (2010). Rereading Fanon for his pedagogy and implications for schooling and education. *Counterpoints, 368*, 1–27.

Fanon, F. (1967). *Black skin, White masks*. Grove Press.

Frankenberg, R. (1993). *White women, race matters: The social construction of Whiteness*. University of Minnesota Press.

Guo, C. (2007). Issues in science learning: An international perspective. In S. K. Abell & N. G. Lederman (Eds.), *Handbook of research on science education* (pp. 227–256). Lawrence Erlbaum Associates.

Holland, D. C., Lachicotte, W., Jr., Skinner, D., & Cain, C. (2001). *Identity and agency in cultural worlds*. Harvard University Press.

Jackson, P., & Seiler, G. (2013). Science identity trajectories of latecomers to science in college. *Journal of Research in Science Teaching, 50*, 826–857. https://doi.org/10.1002/tea.21088

Joseph, J. E. (2010). Identity. In C. Llamas & D. Watt (Eds.), *Language and identities* (pp. 9–17). Edinburgh University Press.

Lemke, J. (1990). *Talking science: Language, learning, and values*. Ablex Publishing.

Moll, L. C., Amanti, C., Neff, D., & Gonzalez, N. (1992). Funds of knowledge for teaching: Using a qualitative approach to connect homes and classrooms. *Theory into Practice, 31*(2), 132–141. https://doi.org/10.1080/00405849209543534

Picower, B. (2009). The unexamined Whiteness of teaching: How White teachers maintain and enact dominant racial ideologies. *Race Ethnicity and Education, 12*(2), 197–215. https://doi.org/10.1080/13613320902995475

Seiler, G. (2011). Becoming a science teacher: Moving toward creolized science and an ethic of cosmopolitanism. *Cultural Studies of Science Education, 6*, 13–32. https://doi.org/10.1007/s11422-009-9240-3

Seiler, G. (2019). Challenging whiteness in science education. In G. Prime (Ed.), *Centering race in the STEM education of African American K-12 learners* (pp. 19–34). Peter Lang.

Seiler, G., & Kwamboka, H. (2019). A neoindigenous perspective on language silencing in science education. *Research in Science Education, 49*(4), 1041–1053. https://doi.org/10.1007/s11165-019-9852-x

Sewell, W. (1992). A theory of structure: Duality, agency, and transformation. *American Journal of Sociology, 98*, 1–29. https://doi.org/10.1086/229967

Sfard, A., & Prusak, A. (2005). Telling identities: In search of an analytic tool for investigating learning as a culturally shaped activity. *Educational Researcher, 34*(4), 14–22. https://doi.org/10.3102/0013189X034004014

Shizha, E. (2010). The interface of neoliberal globalization, science education and Indigenous African knowledges in Africa. *Journal of Alternative Perspectives in the Social Sciences, 2*, 27–58. https://doi.org/10.1007/978-94-6091-561-1_2
Urrieta, L. (2007). Figured worlds and education: An introduction to the special issue. *The Urban Review, 39*(2), 107–116. https://doi.org/10.1007/s11256-007-0051-0

Chapter 11
Understanding Science Teacher Identity Development within the Figured Worlds of Schools

Gail Richmond and Kraig A. Wray

Individuals carry with them diverse beliefs and values about teachers and about teaching into their certification programs (Lortie, 1975; Pajares, 1993), and these serve as filters on the curriculum, learning opportunities (including those situated in the community and in schools), and practice-based expectations they encounter during their preparation for a career in education (Richmond et al., 2011; Sfard & Prusak, 2005). This is the case whether or not the candidate is aware of such beliefs or values (e.g., Schussler et al., 2010) or of the ways they respond either in word or action in the moment or even in the longer term. Some of these representations and encounters may be in alignment with a candidate's values and some may not be, and the extent of alignment may result in a level of consonance—often expressed as confidence, ease, and openness as opposed dissonance, expressed frequently as uncertainty, frustration, and resistance as teacher candidates move through the certification landscape (see, for example, Smagorinsky et al., 2004).

11.1 Challenges for Teachers and Teacher Educators

This experience of consonance or dissonance does not end once certification is gained, but rather moves with the individual as they take on the set of responsibilities expected of a full-time teacher. A beginning teacher who holds values which are in alignment with those that are hallmarks of her certification program may experience dissonance as a result of the sociocultural or sociopolitical climate of the school in which they work or the interactions they have with others in this space (e.g., Wray & Richmond, 2018). An example of dissonance would be experienced

G. Richmond (✉) · K. A. Wray
Michigan State University, East Lansing, MI, USA
e-mail: gailr@msu.edu

H. T. Holmegaard, L. Archer (eds.), *Science Identities*, Contributions from Science Education Research 12, https://doi.org/10.1007/978-3-031-17642-5_11

by a teacher who values instruction responsive to students' needs and their ability to engage fully in scientific practices to develop deep scientific understanding who then finds herself a member of a science department in which teachers are expected not only to implement instructional plans which are didactic in orientation, but to do so in lock-step with one another. In such a scenario, teachers have no freedom to change the content or pace of their instruction in response to any difficulties noted in students' understanding, in the face of unexpected questions or observations, or in pursuit of the goal of handing over increasing responsibility for the learning process to the students themselves.

Not surprisingly, challenges also present themselves to those who help to prepare individuals for the teaching workforce. Teacher educators are enactors of curriculum, supervisors of field-based work, and gatekeepers to successful program completion. Any one of these is a complex role to play, but in many cases, a teacher educator serves in more than one capacity, and it becomes difficult to be an advocate of the program and of the individuals moving through it while at the same time a liaison with mentor teachers, whose own values may or may not be in alignment with those espoused by the program and held by the candidate whom they are mentoring.

11.2 Identity as a Grounding Construct

Being able to name some of the features of identity—as static or dynamic, physical, sociocultural or sociopolitical—is only the beginning of designing strategies to address the problem of teacherfrustration and dissatisfaction, either or both of which can lead to a departure from the classroom. A solution can only be reached if we better understood several things about teachers: what factors or events exert "pushes and pulls" on them, especially early in their careers; how they perceive elements of the worlds they inhabit; and what specific actions they take in response to these perceptions and to their place in that world as they understand it.

For many decades the construct of identity has been invoked in a variety of situations while also being the focus of debate amongst psychologists and sociologists; in more recent years, it has been "borrowed" by educational researchers as a theoretical lens through which to better understand not only the willingness of individuals to take on certain roles and engage in particular practices, but also and more broadly, the ease with which individuals enter and persist in teaching careers. Views about how identity is constructed have historically had either cognitive or sociocultural roots; in the former perspective, identity has been treated as being relatively stable and shaped by internal forces, while scholars in the sociocultural tradition view identity as situationally dependent, as well as dynamically shaped by features and events in the context in which the individual lives and works (for a brief review of these two approaches to understanding identity and its development, see

Richmond, 2015). One specific sociocultural variant is that of professional or teacher identity, which has become a useful tool for our developing understanding of the construct as it unfolds for classroom educators.

Even if we were to agree that a sociocultural view of professional identity is most helpful in understanding how teachers react to the constant challenges they face in these professional roles, it has proved to be a complex problem which, not surprisingly, has been addressed in diverse ways by scholars (e.g., Chu, 2020; Cobb ct al., 2018). In fact, the entries in this volume are an excellent reflection of this diversity of perspectives and framings.

Historically, there has been little "cross-talk" among the multiple theoretical frameworks proposed for the study of teacher identity development, but more recently, several researchers have proposed that a more fruitful approach would be to explore how multiple perspectives might be employed to gain a deeper understanding of this complex process. The work of Varghese and her colleagues is an excellent example of the insights that can be gained and challenges that can be identified through the integration of multiple perspectives (e.g., Varghese et al., 2005; Varghese, 2016). These scholars recognize that identity is a complex construct which is described in various ways by different frameworks. Rather than pit those frameworks against one another, they sought to deepen their understanding of identity by using the strengths of one frame to address the weaknesses of another across disciplines. They argue that there are elements of each framework that are generative and that only through multiple frameworks, such as situated learning, image text, and social identity theory, can a composite framework be created with greater explanatory power. Due to the complexity of identity in multiple contexts, only by considering multiple frameworks can this complexity be maintained and better understood.

We have taken this to heart in the work we have done with those preparing for careers in science teaching, as well as with those who are in the early years of their teaching careers. In some research in this construct domain, agency (which we define more fully in the section which follows) has been considered a second cousin, theoretically speaking, to identity, and the ways in which an individual positions herself with respect to others inhabiting the same community as well as the context itself have only rarely been invoked as factors in shaping this process. In contrast, however, the first author's earlier work on teacher identity told a different story (e.g., Richmond & Muirhead, 2014). That research has served as a foundation for a set of investigations in which the goal has been to understand more deeply the interplay of agency and identity development and how these are shaped by particular and dynamic features of the context (both physical and cultural) in which individuals find themselves by choice or by circumstance. Articulating this relationship has been an intriguing professional and intellectual challenge, and it is to this process of articulation (as aligned with the first and third of our goals for this chapter) that we turn for much of the remainder of this chapter.

11.3 Investigating the Professional Identity of First Year Teachers

Building upon Richmond's prior work, we focused our early studies on the relationship between agency and identity of early-career teachers (Wray & Richmond, 2018). In completing this work, we created a model in which we drew on Bandura's (2001) sociocognitive view of agency, which is centered on intentional acts designed to have a specific outcome, with the sociocultural views of Wertsch et al. (1993), which ground the development of agency as an outgrowth of social interactions individuals and groups have with each other. While both views provide individuals the means to act with intent, together they provide a richer sense of the territory traversed in the development of agency. Bandura's (2001) sociocognitive approach, for example, does not account for the role of the environment in supporting or pushing against agentic action, while Wertsch et al. (1993) sociocultural approach relies heavily on tools available within a particular context to support agency. Additionally, the latter scholars limit agency to the tools immediately available and do not acknowledge a person's past experiences in other contexts as influential in decision-making while this is accounted for in Bandura's (2001) work. It is by assessing elements from these two views of agency that we were able to create a more complete view of action that takes into account the context in which decisions about such actions take place. Figure 11.1 represents our initial model for teacher identity which takes such consideration into account.

In our early framing, we considered professional identity to be the expression of two components: positioning and values. The first of these includes how the teacher positions themself and is also positioned by others in the community they inhabit (e.g., Arvaja, 2016). The second component—values—includes those ideas, goals, or priorities that drive an individual's actions – or a community's actions, for that matter (Richmond et al., 2011). It is worth noting that while every teacher possesses a set of values, those values will not necessarily be prioritized if the context or community does not support actions that align with these values, and it is here that agency is critical; more specifically, in their immediate teaching environment, purposeful action is the result of a teacher recognizing a challenge, developing a plan to address that challenge, and putting that specific plan into action. For the purposes of this study, we saw writing about or speaking about a challenge as *recognition* of that particular challenge. Engaging in actions that align with an individual's values and experiencing some level of success as a result makes it more likely that similarly purposeful actions are subsequently engaged in, that agency develops, and that those initial values continue to be reinforced. When we say "success", what we mean is the outcome of the action is what was intended when it was planned.

Our analyses of classroom observations, journal entries, e-mail correspondence, and interviews allowed us to make claims about these teachers' identities and agency based on what they did, wrote and said and which reflected their primary values, the ways they were positioned by themselves and by others, and the actions they took which were related to those views. Below we share some data to illustrate

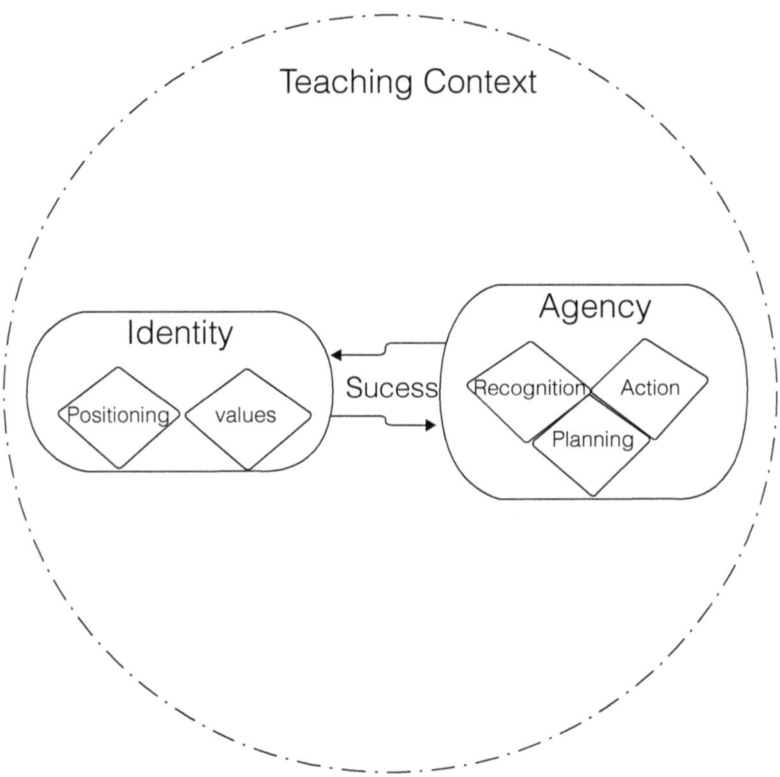

Fig. 11.1 Feedback loop of identity and agency in Pre-service and In-service teaching

the components of our initial model, their relationship, and some issues left unresolved by our analyses.

During her first year of teaching, Nina, one of our participants, repeatedly focused her writing, her talk and her efforts on "... facilitating discussion where students are interacting with each other and not just with me" (Journal, 10/2014). Evidence of this as a core value was reinforced by her referencing the support she felt from the administration and the freedom she had to focus her efforts on developing skills to advance this goal. In an email to Author 2, she stated "... I am still figuring out how to best facilitate small group discussion that is productive and scientific... I want to find a way to encourage students' independence and make them feel confident enough to conduct a conversation with their classmates. I am starting by providing discussion questions for them and having them write their answers on white boards so that we can talk about it as a class" (Journal, 10/2015). In addition to valuing scientific discussion, Nina was positioned by other educators and acted in a way that supported her work in ways that reflected this value. What we were unable to ascertain through our multiple analyses was why she felt a sense of agency, what role the social and physical context as well as other individuals with whom she interacted played in influencing her pedagogical decisions, and the extent of alignment between her choices and those of others around her.

In contrast, as a first-year teacher, William found himself on the other end of the spectrum in regard to support, stating, "…I do sometimes feel like I'm left all alone. And that's not a good feeling; it's kind of scary" (Interview, 1/2016). William's background and experiences in the teacher preparation program informed his primary value of having the students engage in inquiry in a relatively authentic way. However, the lack of support he felt in his school stifled his ability to work toward this goal. During an interview William stated: "I am told to teach microbiology… I've been promised a certain amount of money for this but [the school district] is notorious for not coming through with anything at any point … I am expected to teach high-inquiry science classes but I can't even get paper" (Interview, 6/2016).

Recognizing what little influence he had to get the materials he was promised, William turned his attention to the students as a reason for and potential solution to his inability to teach how he felt was best. William saw what he considered his professional behavior as restricting his ability to gain the respect he felt was necessary to get compliance from the students. "... When I got here, I realized … When you try to be something you're not, these kids can see right through it … they won't trust you, they won't respect you, and they certainly won't do what you tell them to" (Interview, 6/2016). He felt that by behaving in what he perceived was a more natural way, by being more sarcastic rather than formal, the students' respect for him increased and as a result, so did their compliance.

Our analyses of data from these two participants, as well as other beginning science teachers graduating from the same program and working in diverse professional contexts, led us to increasingly wonder what it might be about the school itself which made the experiences different for each of these beginning educators and more importantly, how their interpretation of particular policies, cultural elements, and interactions might shape in some way actions they undertook as classroom educators. We knew that both agency and identity were informed by the school context, but our initial theoretical model did not provide us with access to the dynamic and often subtle ways in which such critical influences developed and often persisted. Having identified this limitation, we sought a way to retain what had proven invaluable as a tool for articulating the interplay between identity and agency but would also permit us to examine and better explain the role that the particular features of the world an individual navigated by privileging that individual's perceptions of the features of that world and retaining the world's inherent complexity.

The limitations of this model arise from the fact that schools are sociocultural institutions or "cultural worlds", and cultural worlds have particular meanings for those who inhabit them. The framework we developed did not account for the nature and dynamics of these worlds in any significant relational way. The limitations of the model all appeared to us to be centered on how decisions affecting action arise and what the consequences of this action are for the individual. That is, the framework, along with many others:

(a) Does not address in any nuanced way, the role that individuals' perceptions of the world in which they work arise;
(b) Does not account for actions taken by individuals as a result of these perceptions,
(c) Does not account for the consequences that taking action has on developing agency and identity.

For example, in the third year of data collection and analysis in the larger project from which the data reported here originate, Nina, through her own actions, was positioned as a valued member of the community and as having increased power and privilege. She was asked to take the lead in developing new curricular pathways for science in the middle grades, a priority for the administration.

We recognized the need for a framework which would feature the interplay between physical, cultural and psychological aspects of the space inhabited by teachers whose decisions and whose lives we were trying to understand more deeply. We also needed a framework that would allow us to privilege teachers' narratives of their experiences. Fortunately, we found a theoretical perspective on identity construction proposed by a group of anthropologists which did just that and which informed our development of a "hybrid" model to guide our work and deepen our understanding about how teachers make decisions and navigate the worlds in which they work.

11.4 Figured Worlds as Identities-in-Practice

In 1998, a group of anthropologists, led by Dorothy Holland, introduced figured worlds as a way of viewing culture and culture-bound identity. These scholars used this framework to explore such identities as beauty in American universities, woman-ness in the caste system in India, and membership in Alcoholics Anonymous, among others. In all of the situations they explored, people enter what they term a "figured world" and make their way by engaging with specific communities within this world and by learning the expectations set forth by community members. It is in these worlds, they claim, that identities are developed and either pushed against or further supported in relation to expectations for action set forth by those within the world who possess relatively greater power (Holland et al., 1998).

Holland and her colleagues (1998) claimed that figured worlds are, by definition, the social and cultural spaces within which people engage in actions in relation to a set of expectations established by the other actors in that world. These worlds are also the space within which identity is developed and refined *in practice*, that is, through perceptions and consequent actions that take place within a particular community. In order to explore how identities are formed through these actions and how those actions are perceived by other members of the community, they identified four features of a figured world—culture, artifacts, community, and power and privilege. In Table 11.1 below, we briefly summarize these four features.

Table 11.1 Features of a figured world

Feature	Summary
Culture	From Holland et al. (1998, p. 54): "The meaning of characters, acts, and events in everyday life was figured against this storyline" In other words... Standard plot of the world against which lived experiences are compared Reflective of the values privileged by those with more power and privilege Can be modified by actions of people participating in the storyline, not necessarily those with greatest power & privilege For example: A school-wide focus on assessment or developing school/home relationships
Artifacts	From Holland et al. (1998, p. 61): "Figured worlds are evinced in practice through the artifacts employed by people in their performances… they are the means by which figured worlds are evoked, collectively developed, individually learned, and made socially and personally powerful" In other words… Physical and discourse-based tools used by participants in a figured world Tools used in actions of those participating within a figured world Include titles and labels given to people that have been historically built and come with required expectations For example: Common departmental assessments or grade-level curriculum
Community	From Holland et al. (1998, p. 192): "...Identity responds to both the imaginary and the embodied communities in which we live" In other words… Social and cultural groups that use artifacts. These groups are characterized by a power dynamic that exists among its members in which certain actions are valued over others Membership in the community is socially developed, and practices or activities take place in historical time. For example: Subject matter departments, specific school committees
Power & Privilege	From Holland et al. (1998, p. 60): "...Gain perspective on such practices and come to identify themselves as actors of more or less influence, more or less privilege, and more or less power in these worlds" In other words… People and actions are compared to one another, naturally ranking them. Being new does not often afford power and privilege, but they can be gained through agentic action in culturally acceptable ways. For example: Differential status among teachers within a school or department based upon years of teaching experience, subjects taught, student populations taught

Holland et al. (1998)

A figured world can be viewed as a space in which an identity is, in part, recognized and developed through the actions in which someone chooses to engage. By participating in activities within a figured world, individuals come to see themselves as participants and can then position themselves as having more or less privilege and power relative to others. Not only does participating in certain practices help a teacher place themselves within the world, but assigning meaning to those practices relative to the "…norms, practices, values, and demands of the setting…" (Carlone et al., 2014) may in turn change their value. It is thus an "identity-in-practice". Our experience in analyzing our accumulated findings is that figured worlds, in conjunction with our earlier model of professional identity are powerful lenses which allow us to more effectively accomplish several goals: to represent the dynamic and "self"-centered process by which individuals accumulate perceptions of each of the multiple worlds they inhabit as well as of their place in these worlds; to consider these perceptions alongside what they value most; and to make decisions about what to do and how to align themselves with others as an outcome of these perceptions. Together they provide powerful insights on the development and refinement of professional identity. In the next section, we describe how we brought these two framings together to create a "hybrid" model which we used to analyze and interpret data from our beginning science teachers.

11.5 A Hybrid Framework for Understanding the Development of Teacher Identity

In more recent investigations we have elaborated on our ideas by focusing more explicitly on the "context" in which the teachers work. We continued to draw upon Richmond and her colleagues' earlier conceptualizations of identity, which include values and positioning at their core (Richmond & Muirhead, 2014; Richmond et al., 2010; Richmond & Wray, 2017; Wray & Richmond, 2018). We married these views with those of Holland et al. (1998) and Carlone et al. (2014) about figured worlds. The result of this synthesis is the model which can be seen in Fig. 11.2. This "hybrid" model brings these views together in what we consider to be a more powerful accounting of the dynamic processes underlying identity development and refinement, one which is positioned with respect to the individual teacher's own perceptions and actions within the figured world they inhabit. In this framework, the four elements of figured worlds – power, culture, community, and artifact – work in concert with the perceptions, actions, positioning, and agency as the grounding for the development of identity. In this model, agency still plays a central role in the development and refinement of science teacher professional identity. Actions that have successful outcomes for students and for the teacher, where these outcomes can be in the cognitive, social, or affective domains and where these outcomes are recognized by those possessing more power and privilege, allow values to continue to be reinforced, agency to grow and the continued successful identity development possible.

An example of this relationship for teachers might be the enactment of an activity sequence for a unit of instruction in which prior students have experienced significant challenges. When such enactment results in greater student engagement and achievement, the teacher feels greater is recognized by more senior colleagues in the department as something they would like to adopt in their own instruction. This success may even reach the "ears" of administrators, who in turn may add their encouragement to the teacher. The positive reactions by both students and colleagues in turn encourage the teacher to engage in similar instructional innovation moving forward.

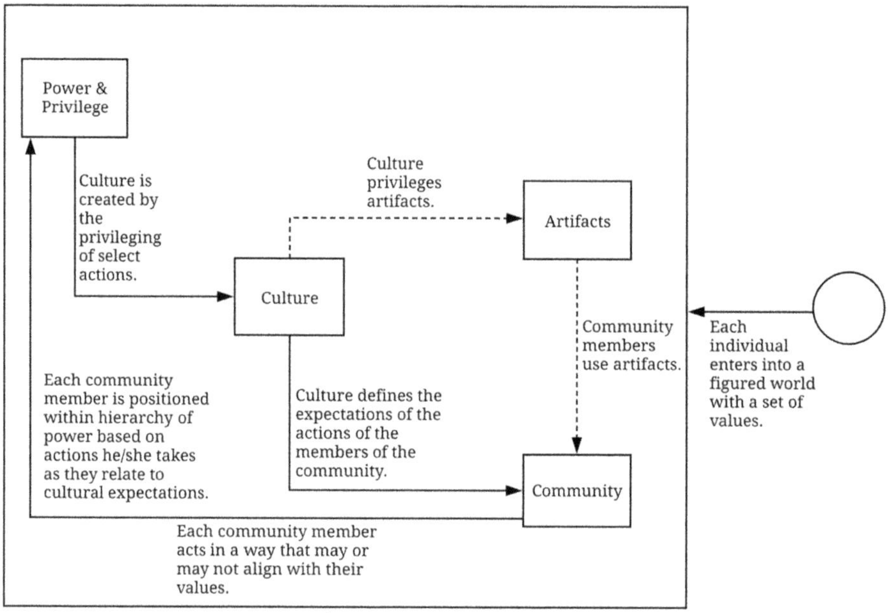

Fig. 11.2 Model of identity development through a figured worlds lens

11.6 Using the Hybrid Model as a Lens on Beginning Teacher Identity Development

We accessed the larger pool of data we had collected from a cohort of educators that extended from their final year of their teacher certification program through their first year of teaching. Using this framework, we leveraged teacher narrative (Richmond et al., 2011), focusing on journal entries, email correspondence, and interviews as our primary data sources. By drawing on written and oral narratives, we were able to develop stories which revealed features of the figured worlds as experienced by the teachers themselves. More specifically, we approached the data using inductive case-based methodology (Glesne & Webb, 1993), modified inductive constant-comparative method (Corbin & Strauss, 2008; Dyson & Genishi, 2005), and by writing analytical memos (e.g., Maxwell, 1996, 2008). The data were initially analyzed for broad themes of identity (values based on recurring themes and positioning in reference to placing or being placed in relation to others) and features of the figured world (Holland, et al., 1998) in which the teachers worked. Data were then re-analyzed for evidence of held values and patterns in prioritized values, instances of positioning, recognition of specific challenges, the identified and pursued solutions and, as applicable, post-action reflections. For each data source, patterns were coded for: primary *values* (the value or issue of most importance), *positioning* (references to interactions with individuals and communities), *agency* (recognizing an issue, planning a solution, and implementing a plan around that issue), and the *figured world* (with specific coding for elements of culture, artifacts, community, and power). Coding was done for teaching-related issues (e.g., teaching practices, relationships, school priorities) focused on by participants and references to interactions and conversations with and about others. References to problems participants had been experiencing (e.g., students not complying with directives, dissociation from the school community) and how this problem may or may not have impacted actionable decisions also was noted. Continued repeated mention of problems or challenges were used to indicate level of priority. Further analysis resulted in more specific codes (i.e., student engagement in science, relationships, respect), which was utilized in a second round of coding. Patterns of codes were then used to draw comparisons *between* cases. Video excerpts were chosen based on what stood out as related to references in journal entries, what was based on submitted lesson plans, and what was relevant to the framework. Lastly, data were triangulated across sources and used to create a more complete profile of each school world and each teacher's professional identity as it developed within it.

Below we briefly present selected data exemplifying three of the four elements of figured worlds as Nina and William experienced them. (Given the participant-generated nature of the data sources to which we had access, our ability to identify significant numbers of artifacts across sources was limited. As a result, we chose to omit treatment of this fourth element from the analyses reported here).

11.6.1 Nina

Nina came to teaching after completing a bachelor's degree in zoology and working for several years as a zookeeper. Having interacted with the public in an educational fashion as part of her duties with the zoo, she came to the teacher preparation program poised to learn how to teach. What she seemed to value in regard to teaching practices in her first year indicated to us that her views of teaching aligned with those of the preparation program. She was then able to carry what she learned and valued into her first year of teaching.

Culture Nina began her teaching career in a public charter school that was housed on the campus of a small university. A charter school in the U.S. is a public school which is created by a group of individuals to meet a perceived need or fill a specific purpose. While it typically has its own independent governing board, it receives public funding in the same way that non-charter public schools do. Several resources made available by this association could be leveraged by the school, including, for example, access to science lab space and equipment and the opportunity for older students to dually enroll in courses and thus earn university credit. However, the school was at the mercy of the university schedule, which limited time and access to classroom and lab space. As a public charter school, its continued existence was dependent upon sufficient student enrollment, and enrollment was a priority for school leaders. It also contributed significantly to Nina's perceptions of what was judged as important in her school's culture (the storyline) and shaped her actions in response to these perceptions. However, the administration generally supported teachers in developing individualized professional growth plans and working with these goals in mind.

> I think because they are really supportive and I think that if like [department chair] and [principal/president] are really supportive of that and like even when [department chair] and I disagree on something like we'll have a discussion about it, but it usually comes down to like it's not a bad thing so just try it. See how it works and I think because the school is brand new, there is still a lot of that, like, "try it to see if it works and if it doesn't work we'll try something different or we'll go back to what we had before." (Interview, 5/2016)

Not only was Nina new to teaching, but she was also in a relatively new school, and one which had as a priority the design and development of curricular pathways aligned with the performance expectations of the United States' new science standards recently adopted by the state (the Next Generation Science Standards or NGSS Lead States, 2013), as well as the expectations of the science courses of the university on whose campus the school resided, where it was located and with which it was partnered.

> We are designing curriculum to challenge all levels of students and develop a variety of skills rather than teaching to any kind of test. We talk about not only supporting our lower students and bringing them up to par with their peers, but also how to create extensions and other opportunities that will challenge our high achieving students as well. Rather than being warned about students, I have heard so many teachers tell me how lucky I am that I will have the students that I have. (Journal, 10/2015)

Community Nina entered into a department of only three sixth- through eighth-grade science teachers, and her department chair was a graduate of the same teacher preparation program. Being a new school with a small science department, lesson designing and teaching were very much in the hands of the teachers. This posed both affordances and constraints; with respect to the former, for example, Nina was free to focus on the pedagogy that she valued and to teach in a way that aligned with her professional identity.

> I really want to focus on developing solid curriculum this year and assessments to go along with the curriculum so that next year I am able to change and perfect the things that I have planned this year and begin adding in even more good stuff that can focus on writing and reading and really prepare students for college. I am really looking forward to working with [author 2] and [department chair] this year to develop some really good units, and I am also looking forward to them helping me to set reasonable goals for this first year. (Journal, 9/2015)

It was this recognition of the *challenge* of developing appropriate curriculum and assessments that contributed to the development of agency. Being a new teacher in a new setting, she also was relatively well supported by those with more power and privilege to reflect on and make changes to her teaching. Her department chair shared many of the values that she had regarding teaching secondary science.

> I really like these meetings as it is a great opportunity to get answers to questions and to run ideas past [principal/president] that have more to do with the big picture and alignment with

> the goals and ideals of the school. He provides a lot of really helpful feedback and encouragement. (Journal, 5/2016)

However, she followed this by sharing the dismay she felt when she was told by the principal that he was no longer going to be able to hold meetings with the teachers in this fashion, essentially removing the support she had grown to expect and appreciate. The pressures placed on the principal to focus on enrollment and recruitment became a priority over his mentorship of teachers.

Over time, Nina positioned herself as someone with leadership capacity in middle school curriculum and assessment. She took on the task of designing the middle-school (6th–eighth grade) science curriculum and was instrumental in co-designing with her department chairperson a performance-based rubric (planning) that was implemented with great success (action) and was not only adopted by the entire department, but also adapted for use in many other departments in the school. This combination of recognition, planning and action, increased Nina's sense of agency. Additionally, her sense of ownership and innovation was instrumental in establishing her as an integral member of the school community and one who could be seen as a catalyst for change.

Power and Privilege Nina was able to develop a sense of power when positioned as the content leader for middle school science. The administration had a vision for the school including a priority to bolster admissions. Also, the department chair had a focus on developing curricular pathways that met the needs of the state standards and was aligned with the expectations of its community college partner. The two primary sources of power, the administration and the department chair, had different primary priorities but were in alignment with a secondary focus; one of supporting teacher development that was responsive to the needs of the individual teachers. This support and resulting freedom helped Nina act in an agentic way that aligned with what she valued. Feeling a sense of success in focusing on high-leverage practices and being recognized by those in power as being successful further confirmed that what she valued was also valued by the community. Additionally, as she acquired responsibility, she was able to work in a way that supported her own values as well as those of the department by aligning the curriculum she was developing with what she felt was good teaching.

> I think that there are lots of things that I have the power to change in my teaching and school. I think the biggest freedom that I have is in my curriculum. There are really no stipulations or guidelines for what I have to create other than we need to follow NGSS. I really like having that freedom to create whatever I want to and to organize the content in whatever manner I feel is best for my students. (Journal, 2/2016)

This provided her with a sense of power that was further acknowledged when a performance based assessment rubric was adopted by multiple departments within the building.

> So like each time I kind of vary the rubric that I use based on what worked and didn't work the last time and so I tried to change my rubric so that I can be more explicit with what it is I'm looking for and communicate that with the kids so they are not surprised when they get

> points off for things like that. So that is definitely one thing. And I've also been talking to the Social Studies and English because they do all similar kinds of projects that we do in Science so I ask them what rubrics do you use. Like can we use a common rubric that way they know in every class what a presentation is going to be graded by and we just put in the content parts later. (Interview, 5/2016)

Through the power she gained, Nina was able to make changes that were taken up by the school, thus re-shaping the storyline that conveyed the school's culture.

11.6.2 William

William decided to pursue a teaching career after having completed a PhD in microbiology. As a result, others had certain expectations of him within the school, and he had expectations of himself as well.

Culture William's first full-time teaching position was in a school in a large urban district in the midst of major upheaval. Due to repeated poor student performance on state-mandated standardized tests, the school was placed under management by an entity directly controlled by the state department of education, and new school leadership was appointed. With this change in leadership came an emphasis on factors that were identified as critical in influencing test scores: school attendance, critical thinking, and reading level. These all too often constitute the storyline of many under-resourced school cultures, and William's school was no different. William stated: "…I am ostensibly a biology teacher but I work in an environment where the ultimate goal that I have for my students is not necessarily to learn biology, but how to look critically at the world" (Journal, 2/2016). While William felt these factors were important, he also felt the execution of plans to create a school culture in response to these factors was making it difficult for him to teach effectively. During an interview he stated: "But the school as a cultural thing you know, we really lack a solid vision as a school, we are working on it….my expectations for my students are not matched by the administration's willingness to hold them accountable for those standards…" (5/2016). William felt that the mission of the school and the ideas privileged by its administrators aligned somewhat with his own views of good teaching and there were indications he may have believed he could achieve his goals of providing students with "hands-on experiences" (Journal, 2/2016), but the upheaval in leadership led to a lack of support, leaving him feeling severely under-supported.

The following year William gained employment in a school within the same district where students had a history of success on standardized exams A clear vision of the school's vision and priorities was evident in his reflections, as when, early in the year, he wrote: "Even when disciplinary actions are taken both the parents and staff are right with you" (Journal, 8/2016).

Community During an interview in January of his first year William mentioned assisting with robotics club and the benefits that potentially affords. He began by discussing that the robotics students are in the twelfth grade and he mostly taught ninth and tenth grade students. But what stands out is when he went on to state "... it's been not only helpful for that [relationship building with students] but because it helps me interact with the staff… I can go two or three days without seeing some of the teachers" (1/2016). Additionally, William came in as a junior member of a science department of two, and shortly after he arrived, the other teacher was pulled into an administrative role, leaving his own classroom with a long-term substitute. When challenges arose for the substitute, planning and student management duties fell to William.

On the rare occasion William was able to interact with a community of teachers, mostly during professional development, the interactions were unhelpful from his perspective.

> So we got teachers that have been there [school 1] for three to four years who I hear side conversations that they're not really adapting to changes, they are not really trying to engage their practice. And I think that hurts the ability for me to set up my learning environment because I am trying to adapt to the changes, I'm trying to do the best I can for my students and they know that. (Interview, 1/2016)

Changing schools in his second year afforded William entry into a new community with a culture which was well established and which had the capacity to provide a set of supports for new teachers which were aligned with his needs. Early in the second year William reflected: "The staff and administration have been amazing about answering questions and helping me get used to the new environment" (Journal, 8/2016).

Power and Privilege

The environment of William's first school presented many challenges. He entered teaching with views reflecting those of his preparation program including that of providing students with hands-on inquiry. He knew that he was coming into a space that was being reorganized for poor student performance However, because of the state's impending take-over, William had little power to pursue what he valued most about science teaching and learning. He also felt vastly under-supported while also feeling immense pressure to help his students to improve. He stated: "…I am dealing with them [administrators] and they're, like, we do what we have to do… we will support you, but then when it comes to the practice there's not the resources there to make it happen" (Interview, 1/2016). Similar comments appeared repeatedly in his journal entries and in interviews. The extreme nature of his situation came out in an interview where he stated: "Sometimes it is like a battle for survival where I'm trying not to compromise my principles, but I still want to make it through the day alive" (Interview, 1/2016).

The transition to a new school in Year 2 was helpful, but not without its challenges. For example, being the junior member of the department meant he did not have his own classroom. “I’m struggling with how best to offer laboratory-related items while roving on a cart… labs can run multiple days or weeks, and I don’t have a ready place to set them up and leave it.” (Journal, 8/2016). William’s *recognition* of his need to provide students with hands-on experiences continued to be challenged, but in different ways. In the first school he was not supported in teaching aligned with his expressed values because of a lack of a community with shared values, insufficient resources, among other factors. In the second school, however, a shared value for teaching science in this way and a larger community more experienced at providing students with explorative learning opportunities allowed him to pursue what he valued, despite large class size, a continued lack of resources, and pressure for higher student performance on standardized tests.

Changing schools meant that William had to navigate a new figured world, sorting out how to work in ways that allowed him to align with that world and teach in ways that respected his values and permitted his continued professional growth. Upon further reflection about his first year of teaching, William found several positives.

> Generally, things are going very well. In many ways, this is like ‘year 1, part 2’, because the environment has been so different for me in terms of teaching practice, environment, and student interaction. If anything, I do have to thank [school 1] for very quickly getting me into a place where I am not willing to tolerate shenanigans from my students. You definitely did learn about management. I believe I can be fair, but that also means establishing and enforcing my role as an adult in the room. Despite what happens “in the moment” when I am forced to do this, it seems that students ultimately respect teachers who hold them accountable. (Journal, 8/2016)

The teachers in this study entered into unique figured worlds, each of which had different but relatively established cultures. The teachers and staff participating in each of these worlds generally worked in ways that were aligned with administrative expectations, and the first-year teachers had to find ways to position themselves with respect to that figured world’s cultural storyline. More experienced teachers generally did their part to provide support for newcomers. In some cases these supports were in opposition to administrative mandates and reflected what veteran teachers valued at that point in time.

Having power and privilege within one figured world does not predict power and privilege in another. Being new generally translates into low levels of power and privilege, and navigating interactions with multiple, often overlapping communities means having to be thoughtful and often strategic about their actions because working in the service of the goals of one world has the potential to push one out of alignment with the goals of another, and this impacts the acquisition of power and privilege in both worlds.

11.7 Implications for Understanding Professional Identity Development and for Supporting Early-Career Science Teachers

This hybrid model has allowed us to uncover the ways in which novice science teachers implement practices within their school contexts. It also has provided us with a sense of what supports might be helpful for preparing teachers to engage in progressive practices (Wray & Richmond, 2018). Ultimately, school communities and school culture more broadly weighed heavily in shaping pedagogical decisions made by the teachers we studied, something rarely addressed within teacher preparation programs and which is even more challenging to address through the typical, largely generic professional supports offered by schools and districts to teachers in the early years of their careers. These observations informed our investigation of school-embedded factors that were most salient in affecting teaching decisions. Our resulting model has been a rich and informative lens through which we have been able to begin articulating how we might not only understand critical elements, perceptions and actions within the figured worlds in which teachers work, but also to consider the means by which we as science teacher educators might better prepare and support early-career science teachers by being responsive to these.

More specifically, there are some activities or features teacher educators can incorporate into methods courses, field experiences, and even as part of induction support, inclusive of professional development efforts. This starts with providing opportunities for teacher candidates and teachers to explicitly name or claim what it is that they value and to explore how it is these values developed—what prior experiences and what individuals, for example, shaped these views about what is important to them personally and professionally. Recent research making use of critical autoethnographic narrative (e.g., Lavina & Lawson, 2019; Yazan, 2019) is promising in this regard.

Using this approach, science teacher educators can provide experiences which allow candidates and beginning teachers to reflect on what is it about the worlds within which they are learning and working and the features within these worlds that align or do not align with those advocated by their preparation and professional support programs. At the same time, experiences might be orchestrated to help teachers develop strategies for gaining expertise aligned with what they value and which can at the same time provide them with a sense of power in the hierarchy which characterize such worlds. If such strategies can also bring into alignment the values of the teacher with those embedded in the school culture, then tensions can be reduced for the individual while a greater sense of agency can develop. An example of this occurred when the rubrics Nina created were taken up by teachers, not only in science, but in other departments in her school.

Agency is an individually developed construct. But power and privilege are gained when an individual is recognized by others as being a leader or expert in some way which is valued by the community. A teacher can be agentic and work in a way they value and yet not gain power and privilege due to the lack of recognition by others. But should others recognize agentic actions and view those actions as aligned with the culture of the figured world they occupy, the teacher may be positioned differently within that figured world. Recognition of what particular actions offer both to the individual and to the school culture contribute to continued and productive professional identity development. Such recognition and upward shifts in power and privilege may also contribute to increased teacher retention, as several of the leading causes of teacher attrition include a sense of powerlessness and a lack of recognition and belonging (Kelchtermans, 2017).

Another aspect of this model that can inform the preparation and support of beginning science teachers is a recognition by teacher educators and professional development providers of what features of these worlds are perceived by teachers to be most important. Such knowledge can be used to help teachers negotiate how best to position themselves with others in the community within the figured world they inhabit while staying true to their professional identity and thus "figure" their place within that world. Acknowledging that both teacher candidates and beginning teachers are entering into spaces that by virtue of the elements that characterize them may align with or operate in opposition to practices that are featured in the preparation program they attended and/or those that are held by the teacher is essential. Actively seeking to discover the primary values of the culture and community the teacher is working in can provide insight into the tensions and therefore potential avenues for agentic action to push against or navigate through the established narrative of a figured world which features prominently in that teacher's life. Similarly, teacher educators can provide guidance to the teachers to be reflective of their practices and to examine reasons those practices are being implemented with fidelity or are posing challenges to enactment. As previously mentioned, if a teacher can gain an awareness of the features of the world that are working in opposition to their pedagogical choices (as well as those which are aligned), solutions can be developed in concert with teacher educators (or with supportive others, including peers and possibly even other professionals within that figured world) to allow the teacher to find success in a way that supports their progressive teaching choices and feelings of agency while also working in parallel with the value of the culture and community of the teaching figured world.

The figured world lens has allowed us to refine our model of teacher identity development in ways which allow us as science education researchers and as teacher educators to gain access to the teaching space our preservice and early-career teachers enter and to understand better how their perceptions guide the way they traverse these spaces. In turn, this allows us the potential for assisting these beginning educators in developing a strong professional identity foundational for continued agentic action in the service of their students' learning. It is through the actions that these beginning educators take that those with more influence in that figured world can begin to recognize novices' expertise and these educators can garner a sense of power and privilege while contributing to the world of which they are a part. This is a critical point because in the research literature, we generally tend to underestimate the potential for novice science teachers to effect change – even reform. And at the same time these teachers are changing their status in the hierarchy of their figured world, they have greater potential to contribute to deeper and more consequential learning of the students they have responsibility for educating.

Acknowledgements This work was made possible by grants from the National Science Foundation (DUE 1340048), Woodrow Wilson Foundation (RC100381) and the Kellogg Foundation to the first author. The views expressed are those of the authors and not those of the funding agencies.

References

Arvaja, M. (2016). Building teacher identity through the process of positioning. *Teaching and Teacher Education, 59*, 392–402. https://doi.org/10.1016/j.tate.2016.07.024

Bandura, A. (2001). Social cognitive theory: An agentic perspective. *Annual Review Psychology, 52*, 1–26. https://doi.org/10.1146/annurev.psych.52.1.1

Carlone, H. B., Scott, C. M., & Lowder, C. (2014). Becoming (less) scientific: A longitudinal study of students' identity work from elementary to middle school science. *Journal of Research in Science Teaching, 51*(7), 836–869.

Chu, Y. (2020). Preservice teachers learning to teach and developing teacher identity in a teacher residency. *Teaching Education, 32*, 1–17. https://doi.org/10.1080/10476210.2020.1724934

Cobb, D. J., Harlow, A., & Clark, L. (2018). Examining the teacher identity-agency relationship through legitimate peripheral participation: A longitudinal investigation. *Asia-Pacific Journal of Teacher Education, 46*(5), 495–510. https://doi.org/10.1080/1359866X.2018.1480010

Corbin, J., & Strauss, A. (2008). *Strategies for qualitative data analysis. Basics of qualitative research. Techniques and procedures for developing grounded theory*. Sage. https://doi.org/10.4135/9781452230153.n4

Dyson, A. H., & Genishi, C. (2005). *On the case* (Vol. 76). Teachers College Press.

Glesne, C., & Webb, R. (1993). Teaching qualitative research: Who does what? *International Journal of Qualitative Studies in Education, 6*(3), 253–266.

Holland, D., Lachicotte, W., Jr., Skinner, D., & Cain, C. (1998). *Identity and agency in cultural worlds*. Harvard University Press.

Kelchtermans, G. (2017). 'Should I stay or should I go?': Unpacking teacher attrition/retention as an educational issue. *Teachers and Teaching, 23*(8), 961–977. https://doi.org/10.1080/13540602.2017.1379793

Lavina, L., & Lawson, F. (2019). Weaving forgotten pieces of place and the personal: Using collaborative auto-ethnography and aesthetic modes of reflection to explore teacher identity development. *International Journal of Education & the Arts, 20*(6), 1–30. https://doi.org/10.18113/P8ijea20n6

Lortie, D. (1975). *Schoolteacher: A sociological study*. University of Chicago Press.

Maxwell, J. A. (1996). *Qualitative research design*. Sage.

Maxwell, J. A. (2008). Designing a qualitative study. In *The Sage handbook of applied social research methods* (pp. 214–253). Sage.

NGSS Lead States. (2013). *Next generation science standards: For states, by states*. National Academies Press.

Pajares, F. (1993). Preservice teachers' beliefs: A focus for teacher education. *Action in Teacher Education, 15*(2), 45–54. https://doi.org/10.1080/01626620.1993.10734409

Richmond, G. (2015). Making sense of the interplay of context, identity, and agency in the development of beginning secondary science teachers in high-poverty schools. In L. Avraamidou (Ed.), *Studying science teacher identity: Theoretical and methodological approaches* (pp. 219–236). Sense Publications.

Richmond, G., & Muirhead, F. (2014, April). *Contextual factors shaping teacher identity and agency among non-traditional science teacher candidates*. Conference presentation. National Association for Research in Science Teaching (NARST).
Richmond, G., & Wray, K. (2017). *Teacher agency as a mediator for professional identity in secondary science teachers* [Conference presentation]. National Association for Research in Science Teaching.
Richmond, G., Parker, J., Gotwals, A., Kang, H., Sato, T., Lark, A., & Anderson, C. W. (2010). *Improving science teacher preparation by studying how knowledge, identity and context affect teaching practices [symposium]*. National Association for Research in Science Teaching.
Richmond, G., Juzwik, M. M., & Steele, M. D. (2011). Trajectories of teacher identity development across institutional contexts: Constructing a narrative approach. *Teachers College Record, 113*(9), 1863–1905. https://doi.org/10.1177/016146811111300907
Schussler, D. L., Stooksberry, L. M., & Bercaw, L. A. (2010). Understanding teacher candidate dispositions: Reflecting to build self-awareness. *Journal of Teacher Education, 61*(4), 350–363. https://doi.org/10.1177/0022487110371377
Sfard, A., & Prusak, A. (2005). In search of an analytic tool for investigating learning as a culturally shaped activity. *Educational Researcher, 34*, 14–22. https://doi.org/10.3102/0013189X034004014
Smagorinsky, P., Cook, L. S., Moore, C., Jackson, A. Y., & Fry, P. G. (2004). Tensions in learning to teach: Accommodation and the development of a teaching identity. *Journal of Teacher Education, 55*(1), 8–24. https://doi.org/10.1177/0022487103260067
Varghese, M. (2016). Language teacher educator identity and language teacher identity: Towards a social justice perspective. In *Reflections on language teacher identity research* (pp. 51–56). Routledge.
Varghese, M., Morgan, B., Johnston, B., & Johnson, K. A. (2005). Theorizing language teacher identity: Three perspectives and beyond. *Journal of language, Identity, and Education, 4*(1), 21–44. https://doi.org/10.1207/s15327701jlie0401_2
Wertsch, J. V., Tulviste, P., & Hagstrom, F. (1993). A sociocultural approach to agency. *Contexts for learning: Sociocultural dynamics in children's development, 23*, 336–356.
Wray, K. A., & Richmond, G. (2018). Factors shaping the agency of beginning science teachers working in high-poverty schools. *Journal of Science Teacher Education, 29*(8), 785–803. https://doi.org/10.1080/1046560X.2018.1514824
Yazan, B. (2019). Toward identity-oriented teacher education: Critical autoethnographic narrative. *TESOL Journal, 10*(1), e00388. https://doi.org/10.1002/tesj.388

Chapter 12
Identities in Action: Opportunities and Risks of Identity Work in Community and Citizen Science

Colin G. Dixon, Emily M. Harris, and Heidi Ballard

In recent years, there has been an expansion of interest in and forms of citizen and community science (CCS) in which members of the public collaborate with professional scientists to conduct scientific investigations or engage in environmental monitoring to generate new knowledge (Bonney et al., 2014; Theobald et al., 2015). With this interest has come a growing body of research about how participation in CCS might support science learning, agency and identity work (Ballard et al., 2017a, b; Bonney et al., 2009, 2016; Brossard et al., 2005). In this chapter, we use two cases – Bryan, a young Mexican-American man, and Diana, a young Black woman, both participating in water quality monitoring projects – to examine the potential of CCS to open up space for young people to see themselves and their possible futures in new ways.

12.1 Bryan

Bryan participated for 4 years in an ongoing out-of-school water quality monitoring and habitat restoration project focused on an urban creek in a major US city. A calm, 15 year old, interested in anime, he had once been, by his own admission, "the one that would slack off." Participants met nearly every day during the summer and on Saturdays in the fall. By Bryan's third year, the lead educator, a Latino field scientist, relied on Bryan to teach new volunteers to take and test water samples, asked

C. G. Dixon (✉) · E. M. Harris
BSCS Science Learning, Colorado Springs, CO, USA
e-mail: cdixon@bscs.org; eharris@bscs.org

H. Ballard
University of California, Davis, CA, USA
e-mail: hballard@ucdavis.edu

H. T. Holmegaard, L. Archer (eds.), *Science Identities*, Contributions from Science Education Research 12, https://doi.org/10.1007/978-3-031-17642-5_12

him to write and present findings, and put him in charge of small crews on clean-up and planting days. The group shared findings with city officials, advocacy groups, neighbors, and scientists at a large geosciences conference, and used what they learned to inform restoration decisions at the creek. The group's research prompted additional monitoring and resources from the city and highlighted the pressures facing urban creek systems for some in the scientific community. The creek vegetation was dominated by invasive plants and the water was affected by nearby encampments, litter, and leaky sewer pipes. Learning all of this, Bryan came to deeply appreciate the creek and believe he could make an "actual impact" on the world, even if small. Though quiet and modest, Bryan came to see himself as a leader in the project and as someone capable of doing science. Bryan also came to see collaborative research as a tool that could address a wide range of environmental and social issues.

12.2 Diana

A participant in a classroom-based water quality project led by the same educator, Diana, 17, loved learning and telling her friends about science – from outer space to how salt forms. She took pride in seeing herself as a "super nerd," but did not plan to become a scientist. In the project, she appreciated being able to go out to collect samples from the lake near her school, but hated inputting and analyzing data. She took the lead on writing for the group's scientific poster and said she really enjoyed presenting at the same geosciences conference as Bryan, where she had a chance to see "what sciency people do." Yet in preparation for the conference, edits that other project scientist-educators made to the poster made her feel that she and her peer's voices had been lost. At the conference, as well as in presentation of data to city representatives, she doubted whether the adults were listening. Unlike Bryan, Diana did not express a sense of ownership for the lake, seeing it as a site of gentrification that did not serve her community. By the project end, she said she still loved science, but felt that she was "not really cut out for the whole science research [world]," and did not feel that science could help her with her goals – to "change the world," or more specifically, to end the oppression of LGBTQ communities and people of color, and "become the first black female world ruler."

As these two cases suggest, identity work – the actions, resources and relationships that people engage with to perform who they want to be and become within cultural worlds (Calabrese Barton et al., 2013; Carlone et al., 2014) – in CCS is complex. Below, we will first describe CCS activities and briefly review prior research on identity development in CCS. We will then unpack four key characteristics of CCS – *contribution to real science, nested purposes, multiple practices, tools and roles,* and *place-based engagement* – that shift the structures of traditional science education. We ask: what forms and functions of agency arise in response to how CCS structures science learning and how does resulting participation support new kinds of identity work in science? We hope the chapter can help educators and researchers gain insight not only into how experiences in CCS can be powerful for science identification, but also when opportunities can fall flat.

12.3 Community and Citizen Science as a Context for Identity Work

Engagement in CCS can involve any step in the scientific production of knowledge that is used for resource management or basic research, from asking questions, to collecting data, to interpreting results and disseminating findings (Shirk et al., 2012). We include "community science," as well as the more recognizable (but problematic) term "citizen science," to acknowledge the importance of community-based approaches and the wide range of people and roles that are critical to how science gets done. CCS participants often produce data, from classifying photos on a computer to analyzing water or air quality samples to identifying birds as they hike. They might also design research questions, conduct data analysis, and disseminate conclusions to researchers or decision-makers (Bonney et al., 2014; Shirk et al., 2012). CCS projects study places as close as neighborhood creeks and school yards, or as distant as galaxies and alpine ecosystems. While some people participate alone, others do so with family, friends or classmates.

CCS is a rich context for learning. As young people do authentic science, they work firsthand with tools and practices of science, and interact with a range of people and places. Activities often involve reasoning about complex scientific and social phenomena: 4th graders argue to defend their bird identifications, watershed groups compare their data to national datasets, and students monitor air around their school and consider local policies. Some CCS projects go even further, connecting research to ongoing community action like habitat restoration, community planning, or political advocacy. Through these types of activities, participants can develop understandings and skills, and become versed in the norms and discourses of scientific communities – critical processes for trajectories in, to and through science (Aikenhead, 1996; Lemke, 1990).

12.4 Prior Research on Identity Development in CCS

A synthesis of research on science learning in informal environments underscores the potential of CCS to foster identification in science, claiming that participation in *authentic science* encourages participants to "[t]hink about themselves as science learners and develop an identity as someone who knows about, uses, and sometimes contributes to science" (NRC, 2009, p. 17). Though published evidence is scant, recent research on adults' identity work in CCS demonstrates this promise: For participants with science backgrounds, CCS can reinforce existing identification as part of a scientific community, while those with very little prior experience in science can develop strong but varied ways of identifying with science over time. Jackson et al. (2015) applied a "community of practice" lens to identity work in an online crowd-sourcing project and found that participants moved toward fuller participation in the project community. Merenlender et al. (2016) found that after a natural history and citizen science training course, more adults could "imagine themselves as scientists" than before. In a longitudinal study of six CCS projects (from tracking precipitation to butterflies and bird identification), Ballard et al. (2017a, b) found that a majority of adult participants strongly identified as "someone who uses and does science" and as "recognized by others as doing so," but rarely identified "as scientists". These findings indicate that work in CCS projects, across a variety of project types, can indeed give participants opportunities to engage in, or re-engage in, science practices for use in local and scientific communities. It also suggests that with adults, rather than thinking about "becoming a scientist," it may be most appropriate to think about how existing identities can be linked to new communities or activities.

For youth, a small but growing evidence base demonstrates that, for those who already identify with science, CCS can offer a way to engage in practices that extend and connect to their identification with science. For those who don't already identify with science, CCS can sometimes be a tool to address other issues they care about, such as conservation or community health. Two ethnographic studies investigated a collaborative CCS project where high school students with significant financial need and/or no family history of college worked to conduct herpetology monitoring and field ecology. In one study, Huffling (2015) found that youth in the project came to see themselves as stewards of the environment and society when they had opportunities to investigate local issues, participate in decision making, and engage in field ecology practices repeatedly. In the other, Carlone et al. (2015) found that gently confronting fears of the outdoors, amphibians and reptiles encouraged young people with limited previous experience outdoors to push comfort zones and engage in identity boundary work. A number of factors were important, such as flexible use of time and space, social support, and regular use of tools that helped youth move out of comfort zones. These two studies suggest that CCS is most valuable when it is collaborative, place-based, and attends to who young people are. Carlone et al. (2015) also challenged educators to go beyond what is familiar to young people and encourage youth "to participate in new communities of practice, engaging in identity work previously unfamiliar to or unusual for them." In a mixed-method study investigating participation in a woodsmoke pollution project, Brickle and Evans-Agnew (2017) found that young people felt empowered as they produced and analyzed air quality data about their homes and told stories using photovoice. Sharing findings with policymakers and community members enabled participants to enact control over their data and become advocates for their personal health and health of their community. In a quasi-experimental study, Wallace and Bodzin (2017) found that young people in a project monitoring plant phenology using mobile technology showed significant changes in their identities as "citizen scientists" and their interest in STEM careers. They felt like they were "helping the environment," and saw science as "more real."

However, outcomes of youth participation in CCS can be mixed. Williams et al. (2021) found that, based on survey responses, classroom-based participation in a citizen science project actually had a slightly negative impact on science identity for middle and high school students, and in an ethnographic study of girls in a milkweed and monarch monitoring project, Brien (2017) found that while some girls demonstrated shifts in science identity, others did not, seeing themselves as just completing repetitive data collection tasks for a scientist. Thus, while existing research provides evidence of CCS' promise for the development of science identity, divergent findings leave many questions about how, when and why young people's experiences in CCS differ.

12.5 Understanding Identity and Agency in CCS

At its best, CCS has the potential to restructure science learning. In many science learning environments, identity work can be limited by historical, social, institutional, or local forces, and entwined with issues of race, class and gender (Carlone et al., 2014). These structures take shape through norms, available and expected roles, and storylines about who people can be and what they are capable of (Gresalfi & Hand, 2019; Holland et al., 1998). In schools, young people have historically been positioned as outsiders to scientific communities and as recipients of scientific knowledge. This limits opportunities for young people to construct knowledge of their worlds, to use the power of science to create change, and to align educational experiences with personal goals and community priorities.

In CCS, science learning happens as part of scientific work, thereby shifting the resources and relations of learning. The local work that young people do is nested within larger initiatives and involves datasets, policies or questions that stretch beyond the social, spatial and temporal boundaries of the classroom (Harris et al., 2020). Students' work and learning also better resembles professional science, distributed across an array of people, skills and tools. By situating science learning in this way, CCS has the potential to:

(a) *Reframe science and science learning as expansive and consequential*, helping young people see their participation not as the obligatory work of schooling, but as part of real science, done within a community comprised not only of teachers and students, but also of scientists, advocates, policy makers, and neighbors (Harris et al., 2020);
(b) *Make new roles and positions available*, diversifying the kinds of practices that count as scientific and allowing young people to participate in ways that make the most of their expertise and best serve their goals (Dixon, 2018; see also Basu et al., 2009; Varelas et al., 2015), and;

(c) *Build connections across school-science communities* such that young people are exposed to a broader range of mentors and role models – scientists, decision makers, neighbors, and community organizations that use or participate in science for many purposes.

To understand how these structures influence identity work in CCS, we analyze identity not as an outcome, but as a process (Carlone et al., 2014; Van Horne & Bell, 2017), using a lens of identities-in-action. We focus on when young people come to see themselves (or not) as community resources, how they imagine their future selves and communities, and how these futures are carried forward through interactions, stories, and artifacts. This approach builds on theories of identities-in-practice that describe how identity work develops through participation in collective endeavors, from basketball to basket weaving, as people engage with practices that are meaningful within particular communities (Nasir & Hand, 2008; Rogoff, 1995). By highlighting identities-in-*action*, we bring attention to activities in which the learning of young people becomes consequential – environments in which the work young people do is valued as a contribution *now*, not solely as preparation for possible contribution in the future. In these settings, it becomes clear that how a person sees themselves is inseparable from what they do with and among others, and depends on whether they believe what they do matters.

Identity work is therefore inextricably linked to *agency*, the capacity of individuals or groups to act upon, modify, and give significance to the world in purposeful ways, with the aim of transforming themselves and/or impacting the conditions of their lives (Holland et al., 1998). While identity is shaped through participation, participation is shaped by how people take up, transform, or resist the practices, roles and resources available to them (Varelas et al., 2015). Whether in science classroom, lab or field, young people make some positions and constraints visible and submerge others; they ignore some ideas and tasks, while embracing and introducing others; they listen, write, analyze, ask and answer questions, but also joke, talk to friends, and skip school. In these ways young people negotiate with their peers and educators the norms and narratives of their classrooms – norms and narratives that carry status, relationships, and expectations for what is possible and who belongs in science (Patterson, 2019).

Attending to agency makes visible how young people find ways through, and even reshape, learning situations to reflect their experiences, goals and concepts of self (Calabrese Barton & Tan, 2010; Carlone et al., 2014; Cobb & Hodge, 2011). Theories of agency help us think critically about science learning in the context of social structures and local action, looking not just for where learners take up recognizable science identities, but also for moments of improvisation and resistance (Holland et al., 1998). Identifying these moments is critical to understanding when shifts to the historical patterns and relations of science education – like the shifts possible with CCS – offer up meaningful resources for identity work, and to whom.

Adapted from Basu et al.'s (2009) concept of critical science agency, the construct of *environmental science agency* (ESA) provides an additional tool with which we examine how identities-in-action get linked to science and science learning in CCS. ESA is the ability to use experiences in environmental science to make positive changes in one's life, landscape and community. ESA brings our attention to a cycle of learning in which young people take up practices and roles within scientific work, build expertise, and take action in their communities or come to see themselves in new ways (Ballard et al., 2017a, b). These positive changes might occur in small, discrete ways, such as carrying a practice from the CCS project to an extracurricular program, or in ways that reflect a deeper, enduring shift, like joining a new program or class, expressing newfound confidence, or describing new possibilities for one's future (Ballard et al., 2017a, b).

12.6 Examining Identity Work in CCS at Two Levels

To better understand when identities-in-action become meaningful – and when they do not – we must examine identity work and agency at both the *activity* level and *individual* level, looking at what practices and tools young people are putting to work in the world, as well as what worlds young people see themselves in (Varelas et al., 2015). At the individual level, we examine identities through narrative identities, or stories people tell about themselves, their world, and who they want to be (Sfard & Prusak, 2005). Analysis of narrative identities foregrounds how young people make meaning of their experiences and project themselves in a given context through their self-concept and narratives about others. The stories young people tell can also make visible where they see themselves stepping up and stepping away, moments of agency that might be hard to recognize as an observer. In our research, we accomplish this through semi-structured interviews, both early in a CCS program and after a program has ended, as well as, in observation, through "small stories" (Bamberg, 2004) and histories that young people surface or imply through their interactions with peers and adults. These narratives can help analysts understand how young people see, or frame, what they are doing and why (Dixon & Martin, 2017; Engle et al., 2012). A narrative approach also helps to foreground experiences of young people who have been historically marginalized (Solórzano & Yosso, 2001).

Alongside analysis through this individual lens, we also use an activity-centered lens on identities-in-action. We draw on observations, artifacts, and audio recordings of interactions to capture the dynamics and consequences of identity work, as well as the ways that power and history come into play. With an activity-focused lens, researchers can better recognize how young people author identities as they take up, ignore, or transform the practices and labels of science, depending on where, with whom, and for what they are doing science (Tan & Calabrese Barton, 2008). Focusing on activity can also help make power and privilege visible within CCS environments because it helps us see what kinds of resources for expression, contribution, and identity work are legitimized or undermined through moment-to-moment interactions (Nasir & Hand, 2008).

Analysis that uses both individual- and activity-centered lenses enables CCS researchers to look for shifts not just across time, but across contexts and tasks, and to attend to where young people themselves see change. This means:

(a) Documenting how young people's status, behavior, ways of talking and working might change as they move from a classroom to a scientific conference; and
(b) Unpacking the meaning they make of their work in each setting – what practices they highlight and what roles they claim (Dixon, 2018; Gresalfi & Hand, 2019).

It also means considering when participants' perspectives on what they are doing changes, even when the setting or task itself does not (Beach, 1999), such as when they start to talk about their work not as "doing school," but as contributing to community well-being or scientific research. Looking for places of alignment and resistance across these accounts can reveal where young people are doing important work to reconcile and create their worlds.

To investigate CCS as a context for agency and identity work, we studied eight CCS programs working with more than 135 young people, ages 12–18 (see Ballard et al., 2017a, b). These included four in-school and four community-based programs, and ranged from short-term (2 weeks) to long-term (six plus months), with some young people like Bryan participating over multiple years. Young people investigated a variety of organisms and ecosystems, from monarch larvae and milkweed plants in restoration areas, to water quality in neighborhood creeks and lakes, to crab abundance at regional beaches. Users and audiences for the data and findings produced by the youth participants included university scientists, researchers and managers at a National Marine Protected Area, local environmental and advocacy groups, city environmental agencies, and school administrators. Participation was demographically diverse in terms of race, ethnicity, and gender, and reflected varied levels of experience and interest in science. Four programs, including Bryan's and Diana's, deliberately worked with students of color and/or students with significant financial need.

Using a case study design (Yin, 2013), we collected observational field notes, reviewed program and student-produced artifacts (Bowen, 2009), and conducted semi-structured interviews (Patton, 2002) with educators and 6–10 focal youth participants per program (54 total), before and after their CCS experiences. Narrative and activity analysis of these cases provided a close-up look at the kinds of interaction, learning, and work that took place during CCS experiences. The four themes discussed below emerged from analyses across the multiple projects and reflect ways that we saw CCS structure science learning and opportunities for identity work (for more details, see Ballard et al., 2017a, b; Harris et al., 2020). In this chapter, we analyzed cases of individual participants in order to look more closely at these four aspects of CCS and to understand, through an identities-in-action lens, how they shape agency and identity work (Becker, 2014).

We purposefully selected the cases of Bryan and Diana to help us explore a range of experiences in CCS and when these key aspects of CCS have different impacts given different situations. Both individuals participated in water quality monitoring, led by the same scientist-educator, with the scientific data they produced used to inform local governmental and community actions. However, the programs differed in many respects. Bryan participated during the summer and weekends, in a small group of four to eight youth participants, depending on the day. Diana participated as part of an elective science course at her high school, with approximately 20 other students. Of course, their experiences reflect other factors as well – both unique to them as young people and intersecting with constructions of gender, race, sexuality and economics in science and society. The cases are not meant to be representative, but are reflective of trends we saw across multiple projects and participants. Though they are limited in generalizability, we believe they can be generative for research and relevant to practice, providing insight that can help better understand the varied potential for identity work in science learning environments.

12.7 Key Aspects of CCS

Though trajectories through CCS are unique, across our analyses of the eight case studies we identified four key aspects of CCS participation that reconfigure participation in ways that provide opportunities for agency and identity work in science:

(a) Contribution to "real science" and scientific communities (Ballard et al., 2017a, b);
(b) Nested purposes Harris et al., 2020;
(c) The availability of multiple practices, tools and roles (Dixon, 2018; Harris & Ballard, 2021), and;
(d) Place-based engagement (Ballard et al., 2017a, b; Dixon et al., 2021).

Below, we discuss these aspects, returning to the stories of Bryan and Diana to illustrate how they can influence development of science-linked identities-in-action.

12.7.1 Contributions to "Real Science" and Scientific Communities

Participation in authentic science can be a powerful motivator of learning and identity work. Though it is important to consider what counts as "real" and to whom (Rahm et al., 2003), one form of authenticity is recognition and use of one's work by others. In CCS, this emerges from participants' contribution to institutional science, typically through data that informs basic research or environmental management (Bonney et al., 2014). Through this production, young people have the opportunity to act and be seen as experts, creating a pathway to move from the far peripheries of science – a distant planet of textbooks and worksheets to which many students feel they have been sent – to more central positions in science that are recognizable and valued.

Part of the value of production – whether of physical objects, digital artifacts, or knowledge shared with and used by scientists, managers, or local decision makers – comes from seeing "a little bit of me in the world," a process of productive agency with unique affordances for disciplinary and identity-linked learning (Engle et al., 2012; Okita & Schwartz, 2013). Further, when CCS work involves direct interaction with scientists, youth participants may develop new understandings of scientific communities and be better able to envision possible selves in science.

Bryan By his third year, Bryan came to appreciate his group's water quality research at the creek, seeing it as something "bigger" that reflected his ability in science better than his grades did. He saw the group's data production – testing water to track spikes in ammonia, nitrates and phosphates, and identifying possible pollution sources – as critical to their work restoring native plant species, cleaning the creek, and reporting results to the city government and other stakeholders to prompt action. Presenting findings at a large scientific conference also became very important to Bryan. He talked about how, in his first year, his brother taught him to tie a tie to wear to the conference. After 3 years of presenting, Bryan was comfortable speaking at the conference. He felt challenged but not intimidated by questions from scientists visiting his poster, and saw himself as a coach for his peers, gently prompting them when they forgot to communicate something important. Attending the conference, which took place mid-week, also forced Bryan to tell his teacher and principal that he did science outside of school, something they did not know despite his years-long participation. Bryan reflected on handing his teacher the permission slip for the conference, saying:

> I told [my biology teacher] what I did and she called me a nerd – but she didn't mean it in an offensive way. She meant it as a good way. That's how it came out. I was like, 'Of course I'm a nerd'... She told me that she was a nerd too when she was my age. So yeah, she was pretty proud.

Through his contributions to science and the participation in scientific communities these contributions afforded, Bryan came to understand and appropriate the norms and performance of institutional science. This identity work in CCS eventually extended to his work in school, creating an opportunity for him to be recognized by his teachers in a new way. For young people like Bryan, science that feels "real" can form a foundation of collective action, providing legitimacy to act in new ways and new spheres.

Diana Diana appreciated being outdoors at the lake during the CCS project, learning about environmental impacts, and collecting real water quality data rather than working from textbooks. To Diana, getting good data was really important, but she saw how her classmates were "not really taking it seriously," and reported to us that "that really bothered me the whole time because I was like, 'this is inaccurate information. We can't present this!'" Despite this concern, Diana was proud to present at the scientific conference. Though at first skeptical, she ended up loving being at the conference. She thought it was "cool to see what sciencey people do with their work... and really cool just being surrounded by a ton of nerds." However, Diana revealed after the program ended that she felt "slightly insulted" when a scientist helping the group prepare for the conference changed the text on their poster:

> [An assistant instructor] sent all our stuff to his boss or some other guy who has apparently done a lot of science posters, and he changed basically everything. So we didn't really make it [the poster]. They just kind of told us we did.... I'm not super insulted because it *was* better, but at the same time, we all tried really, really hard and did our best.... I feel like the poster that we came up with doesn't quite display our voices.

At the conference itself, Diana also watched one of her peers get rebuked by a female scientist visiting their poster, because he invited her to the poster by calling her "gorgeous." For Diana, this exchange stuck out. She found it funny, but also sympathized with her classmate who reflected on the moment to us, saying:

> I mean, I can see how I faulted at it. But you know, she faulted as well. She didn't have to react the way she did... Just because you're a professor doesn't make you any more of a person than me. I mean, that's how I greet people. That's how I talk. That's just me as a person.

As Diana's experience demonstrates, participation in and contribution to professional scientific communities can be a double-edged sword. Diana valued the chance to be among fellow "nerds," but also became frustrated with the norms of the scientific community, feeling that her voice was written out of the group's work and doubting the value of her group's data and findings – their product. Science was an interesting but challenging and sometimes unwelcoming place. Though Diana exercised agency to develop a meaningful and personal role in project work, one she felt reflected her strengths and identity, seeing her contributions to the poster undercut seemed to in turn undercut greater identification with science.

12.7.2 Nested Purposes

While producing data for professional scientists is one form of participation in science, multiple forms of authenticity emerge from the relationship between participants and the social and ecological systems in which they act. CCS projects where there are additional purposes for the participants' work – such as advocacy, restoration, and individual learning – provide more robust opportunities for identity work and agency. We refer to these as nested purposes (Harris et al., 2020) – ways of working that situate, or nest, the young person's activity within communities beyond the classroom, whether these communities are geographically, politically, or professionally defined.

In cases like Bryan's experience, data and findings not only contribute to larger scientific endeavors as previously described, but also inform valued audiences, including peers, school administrations, neighbors and family members, and motivate direct action, such as restoration, policy change, or community organizing.

These multiple uses allow learner goals, such as social relationship building, skill development, and community impact, to sit alongside the scientific value of CCS work. Establishing multiple purposes can motivate participation and a sense of efficacy, and help shift the way young people are seen and see themselves in communities that matter to them.

Sharing scientific work with community members, such as family members, peers, or the general public, positions young people as knowledgeable and valuable to their communities (Calabrese Barton & Tan, 2010), and when CCS work results in local community action, young people can envision how science is useful in their daily lives and develop new narratives about science and themselves (Stepenuck & Green, 2015).

Bryan By the end of Bryan's third year, he saw community science as a way to create real, if small, change at the creek, and as a strategy that could be employed across a range of problems that matter to him, like violence, poverty, and police brutality.

> This [community research and action] could go further. It could go to, I'd say, violence… people, they could prevent violence; they could report that type of thing. You could just help many places, not only with our small program but through the whole [city]…. Science isn't only for the laboratory. Science is everywhere, even in your community, even when you don't know it…. Science explains what's wrong, why – and how – it's wrong. Science can also say how to fix a problem and how to make a better future.

Bryan also saw that the group's work began to shift how people around the creek used the space. Neighbors – young and old – walked or played near the creek more frequently, and many of them thanked Bryan for being out there. As Bryan came to see his work through the interest and appreciation of the creek neighbors, the group's restoration and advocacy work – "an actual impact to the community" – became the most important part of the CCS project to Bryan.

While scientific contribution remained an important purpose for Bryan, restoration work and community engagement became the heart of the project. Alongside the scientific products, these community purposes helped Bryan see himself as a valuable member of his neighborhood, and through his CCS work, Bryan also was recognized as a legitimate expert not only by scientists, but also by neighbors.

Diana Unable within the timescale of participation to see the social and ecological results of her group's work, Diana was skeptical about whether their work mattered, even to the people who said they cared. She explained,

> [One lady who works with the city] came to class one day and we presented it to her…. So, I guess she took the information back – maybe if she actually listened to us…. in my experience people sometimes don't. Adults sometimes don't listen to high schoolers, but sometimes they do. I'm not sure.

She reflected that she wished she better understood exactly why the group was doing water testing, and she had a hard time connecting environmental science to issues, communities and goals she cared about: Diana wanted to graduate, go to college, make videos that tell people's stories, especially those of LBTQ community and women of color, and, ultimately, "to dismantle systems of oppression."

Diana's experience illustrates how purposes of CCS work can be unclear, especially for young people involved in projects for which direct results are not visible. Diana's work outside the CCS project exemplifies the strong link between one's identity and one's action: she used what she wanted to accomplish—to create a more just world—to say much about who she was. Yet, she found it difficult to connect her work in CCS to the purposes she had for school and community, and to her vision of herself. Her experiences as a young Black woman may have also inclined her to doubt that authorities truly listened when she and her peers spoke. To her, the data she produced ended up having only one purpose: to help her "get an A".

12.7.3 Availability of Diverse Practices, Tools and Roles

In CCS, producing high quality data and analysis requires teamwork, which can mean that a wider range of practices, tools and roles are available for young people to take up than are commonly valued in school science. In school, the scientific process is often defined in terms of a narrow set of practices and personas, dominated by the archetype of a lab-coated White man. However, science happens across an array of institutions and requires a multitude of tools, from cameras to DNA splicers to shovels, and social practices, from presenting a talk to teaching students to editing a manuscript (Pickering, 1995). In a CSS project, young people might have the chance to write up analysis, lead monitoring, organize new volunteers, research important concepts, or create graphs and graphics. In these roles, young people can leverage existing areas of expertise, find pathways from initial entry points into new practices and positions, and tailor their ways of being in science (Dixon, 2018; Varelas et al., 2015). CCS participants are able to experiment with new tools, new roles, and, ultimately, new ways of seeing themselves and being recognized by others. This opportunity for experimentation and movement is critical for expansive identity work (Carlone et al., 2015; Carlone & Johnson, 2007; Holland et al., 1998; Nasir & Hand, 2008).

Bryan Bryan admitted to us that in his first year of participation, he primarily did water quality sampling and testing, but was "the one that would slack off sometimes." When asked about his role now, he said modestly and hesitantly that he considered himself a leader, qualifying that leader meant being a "team player". Bryan taught new participants how to sample water from the creek, "re-explained" ideas and instruction to newer participants, and got into the weeds to do trash cleanup with his team. His role in the group diversified and expanded as he moved from creek to conference to computer room, and across jobs that needed to be done. As his experience and confidence grew and he became interested in new areas,

Bryan moved into aspects of the project with greater accountability, such as entering the group's data into the years-long log and identifying invasive species for removal. In his third year, he was tasked with drafting the group's scientific poster, describing the task this way:

> That was really challenging since I have a bit of a hard time writing. [The lead educator], he still believes in me. You guys both helped me a lot and I got it finished along with the help of [the program director] and his edits.... I'm not the best writer. But I'm not the worst.

Diana Diana expressed clearly that she "hated" inputting data and found it "really annoying" to work with spreadsheets. However, she said she took the lead "when we were writing the poster and stuff." Recognized by the assistant instructor as a good writer, Diana drew on her experience with film scripts to step into this role. She also drew on her experience with an advocacy group to step up and speak at the scientific conference, where she felt she did well as one of the few students who communicated the group's findings and story.

In their engagement with the practices and roles of science, Bryan and Diana provide two complementary trajectories that we saw across projects. Bryan tried out and became proficient in new aspects of the project over time and expanded his engagement with new practices, like writing. These experiences enabled him to see that he could be and do a wide range of things in science, including lead. Similarly, Diana was able to connect an existing area of expertise and identity, as a storyteller and advocate, to science in ways she did not expect. She expressed pride in this work, even if ultimately these roles felt temporary.

12.7.4 Place-Based Engagement

The importance of place in science and environmental education is well established (Kudryavtsev et al., 2012; Lim & Calabrese Barton, 2006). In CCS, repeated work in familiar places allows participants to draw on lived experiences and understand scientific work in relation to community members' priorities and assets (Haywood, 2014). This situates the environmental issues they encounter in the social systems they navigate and within which identities are deeply entwined (Calabrese Barton & Tan, 2010; Carlone et al., 2015).

Bryan When Bryan began the CCS project, he knew almost nothing about the creek that ran near his house. By the end, the creek and the people around it were important to him, from young kids playing nearby to the retired neighbor who stored tools for the group. Work in the project changed how he saw the creek itself, from believing "all green is good" to understanding that the type of vegetation matters, and realizing that water can be clear but unhealthy. It also changed how he saw people in his community, from believing that they did not care about the city to seeing their appreciation for his work and their eagerness to improve things.

> I guess [being at the CCS site] kind of changed my perspective. People do care. A lady yesterday, she was just walking her dog and then she saw what we were doing and she was really, really interested. She wanted to be part of it. She was like, 'Wow, yeah. This is my place. This is where I've been in my life, my whole life'... So, I was like, "Whoa, okay. Strangers, it's nice that you care".

Importantly, his work at the creek changed how people saw him. Neighbors often assumed that young people – young men of color in particular – at the creek were there to cause or escape trouble. Yet they greeted and thanked Bryan when he was out there. They saw the positive change in their neighborhood as the group's work improved the landscape around the creek.

Yet work in CCS also brought Bryan into contact with contrasting perceptions of his place and experience. Bryan recounted a memorable interaction in which he explained to a scientist who stopped by his conference poster that "there's people defecating and urinating in the creek." Bryan reflected with a laugh that "it came to him as a shock and surprise.... I was like, 'that's normal. That's just how it is.' I guess for this man, no. He has different experiences with creeks".

As he learned to identify invasive species, locate garbage dumps, and gauge sources of contamination, Bryan learned that the creek held a history of the people around it – a history that included him and that he could help write. As his case shows, place-based investigation of social and environmental phenomena can transform a learner's perception and narrative of a place, as well as how they are recognized by community members.

Diana Outside the CCS project, Diana did not spend a lot of time outdoors and did not spend a lot of time at the lake, despite it being just a block away from school. Yet she enjoyed being there during project work sessions, investigating the real world, and noticing new things about the lake. However, Diana also wondered why the class did not talk about other aspects of the lake and city, such as gentrification, which she saw as stemming from a major city-led residential and recreational development around the shores where she sampled water quality.

> Now I'm learning how [gentrification] affects people.... They're basically just kicking out a lot of these poor people, who are mainly people of color.... there's tons of big high rises and stuff being built [around the lake], and hella expensive... like $3,000 a month. Who can afford that? Not Black people.

With these experiences in mind, Diana at times questioned the impact and value of monitoring, and did not see herself as a steward of the lake. Where Bryan was able to laugh off the scientist's view of the creek, putting aside conflicting perceptions of the lake was more difficult for Diana: the city's view of the lake as a site of economic development forced confrontation with her experience of gentrification around the school and across the city. She did come to better understand the lake through a scientific lens, but the place did not feel like hers.

12.8 Summary of Cases

Analyzing these key aspects of CCS helps us to understand what kind of identity work and agency arise when engagement with science is not just in service of learning, but of action. Work in CCS can disrupt traditional narratives about who is and can be successful in school science by expanding the norms, roles, and practices of science education to better reflect professional science, while also allowing a more diverse range of youth experiences and expertise to become relevant (see Gresalfi & Hand, 2019). For Bryan, aspects of CCS opened space for him to take up a position as contributor to consequential science and to align his work in science with his sense of self. From these new positions and practices, he was able to see and strengthen an expanded idea of what was possible for himself and his community. This took time. Bryan was a third-year participant at the time we joined his project. Long term participation offers stronger relationships to educators, opportunity to gain varied expertise and step into new roles and time to see possible impacts of scientific work on the environment, policy, and/or community action. Though we saw many cases of transformative identity work in our larger study, including in short term programs, the importance of time cannot be overlooked.

Diana embraced the opportunity to be a nerd and still loved science, but by the end of participation, did not feel any closer to it. Across the projects we studied, Diana's experience was neither common nor unique; many students struggled to figure out whether their work and who they were really mattered in "a scientist's science" (Moss et al., 1998; see also Brien, 2017). Her experience demonstrates how aspects of CCS that can restructure science learning – interactions with scientists and community decision makers, learning and engagement rooted in familiar places – can also undermine the potential for transformative identity work in relation to "inbound" science trajectories. In this case, interactions reinscribed Diana's position as outsider, and strengthened her perception that the values and priorities of scientific communities were contrary to her own and those she held for her community. Diana's existing narratives of science as hard, disconnected from her social justice goals, and separate from the issues affecting Brown and LGBTQ people, became resources that worked against realignment of her identity with science. Yet, asserting her priorities and expertise were still expressions of agency and important forms of identity work (Benson, 2003) that Diana did within the project. We saw evidence that her participation in CCS (as well as in reflective interviews that were a part of research) provided valuable opportunities to further develop who she wants to be and what she wants to do. This reminds us that as researchers and educators, we must value the many possible selves and futures that young people are constructing, regardless of their relationship to science or science education.

Looking at the experiences of Bryan and Diana, it is clear that identity work is not and cannot be done in isolation: how young people see themselves is inextricably linked to what actions they take and how they perceive the purpose and consequence of those actions in relation to their lived experiences. While CCS can transform the conditions of identity work, young people bring with them existing histories and goals, and it is from these that affordances for science-linked identities-in-action arise. Unrecognized histories and inattention to larger scale social structures can undermine potential openings and resources, particularly for young people of color (Tokunow, 2016; Tzou et al., 2010).

12.9 Implications for Practice and Research

While engaging classrooms and groups in CCS can provide a mechanism to open space for new kinds of identities-in-action, this must happen throughout young people's interactions in CCS, not just in project planning and structure. Based on our research in CCS we have identified heuristics that can help educators design and facilitate learning environments to support expansive identity work in science (Ballard et al., 2017a, b). Here we highlight two of these, connecting with multiple stakeholders and engaging complex socio-ecological issues, that are relevant to the experiences of Bryan and Diana and important to thinking about identity work in many science learning environments. Translating the analytical lenses of narrative and activity to practice, these heuristics prompt opportunities for young people to explicitly author who they want to be and to put scientific practices to use as ways to act and identify in their communities.

12.10 Facilitate Connection with Multiple Stakeholders

When young people see others make use of their data or take action based on their analysis, they can feel a part of the scientific community and gain a sense of connection to their scientific work (Ballard et al., 2017a, b; Harris et al., 2020). While involving scientists can be valuable, as it was for Bryan, connections with local decision makers, neighbors, or family members may offer opportunities that better fit some students' goals and priorities. Interactions with these community stakeholders position young people to be recognized as community science experts (Calabrese Barton & Tan, 2010), making valued contributions to their communities. However, educators cannot assume that bringing learners *to* a community means bringing them *into* a community: Engaging audiences that young people do not trust or care about, or that cannot make real use of the data or findings, can have unexpected negative consequences, as it did for Diana. Educators must attend to ways that new audiences and practices may be in conflict with young people's narratives and look for opportunities for young people to share their experiences and questions, not just their data.

12.11 Engage with Complex Socio-Ecological Issues

Through both scientific protocols and informal interactions, educators can ask young people to look closely at natural *and* social worlds (Ballard et al., 2017a, b). CCS work can risk focusing primarily on the natural world and making young people feel like automatons collecting data for scientists (Moss et al., 1998). The connections Bryan saw between the CCS project and ways of addressing other social issues helped him see how he, as a young person with little political or economic power, could impact his community. For Diana, lack of space in the project to engage the social systems that contributed to ecological conditions the group was researching made her believe these social systems were outside the scope of scientific inquiry. When designing projects and facilitating research in the field, educators and other practitioners should create space in protocols to observe phenomena beyond the project foci, including where and how human activity is influencing what they notice in nature. Discussions can go beyond science content required for CCS protocols and allow young people to guide the group's learning (Stroupe, 2017) and draw on funds of knowledge about places and socio-scientific issues (Bang & Medin, 2010). It is important to make clear that human activity is an integral part of ecological systems, with both positive and negative effects; science is not a process of figuring out right and wrong answers, but of using and building understandings about the complex social-ecological systems in which we live. Beyond lesson or program design, this requires that teachers be ready to soften control over topics and purposes of classroom activity, and be able to facilitate discussions that lead in unexpected directions. Doing so can help surface underlying issues of power, race and class that, if left submerged, can limit forms of agency and identity work (Sheth, 2019).

12.12 Research Questions and Recommendations

With its connection to larger scientific and/or social projects, CCS is uniquely situated for studying, through the identities-in-action lens, how learning happens across both individual and group activity. Long-term studies can help document how identity work in CCS – through narrative and activity – influence young people's trajectories far beyond the project, including into forms of civic participation. We also have questions about how insights gained from studying field-based environmental CCS might apply to online CCS work and communities – a common form of CCS participation – and to diverse disciplines, such as astronomy, biomedicine or social sciences. Further, research on CCS needs to interrogate assumptions and goals underlying what education is for: we need to deliberately investigate how changes in identity and agency may not simply influence individual trajectories, but how they impact the organization and resilience of communities. CCS presents a chance for democratization and disruption of science that could lead intentionally away from conventional conceptions of science and science identity. What can and should this look like — and how can young people be the ones to decide?

12.13 Conclusion

In the context of CCS, we have argued the importance of looking at identities-in-action: understanding identity through how young people take action in the world. We presented a theory of agency that links development of practices and identities within a community, field or discipline, to the roles young people take within those communities and what they might carry beyond them. From this perspective we discussed four aspects of CCS:

(a) Contribution to scientific work;
(b) Nested purposes;
(c) Engagement with multiple practices, tools and roles, and;
(d) Place-based engagement.

These aspects can support identification with science, sometimes in divergent ways for different young people. Based on this and earlier empirical research, we posed implications for practice that can help CCS scientists, designers and educators best support young people's identity work in science.

While CCS is unique in the ways it can support young people's identification with science, we see it as an example of a larger class of learning environments in which learners are engaged in *real*, not just *realistic*, work (Heath, 2004). We believe that some combination of these affordances for identity work can be found across a range of settings in which young people produce something to share beyond program walls, from bike shops to media programs to advocacy organizations to makerspaces. However, the strengths of these environments – public, boundary-crossing, collaborative – also create vulnerabilities, as they did for Diana. With this in mind, the four aspects of CCS we highlight shed light on the intersubjective processes through which identities are formed across a range of settings and disciplines, not just in science. They bring attention to ways that young people's experiences can diverge, but also remind us of ways that we, as educators, researchers, and scientists, can join young people in authoring identities, as individuals and communities, through shared work in the world.

References

Aikenhead, G. S. (1996). Science education: Border crossing into the subculture of science. *Studies in Science Education, 27*(1), 1–52. https://doi.org/10.1080/03057269608560077

Ballard, H. L., Dixon, C., & Harris, E. M. (2017a). Youth-focused citizen science: Examining the role of environmental science learning and agency for conservation. *Biological Conservation, 208*, 65–75. https://doi.org/10.1016/j.biocon.2016.05.024

Ballard, H. L., Yamashita, L., Phillips, T. B., & Bonney, R. (2017b). *Examining the influence of participation in citizen science projects on participants' identity with respect to science: A study of 70 people across 6 projects in the U.S.* Conference presentation. National Association for Research in Science Teaching, San Antonio, USA.

Bamberg, M. (2004). Form and functions of 'slut bashing' in male identity constructions in 15 year-olds. *Human Development, 47*, 331–353. https://doi.org/10.1159/000081036

Bang, M., & Medin, D. (2010). Cultural processes in science education: Supporting the navigation of multiple epistemologies. *Science Education, 94*(6), 1008–1026. https://doi.org/10.1002/sce.20392

Basu, S. J., Calabrese Barton, A., Clairmont, N., & Locke, D. (2009). Developing a framework for critical science agency through case study in a conceptual physics context. *Cultural Studies of Science Education, 4*(2), 345–371. https://doi.org/10.1007/s11422-008-9135-8

Beach, K. (1999). Consequential transitions: A sociocultural expedition beyond transfer in education. *Review of Research in Education, 24*, 101–113. https://doi.org/10.3102/2F0091732X024001101

Becker, H. S. (2014). *What about Mozart? What about murder? Reasoning from cases*. University of Chicago Press.

Benson, C. (2003). The unthinkable boundaries of self: The role of negative emotional boundaries for the formation, maintenance, and transformation of identities. In R. Harré & F. Moghaddam (Eds.), *The self and others: Positioning individuals and groups in personal, political, and cultural contexts* (pp. 61–84). Praeger.

Bonney, R., Cooper, C. B., Dickinson, J., Kelling, S., Phillips, T., Rosenberg, K. V., & Shirk, J. (2009). Citizen science: A developing tool for expanding science knowledge and scientific literacy. *Bio Science, 59*(11), 977–984. https://doi.org/10.1525/bio.2009.59.11.9

Bonney, R., Shirk, J. L., Phillips, T. B., Wiggins, A., Ballard, H. L., Miller-Rushing, A. J., & Parrish, J. K. (2014). Next steps for citizen science. *Science, 343*(6178), 1436–1437.

Bonney, R., Phillips, T. B., Ballard, H. L., & Enck, J. (2016). Can citizen science enhance public understanding of science? *Public Understanding of Science, 25*(1), 2–16. https://doi.org/10.1177/2F0963662515607406

Bowen, G. S. (2009). Document analysis as a qualitative research method. *Qualitative Research Journal, 9*(2), 27–40. https://doi.org/10.3316/QRJ0902027

Brickle, M. B., & Evans-Agnew, R. (2017). Photo voice and youth empowerment in environmental justice research: A pilot study examining wood smoke pollution in a Pacific Northwest Community. *Journal of Community Health Nursing, 34*(2), 89–101. https://doi.org/10.1080/07370016.2017.1304148

Brien, S. C. (2017). *Citizen science: Opportunities for girls' development of science identity*. Masters dissertation, University of California.

Brossard, D., Lewenstein, B. V., & Bonney, R. (2005). Scientific knowledge and attitude change: The impact of a citizen science project. *International Journal of Science Education, 27*(9), 1099–1121. https://doi.org/10.1080/09500690500069483

Calabrese Barton, A. C., Kang, H., Tan, E., O'Neill, T. B., Bautista-Guerra, J., & Brecklin, C. (2013). Crafting a future in science: Tracing middle school girls' identity work over time and space. *American Educational Research Journal, 50*(37). https://doi.org/10.3102/2F0002831212458142

Calabrese Barton, A. C., & Tan, E. (2010). We be burnin'!: Agency, identity and science learning. *Journal of the Learning Sciences, 19*(2), 187–229. https://doi.org/10.1080/10508400903530044

Carlone, H. B., & Johnson, A. (2007). Understanding the science experiences of successful women of color: Science identity as an analytic lens. *Journal of Research in Science Teaching, 44*(8), 1187–1218. https://doi.org/10.1002/tea.20237

Carlone, H. B., Scott, C. M., & Lowder, C. (2014). Becoming (less) scientific: A longitudinal study of students' identity work from elementary to middle school science. *Journal of Research in Science Teaching, 51*, 836–869. https://doi.org/10.1002/tea.21150

Carlone, H. B., Huffling, L. D., Tomasek, T., Hegedus, T. A., Matthews, C. E., Allen, M. H., & Ash, M. C. (2015). 'Unthinkable' selves: Identity boundary work in a summer field ecology enrichment program for diverse youth. *International Journal of Science Education, 37*(10), 1524–1546. https://doi.org/10.1080/09500693.2015.1033776

Cobb, P., & Hodge, L. L. (2011). Culture, identity, and equity in the mathematics classroom. In A. Sfard, K. Gravemeijer, & E. Yackel (Eds.), *A journey in mathematics education research* (Vol. 48, pp. 179–195). Springer.

Dixon, C. (2018). *Stepping up: Tailoring contribution and learning in youth-based citizen and community science*. Doctoral dissertation. University of California.

Dixon, C., & Martin, L. (2017). Make to relate: Analyzing narratives of community practice. *Cognition and Instruction, 35*(2), 103–124. https://doi.org/10.1080/07370008.2017.1282484

Dixon, C., Harris, E. M., & Ballard, H. L. (2021, April 8) *Bravery & vulnerability in youth community science*. Conference paper. Annual Meeting of the American Educational Research Association.

Engle, R. A., Lam, D. P., Meyer, W. S., & Nix, S. E. (2012). How does expansive framing promote transfer? Several proposed explanations and a research agenda for investigating them. *Educational Psychologist, 47*(3), 215–223. https://doi.org/10.1080/00461520.2012.695678

Gresalfi, M., & Hand, V. M. (2019). Coordinating situated identities in mathematics classrooms with sociohistorical narratives: A consideration for design. *ZDM, 51*(3), 493–504. https://doi.org/10.1007/s11858-019-01034-y

Harris, E. M., & Ballard, H. L. (2021). Examining student environmental science agency across school science contexts. *Journal of Research in Science Teaching, 58*(6), 906–934. https://doi.org/10.1002/tea.21685

Harris, E. M., Dixon, C., Bird, E. B., & Ballard, H. L. (2020). For science and self: Youth interactions with data in community and citizen science. *Journal of the Learning Sciences, 29*(2), 224–263. https://doi.org/10.1080/10508406.2019.1693379

Haywood, B. K. (2014). A "sense of place" in public participation in scientific research. *Science Education, 98*, 64–83. https://doi.org/10.1002/sce.21087

Heath, S. B. (2004). Risks, rules, and roles: Youth perspectives on the work of learning for community development. In A. N. Perret-Clermont, C. Pontecorvo, L. B. Resnick, T. Zittoun, & B. Burge (Eds.), *Joining society: Social interaction and learning in adolescence and youth* (pp. 41–70). Cambridge University Press.

Holland, D., Lachicotte, W., Skinner, D., & Cain, C. (1998). *Identity and agency in cultural worlds*. Harvard University Press.

Huffling, L. D. (2015). *Critical environmental agency in a field ecology program*. Doctoral dissertation. Michigan State University.

Jackson, C., Osterlund, C., Crowston, K., Mugar, G., & Hassman, K. D. (2015). Motivations for sustained participation in citizen science: Case studies on the role of talk. *International Conference on System Sciences, 48*, 1624–1634. https://doi.org/10.1109/HICSS.2015.196

Kudryavtsev, A., Stedman, R. C., & Krasny, M. E. (2012). Sense of place in environmental education. *Environmental Education Research, 18*(2), 229–250. https://doi.org/10.1080/13504622.2011.609615

Lemke, J. L. (1990). *Talking science: Language, learning, and values*. Ablex Publishing Corporation.

Lim, M., & Barton, A. C. (2006). Science learning and a sense of place in an urban middle school. *Cultural Studies of Science Education, 1*(1), 107–142. https://doi.org/10.1007/s11422-005-9002-9

Merenlender, A. M., Crall, A. W., Drill, S., Prysby, M., & Ballard, H. (2016). Evaluating environmental education, citizen science, and stewardship through naturalist programs. *Conservation Biology, 30*(6), 1255–1265. https://doi.org/10.1111/cobi.12737

Moss, D. M., Abrams, E. D., & Kull, J. A. (1998). Can we be scientists too? Secondary students' perceptions of scientific research from a project-based classroom. *Journal of Science Education and Technology, 7*(2), 149–161. https://doi.org/10.1023/A:1022564507639

Nasir, N. I. S., & Hand, V. (2008). From the court to the classroom: Opportunities for engagement, learning, and identity in basketball and classroom mathematics. *The Journal of the Learning Sciences, 17*(2), 143–179. https://doi.org/10.1080/10508400801986108

National Research Council. (2009). *Learning science in informal environments: People, places, and pursuits*. The National Academies Press.
Okita, S. Y., & Schwartz, D. L. (2013). Learning by teaching human pupils and teachable agents: The importance of recursive feedback. *Journal of the Learning Sciences, 22*(3), 375–412. https://doi.org/10.1080/10508406.2013.807263
Patterson, A. D. (2019). Equity in groupwork: The social process of creating justice in a science classroom. *Cultural Studies of Science Education, 14*, 361–381.
Patton, M. Q. (2002). *Qualitative evaluation and research methods* (3rd ed.). Sage.
Pickering, A. (1995). *The mangle of practice: Time, agency, and science*. University of Chicago Press.
Rahm, J., Miller, H. C., Hartley, L., & Moore, J. C. (2003). The value of an emergent notion of authenticity: Examples from two student/teacher-scientist partnership programs. *Journal of Research in Science Teaching, 40*(8), 737–756. https://doi.org/10.1002/tea.10109
Rogoff, B. (1995). Observing sociocultural activity on three planes: Participatory appropriation, guided participation, and apprenticeship. In J. Wertsch, P. Rio, & A. Alvarez (Eds.), *Sociocultural studies of mind: Learning in doing: Social, cognitive and computational perspectives* (pp. 139–164). Cambridge University Press.
Sfard, A., & Prusak, A. (2005). Telling identities: In search of an analytic tool for investigating learning as a culturally shaped activity. *Educational Researcher, 34*, 14–22. https://doi.org/10.3102/2F0013189X034004014
Sheth, M. J. (2019). Grappling with racism as foundational practice of science teaching. *Science Education, 103*, 37–60. https://doi.org/10.1002/sce.21450
Shirk, J. L., Ballard, H. L., Wilderman, C. C., Phillips, T., Wiggins, A., Jordan, R. C., McCallie, E., Minarchek, M., Lewenstein, B. V., Krasny, M. E., & Bonney, R. (2012). Public participation in scientific research: A framework for deliberate design. *Ecology and Society, 17*(2), 29–48.
Solórzano, D. G., & Yosso, T. J. (2001). Critical race and LatCrit theory and method: Counter-storytelling. *International Journal of Qualitative Studies in Education, 14*(4), 471–495. https://doi.org/10.1080/09518390110063365
Stepenuck, K., & Green, L. (2015). Individual-and community-level impacts of volunteer environmental monitoring: A synthesis of peer-reviewed literature. *Ecology and Society, 20*(3), 19.
Stroupe, D. (2017). *Reframing science teaching and learning: Students and educators co-developing science practices in and out of school*. Routledge.
Tan, E., & Calabrese Barton, A. (2008). Unpacking science for all through the lens of identities-in-practice: The stories of Amelia and Ginny. *Cultural Studies of Science Education, 3*(1), 43–71. https://doi.org/10.1007/s11422-007-9076-7
Theobald, E. J., Ettinger, A. K., Burgess, H. K., DeBey, L. B., Schmidt, N. R., Froehlich, H. E., Wagner, C., Hille Ris Lambers, J., Tewksbury, J., Harsch, M. A., & Parrish, J. K. (2015). Global change and local solutions: Tapping the unrealized potential of citizen science for biodiversity research. *Biological Conservation, 181*, 236–244. https://doi.org/10.1016/j.biocon.2014.10.021
Tokunow, M. (2016). *Re:Coding conservation: A field project by Miles Tokunow*. Unpublished report. https://www.academia.edu/29398304/Re_Coding_Conservation_A_Field_Project_by_Miles_Tokunow
Tzou, C., Scalone, G., & Bell, P. (2010). The role of environmental narratives and social positioning in how place gets constructed for and by youth. *Equity & Excellence in Education, 43*(1), 105–119. https://doi.org/10.1080/10665680903489338
Van Horne, K., & Bell, P. (2017). Youth disciplinary identification during participation in contemporary project-based science investigations in school. *Journal of the Learning Sciences, 26*(3), 437–476. https://doi.org/10.1080/10508406.2017.1330689
Varelas, M., Tucker-Raymond, E., & Richards, K. (2015). A structure-agency perspective on young children's engagement in school science: Carlos's performance and narrative. *Journal of Research in Science Teaching, 52*(4), 516–529. https://doi.org/10.1002/tea.21211

Wallace, D. E., & Bodzin, A. M. (2017). Developing scientific citizenship identity using mobile learning and authentic practice. *Electronic Journal of Science Education, 21*(6), 46–71.
Williams, K. A., Hall, T. E., & O'Connell K. (2021). Classroom-based citizen science: impacts on students' science identity, nature connectedness, and curricular knowledge. *Environmental Education Research, 27*(7), 1037–1053. https://doi.org/10.1080/13504622.2021.1927990
Yin, R. K. (2013). *Case study research: Design and methods* (5th ed.). Sage.

Part V
Multi-layered Methodological Approaches to Science Identities

Chapter 13
Using Qualitative Metasynthesis to Understand the Factors That Contribute to Science Identity Development Across Contexts in Secondary and Post-Secondary Students from Underrepresented Groups

Sylvia M. James Butterfield and Karen Benn Marshall

13.1 Introduction

The literature on student identity has expanded considerably over the last 20 years to include a multiplicity of meta-analyses, literature reviews, metastudies, and other types of syntheses focused on identity and academic achievement. It is no coincidence that studies of science learning, achievement, and persistence, especially those focused on groups underrepresented and underserved in "science, technology, engineering, and mathematics" or STEM, now take into account the role of non-cognitive factors such as interest, motivation, and identity. In an attempt to understand the persistent achievement gap in K-12 schools, as well as the decrease in the number of women and individuals from underrepresented groups that are earning STEM degrees in some fields, syntheses provide insight into the complex array of factors that come into play. The majority of studies, however, focus primarily on the synthesis of quantitative research.

Over the last 20 years there has also been a proliferation in the number of studies focused on understanding the role of identity in science. These studies help to highlight the important role of personal and social identity or how multiple identities intersect, when one is entering a field that has long been portrayed as the domain of men and not women or where few well-known role models exist for young people of color. There is still a lack of research that looks beyond quantitative studies to understand how to cultivate an identity that enables students to enter into science

S. M. James Butterfield (✉)
National Science Foundation, Alexandria, VA, USA
e-mail: smbutter@nsf.gov

K. B. Marshall
Oakwood University, Huntsville, AL, USA

H. T. Holmegaard, L. Archer (eds.), *Science Identities*, Contributions from Science Education Research 12, https://doi.org/10.1007/978-3-031-17642-5_13

communities of practice in secondary school, persist to a point that enables successful navigation of undergraduate and graduate environments, and gain entry to academia, research, or other STEM rich careers. A qualitative metasynthesis is an ideal approach to delve more deeply into studies on science identity because this type of scientific inquiry uses the findings from qualitative research studies as primary data, which may result from a range of qualitative methodologies. The research findings comprise the metasynthesis "sample" and the language in the findings is intensely reviewed, integrated, and then re-interpreted in a manner that offers new perspectives on the findings as a whole (Sandelowski & Barroso, 2007).

The number of qualitative studies that examine the role of science identity in students from underrepresented groups is limited but the rich findings allude to quintessential aspects of identity development that are not always apparent in singular quantitative studies. Qualitative researchers utilize multiple approaches to study science identity development that reveal aspects of competence, performance, and recognition in science using ethnography, phenomenology, interviews, and case studies (Carlone & Johnson, 2007). It is not unusual to see studies that use STEM as a proxy for "science" and address multiple disciplines (Adams et al., 2014; García et al., 2019). Science identity may emerge in the findings even if it was not the primary focus of the research questions (Brand et al., 2006). Also, mixed-methods studies include designs that are qualitative dominant to allow for a closer examination of science identities (Archer et al., 2015). While this metasynthesis necessarily defines the parameters for inclusion, those parameters take into account the variability that is represented in qualitative studies that examine science identity.

Accordingly, the study aim is to explore the findings from qualitative studies that address science identity in secondary and postsecondary students from underrepresented groups across multiple science learning contexts, including those from African American, Black, Hispanic, Latinx, and Native American backgrounds. This qualitative metasynthesis seeks to answer the following research questions:

(a) What is science identity and what are the factors that impact science identity development in secondary and post-secondary students from underrepresented groups in varied learning contexts?
(b) How is science learning helped or hindered by science identity development in varied learning contexts?

13.2 Theoretical Framework

In this chapter, the framework is informed by identity and social identity theory, science identity, and intersectionality. Several attempts have been made to operationalize the term identity which addresses the role of the individual in relationship to others. Gee (2001) describes identity as "the kind of person one is recognized as being," which emphasizes the importance of the social context and the opinion of other individuals in shaping identity (p. 99). The theoretical framework for the qualitative metasynthesis integrates the social contexts in which science identity develops and includes three main components: social identity theory, science identity, and intersectionality.

13.3 Science Identity as a Social Identity

Social identity theory has its origins in the work of Henri Tajfel (1974) who defined social identity as an individual's "knowledge of his membership as part of a social group (or groups) together with the emotional significance attached to that membership" (p. 69).

Social identity theory is based in social psychology and explores the self in relationship to group membership and group behavior (Hogg, 2003; Wellman, 1971). Categories of group membership traditionally included in social identity may include racial, ethnic, religious, political, gender, and personal affiliations, as well as vocations and avocations (Kernis & Goldman, 2003). Science identity is considered a type of social identity because it represents a group membership and affiliation with those who engage in or seek to engage in science study, practice science, or work in a science or STEM field (Brickhouse, 2001; Brickhouse & Potter, 2001; Stets et al., 2017).

According to Tajfel (1974), social identity and the desire to be associated with a particular group, is always considered in comparison to other groups that are similarly desirable in status or alternatively, less attractive. Group membership may have a positive impact on an individual and if so, they will elect to remain part of the group or if dissatisfaction ensues, then the individual will leave the social group. If leaving is too difficult, then one may accept the status quo, redefine the group status, or seek to change it (Tajfel, 1974; Tajfel & Turner, 2004). Social identity theory also infers that membership or affiliation with a particular social group has a major influence on the behavior of the individuals that are part of the group (Hogg, 2003; Tajfel, 1974).

Because of the prominent role of race and culture in identity construction, and the emphasis on science learning in group settings, this broader, more inclusive view of identity that acknowledges the impact of multiple social groups is a fitting framework. Also, social identity accommodates the postmodern view which holds that identity is dynamic and malleable as opposed to the traditional view that the "final identity" is set at the end of adolescence (Erikson, 1968). The postmodern view purports that identity is not stagnant, but constantly evolving. Individuals have multiple identities that are constructed in relationship to social interactions associated with people, practices, and institutions. In this sense, science identity, like all social identities, is dependent on the social, cultural, and historical context, and one may have different identities that are congruent with diverse sociocultural contexts (Nasir & Saxe, 2003).

13.4 Dimensions of Science Identity

Science identity develops in a variety of social contexts such as classrooms, labs, study groups, professional meetings, and job settings. It is nurtured (or hindered) by peers, friends, family, teachers, and "important science others" (Remich et al., 2016). Carlone and Johnson (2007) provide a much-used model for defining science identity, which is an essential component of the theoretical framework. This model consists of three dimensions which are used to describe science identity. The first dimension is competence, which refers to science content knowledge and understanding of concepts and processes. The second dimension is performance or the ability to do science or engage in science practices using tools and language in appropriate social settings such as classrooms, laboratories, and professional environments. Finally, recognition, the third and most important dimension of the science identity model, refers to acknowledgement or validation by others. Validation may come from teachers, peers, or other science professionals, confirming that one is indeed a "science person" as well as self-recognition that one is part of the science enterprise.

While the authors acknowledge that there is some overlap among the three dimensions, they contend that the value of this model is that it recognizes that science identity is not developed in isolation of other social identities such as race, ethnicity, or gender, but concurrently with other social identity groupings which in turn impact science identity development. Additionally, this model defines science identity broadly to include life, physical, and social sciences and consequently is often used to support studies of science and/or technology, engineering, and mathematics identity. The recognition that science identity interacts with other social identities is significant and also the reason that intersectionality has garnered so much attention in recent years.

13.5 Intersecting Identities: Power Struggle or Opportunity for Agency?

Intersectionality, the final component of the theoretical framework, is frequently used to address the consequences that result when multiple social identities are deemed inferior (Crenshaw, 1991). Intersectionality has its origins in a landmark case which focused on racial and gender discrimination against Black women. This framework has been embraced across various disciplines to explain how multiple social identities overlap to intensify discrimination against women and racial and ethnic minorities. The term intersectionality therefore embraces systems of power and identity and aptly describes how discrimination and inequality can be compounded and negatively impact individuals who possess multiple social identities that are targets of discrimination. Crenshaw (1991) notes that this extends to "structural, political, and representational" experiences as well. Structural intersectionality refers to organizational structures, while representational refers to cultural representation. Political intersectionality highlights the lack of representation of women of color in political agendas that espouse antiracism, feminism, or both. Science identity, however, frequently incorporates both the structural and representational aspects of intersectionality.

Next, two examples of how intersectionality can enhance studies of social identity and science identity in particular are provided. First, in the book, *Cultural constructions of identity, meta-ethnography and theory*, Urrieta and Noblit (2018) discuss how identity and intersectionality are linked in multiple types of identity, as well as, whether or not intersectionality is intentionally addressed as a critical aspect of an identity study. These authors note that identity is ultimately about power. Consequently, even if science identity is defined as competence, recognition, and performance (Carlone & Johnson, 2007), generally, it is someone from the racial majority who has the power to *judge* competence, *bestow* recognition, or *provide* access to classes or labs that enable students to perform or carry out scientific activities. The authors argue that an intersectional approach "disrupts" the binary or oppositional identity approach, and incorporates other ways of knowing related to racial, cultural, and gender identities. Traditional identity theory forces a choice between either/or, Black or White, scientist or non-scientist, competent or incompetent, who does science or who does not do science, since this approach says *you are who the other is not.*

Intersectionality, on the other hand, requires an examination of complex identities and relationships of power (Urrieta & Noblit, 2018). This perspective suggests that some level of agency is restored when an intersectional approach is applied, which in the case of underrepresented students, enables them to choose which social identities they wish to embrace rather than just accepting a label that has been imposed by society. Some scholars also contend that in addition to agency, one must also be attentive to social structures which also intersect and are inherently based in culture (Phoenix, 2006).

In the second example, Sparks (2017) uses an intersectional lens to highlight the importance of applying this analytical approach to understand the complex identity relationships that ensue when female students of color are engaged in science studies. He cautions that students may become victims of what he refers to as the intersectional trap. The trap occurs when students feel compelled to conform to behaviors associated with majority groups (e.g., in science or mathematics classes, labs, or clubs) or when others make assumptions about Black women and girls when they attempt to join science groups. The impact of the intersectional trap may be that Black women may feel unwelcome, wholly adopt the behaviors of the science majority group, or reject or suspend their other social identities, including gender, racial, ethnic, or cultural identity to fit into the group (Sparks, 2017). While Sparks does not contend that all diverse racial and ethnic groups are subject to the intersectional trap, the early studies on social identity suggest that the desire for group membership can generate similar behaviors (Tajfel, 1974; Tajfel & Turner, 2004). Sparks' (2017) intersectional trap and Urrieta and Noblit's (2018) explanation of the relationship between identity, power, and agency, emphasize the benefit of incorporating intersectionality as part of the framework for an examination of science identity to fully understand the role of the social context in which science identity develops.

13.6 Theoretical Framework Summary

This qualitative metasynthesis is designed to provide a deeper understanding of how science identity development occurs in secondary and post-secondary students from underrepresented groups in multiple learning contexts. The theoretical framework, composed of social identity theory, Carlone and Johnson's (2007) science identity model, combined with an intersectionality lens, will examine the multiple factors which appear to impact science identity development. By grounding the study in social identity, science identity is better understood as one type of group membership that impacts the behavior of secondary and post-secondary students from underrepresented groups. Underrepresented students may be positively impacted by science identity, in which case they may chose to remain in the social settings that support its development (e.g., science classes, study groups, cohorts, undergraduate or graduate studies). However, choosing to stay or even choosing to leave may result in behavioral changes, not all of which are positive, such as the "intersectional trap" which may force a choice between science identity and racial and ethnic identity (Sparks, 2017; Tajfel, 1974; Tajfel & Turner, 2004). Intersectionality is a reminder that social identities are rarely considered in isolation and are never entirely neutral for women and people of color. The inclusion of intersectionality in the framework ensures that agency, the power and the powerlessness engendered by science settings, as well as the associated cultural and organizational structures, are considered in the interpretation of the metasynthesis results (Crenshaw, 1991; Phoenix, 2006; Urrieta & Noblit, 2018). The findings will underscore the significance of science identity as an essential component of social identity for students from underrepresented groups in science and how science identity development is facilitated or impeded.

13.7 Methodology

Conducting a qualitative metasynthesis is a systematic and thorough process that results in "the integration of research findings" allowing for a deeper understanding of a collection of studies (Sandelowski & Barroso, 2007). Erwin et al. (2011) outlined six steps for conducting a metasynthesis:

(a) Identification of the research question;
(b) Completion of a literature review;

(c) Development of inclusion criteria;
(d) Identification of the metasynthesis approach;
(e) Presentation of findings, and;
(f) Researcher self-reflection.

The next section describes each step and the associated outcomes.

13.8 Research Questions and the Literature Review

The first step is the identification of the research questions, along with a rationale for undertaking these questions. To summarize, the aim of the study is to synthesize the results from qualitative studies that address science identity in secondary and postsecondary students from underrepresented groups in a variety of science learning contexts, including academic and non-academic settings. The first research question seeks to understand how science identity is defined and identify the factors that impact science identity development in secondary and post-secondary students from underrepresented groups. The second question focuses on how science learning is helped or hindered by science identity development.

For the second step in the metasynthesis, a comprehensive search of the literature was conducted to identify qualitative studies that address identity across multiple science learning contexts including secondary and post-secondary classrooms, laboratories, study groups, and enrichment programs conducted in academic and community settings. Enrichment programs in community settings generally include after-school and summer activities taking place in museums, science centers, community centers, and other informal science learning settings. The search process took into account the fact that the number of qualitative studies focusing on science identity development in racially and ethnically diverse populations is limited. Studies of secondary and post-secondary students from underrepresented groups as a subset of an already sparse pool, if synthesized, may offer insights on the importance of science identity development in this population and ideas for future studies to grow the knowledge base.

The literature search used the following terms either individually or in various combinations: identity, science identity, STEM identity and underrepresented students and/or science, technology, engineering, and mathematics. Searches using different terms to describe students from groups that are underrepresented in STEM were conducted. This included terms such as *minority* and *diverse* and terms encompassing various racial and ethnic groups that are denoted as being statistically underrepresented in STEM (e.g., Black, African American, Latino, Hispanic, Native American, girls, female). Terms for the various educational levels included high school, secondary, post-secondary, and undergraduate. Psychological terms such as motivation, interest, engagement, and persistence were also used, because some qualitative studies examine multiple variables associated with the participation in science by students from underrepresented groups. The electronic databases used included JSTOR, PsychInfo, ERIC, Wiley, and NIH Public Access. Secondary manual searches were conducted using the reference sections of articles including literature reviews and synthesis papers, as well as dissertations, citations, footnotes, and author searches. A screening form was developed to assist with the identification of the criteria that were used to include or exclude studies from the metasynthesis, as well as to assess study quality.

13.9 Identification of Inclusion Criteria

The third step in the metasynthesis process, development of the inclusion criteria, was informed by the research questions, the focus on science identity development, and the literature review (Erwin et al., 2011). The screening form categorized information on the study goals, research questions, science subjects, conceptual or theoretical framework, research design, unit of analysis, methods, data analysis, and results. Inclusion criteria, as indicated on the screening form, included primary literature, consisting of peer reviewed studies published in 2000 or later, focused on science identity development in secondary or post-secondary students from underrepresented groups. Additionally, all studies had to use validated qualitative research methods. Due to the paucity of qualitative studies in this area, a broad definition of science identity was used for inclusion. For example, in the course of the literature review, it became apparent that in many studies, the term "science" is often used interchangeably with STEM. Only studies that specified the science discipline or STEM content that students were studying were included in the metasynthesis. Also, "identity development" associated with science learning may not always be clearly defined as science identity, although studies commonly refer to science identity development in the literature review, results, and discussion, and may also incorporate multiple disciplines.

Studies had to target students from underrepresented groups or include students in a secondary school that served primarily racially and ethnically diverse populations in a community education (such as afterschool enrichment programs) or post-secondary educational setting. Exclusion criteria consisted of quantitative studies and mixed-methods studies with quantitative dominant designs. As previously noted, while there is a plethora of quantitative studies on identity, these authors believed that it was beneficial to examine findings in qualitative studies to better understand elements of science identity development. Additionally, critical reviews, literature reviews, position papers, dissertations, and synthesis papers were excluded, as well as policy documents and book reviews.

This process resulted in 26 articles of which 18 were qualitative studies which met the inclusion criteria after being further assessed for study quality drawing on established criteria for credibility (Lincoln & Guba, 1985; Shenton, 2004). In addition to using peer review as a criterion for study quality, the study design, methods, and the credibility of the data and findings were also considered. All studies met key standards for credibility such as the use of valid qualitative research methods, in addition to employing triangulation, member checks, critical self-reflection, detailed audit trails, and peer review to ensure validity and reliability (Lincoln & Guba, 1985; Merriam, 2002; Shenton, 2004). In total, the articles represent over 360 participants ($n = 364$).

13.10 Analysis

The researchers utilized the screening form to develop a typology of the 18 qualitative studies which highlights and summarizes key study components, as well as subject demographics (as described in the qualitative studies themselves) and research questions (see Table 13.1). All studies were read at least two times. The initial reading was used to complete the screening form and determine if basic inclusion criteria were met. A second reading along with any subsequent readings, was used to ensure that the key components were accurately depicted in the typology and to identify concepts and categories for coding. The researchers also discussed each article and had to agree on the definition of identity and findings related to science identity development in order for it to be included.

The metasynthesis approach, which is the fourth step, outlined by Erwin et al. (2011), utilizes modified open coding techniques adapted from grounded theory to identify overarching concepts from the findings in each of the 18 studies using an inductive approach (Charmaz, 2006). To obtain the major concepts, the findings were first grouped according to similarity in topics. The articles were subsequently mined for participant quotes which were also grouped according to similarity in topics that resulted from the groups of study findings. The following 11 categories resulted from grouping the findings and quotes: stereotypes, systems, and policies, microaggressions, social group rejection, persistence, competence, performance, recognition, role models, novice/emerging science identity, social group inclusion, and expert. After this process, three overarching concepts emerged from the coding (see Figs. 13.1, 13.2, and 13.3):

(a) Support for science identity development and science learning;
(b) Barriers/obstacles to science identity development and science learning, and;
(c) Evolving science identity trajectories.

Table 13.1 Characteristics of studies used in the qualitative metasynthesis

Author(s)		Year	Race/Ethnicity	Educational level	N	Methods	Study questions/Goal
1.	Adams, Gupta & Cotumaccio	2014	African American Latina Southeast Asian	Secondary	$n = 6$	Focus groups and interviews	In what ways does long-term participation in OST science programs shape the interest, motivation, and ability of young women of color to pursue and persist in STEM majors?
2.	Adjapong, Levy & Edmin	2016	Haitian American and Latina	Secondary	$n = 3$	Interviews and journals	Does the participation in an authentic science internship affect the science identity of girls of color? Does the participation in an authentic science internship influence the pursuit of careers in STEM of girls of color?
3.	Archer, Dewitt & Osborne	2015	African/Caribbean	Secondary	$n = 10$	Interviews	Why are black students less likely to express aspirations for careers in science? What makes science careers less "thinkable" (conceivable and achievable)? And what enables some students to buck the trend and "aspire otherwise"?
4.	Brand, Glasson & Green	2006	African American	Secondary	$n = 5$	Interviews	How do African American students define their experiences in science and mathematics classes?
5.	Brown	2004	African American Hispanic Vietnamese	Secondary	$n = 35$	Ethnography	What types of discursive identities are constructed as students attempt to assimilate into the culture of science classrooms?

6.	Burt, Williams & Smith	2018	Black	Graduate	$n = 21$	Semi-structured interviews	What ecological and sociological barriers within a College of Engineering promote non-normative student role strain for black male students pursuing graduate engineering degrees? How do these ecological and sociological barriers influence black males' performance in engineering graduate programs?
7.	Carlone & Johnson	2007	Latina, Black & Native American	Under-graduate	$n = 15$	Ethnography	How do successful women of color negotiate and make meaning of their science experiences? How do women of color develop and sustain their science identities throughout their undergraduate and early science careers? What is the relationship between the women's science identities and racial, ethnic, and gender identities?
8.	Chapman & Feldman	2017	Hispanic, Black White Asian	Secondary	$n = 12$	Interviews and journals	Did students perceive their experience as one of authentic science? How did participation of students in an authentic science experience affect their identities as scientists and perceptions about scientists?

(continued)

Table 13.1 (continued)

Author(s)		Year	Race/Ethnicity	Educational level	N	Methods	Study questions/Goal
9.	Gazley, Remich, Naffziger-Hirsch, Keller, Campbell & McGee	2014	Black, Latina, Native American	Graduate	$n = 52$	Semi-structured interviews	Why do some college graduates from underrepresented populations who aspire to biomedical research careers enter a PREP program rather than go to graduate programs directly from college? What do these entering PREP scholars expect from their postbaccalaureate experience to facilitate their education and career decision making?
10.	Gibau	2015	African American	Graduate	$n = 18$	Phenomenology	What were the experiences of the underrepresented minority (URM) students? Which experiences were most critical to their persistence in their programs? And how can these experiences potentially inform and enhance intervention strategies?
11.	Gudyanga	2016	Zimbabwean	Secondary	$n = 9$	Semi-structured interviews; classroom observations	What factors of identity formation could be considered as contributing to developing an orientation to physics by female students?
12.	Haun-Frank	2011	African American	Secondary	$n = 14$	Semi-structured interviews, informal interviews, and focus groups	What meanings of science, self, and science careers did students construct as they traversed various spaces along their trajectories? What critical spaces for identity work were evident in students' narratives? What role did these spaces play in their science trajectories?

13.	Malone & Barabino	2009	African American Caribbean African Other	Under-graduate & Graduate	$n = 24$	Focus groups and interviews	How are minority students positioned within interactions in terms of race and science community?
14.	McGee, Griffith & Houston	2019	African American	Doctoral and post-doctoral students	$n = 48$	Focus groups and interviews	How do black engineering and computing doctoral students and postdoctoral researchers describe and make sense of the stressors and strains in their training programs? How do black engineering and computing doctoral students and postdoctoral researchers cope with the stressors and strains in their training programs?
15.	Remich, Naffziger-Hirsch, Gazley & McGee	2016	African American, Hispanic/ Latino Native American Asian White/ non-Hispanic	Under- graduate, post-baccalaureate, graduate	$n = 48$	Interviews	To understand how identity as a graduate student and future scientist develops.
16.	Sayman	2013	Latinas	Secondary & post-secondary	$n = 10$	Focus groups and interviews	What are the experiences of Latinas in state-supported residential STEM schools? What factors are involved in recruitment and retention of this population?
17.	Sriram, Rishi & Diaz	2016	Hispanic, African American, East Asian, Southeast Asian	Under-graduate	$n = 8$	Phenomenological case study	To address gaps in the literature in living learning programs; lack of studies on students of color, how LLPs can promote success in postsecondary students
18.	Villa, Wandermurem, Hampton, & Esquinca	2016	Hispanic	Under-graduate	$n = 26$	Semi-structured interviews	What is the relationship among identity, resilience, and persistence in Latinas in computer science and engineering?

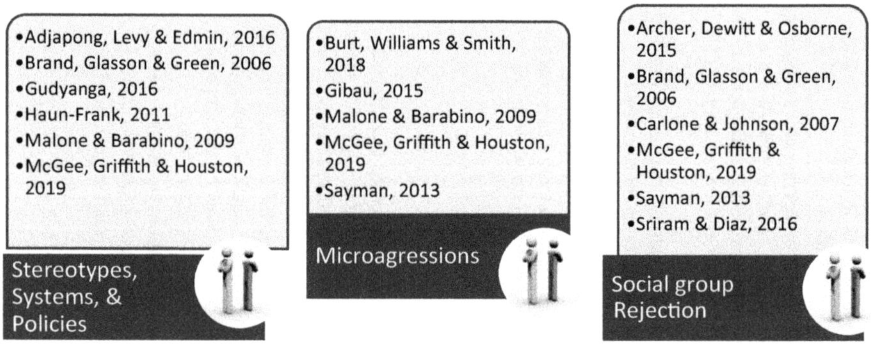

Fig. 13.1 Studies that address barriers to science identity development

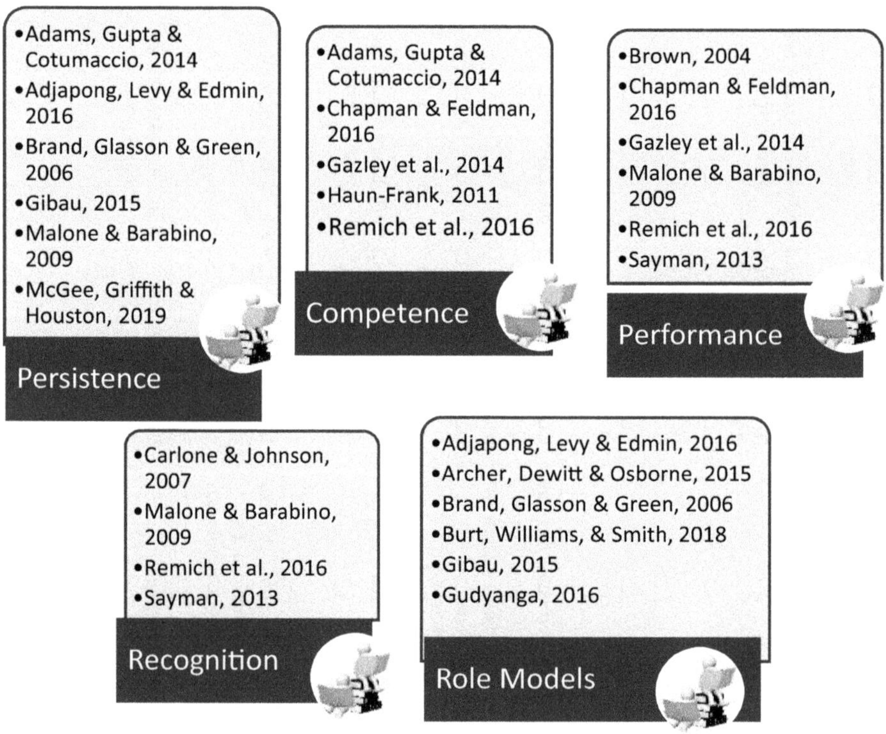

Fig. 13.2 studies that address support for science identity development and learning

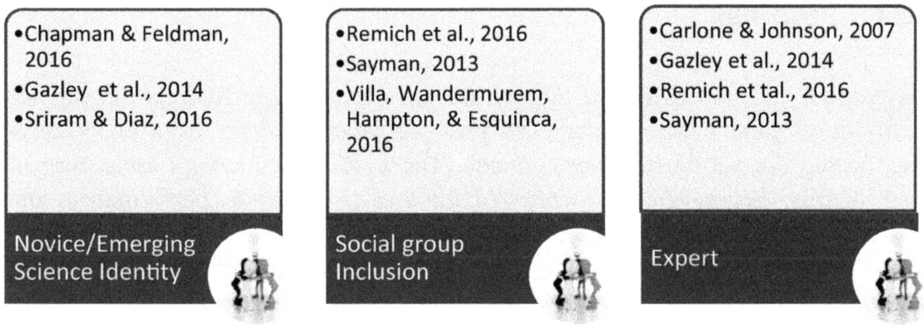

Fig. 13.3 Studies that address evolving science identity trajectories

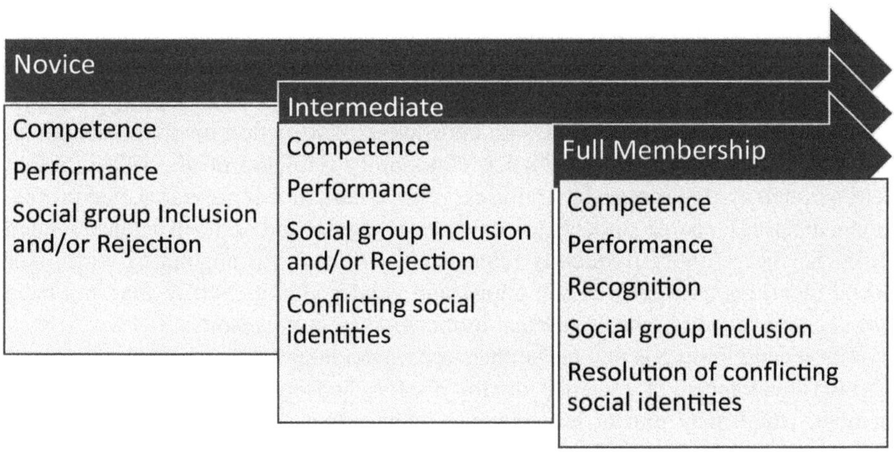

Fig. 13.4 Longitudinal science identity development trajectory

The overarching concepts align with the research questions by providing insight into the critical factors that impact science identity development for students from underrepresented groups and the manner in which critical factors influence learning in secondary and postsecondary settings. Finally, a science identity development trajectory (see Fig. 13.4) emerged from the studies which showed a possible pathway from novice to identity as a future scientist. The presence of a sequential longitudinal trajectory of important events during the process of science identity development is a novel and significant finding that requires further examination through an intersectional lens. It is probable that these findings can provide an understanding of how science identities develop over the course of secondary and postsecondary education and clarify strategies to avoid trajectories that impede underrepresented students' progress in STEM careers.

13.11 Findings

This purpose of this qualitative metasynthesis is to integrate findings on science identity development in secondary and post-secondary students from underrepresented groups in varied learning contexts. The guiding framework consisting of social identity theory, science identity (defined as competence, performance, and recognition), and intersectionality, also provided a mechanism for coding the findings (Carlone & Johnson, 2007; Crenshaw, 1991; Tajfel, 1974). The fifth step in the metasynthesis process is the presentation of the findings (Erwin et al., 2011). When looking across the set of qualitative studies included in the metasynthesis at the role of identity in STEM students' academic trajectories, three significant findings emerged: approaches to defining science identity, factors impacting science identity development, and longitudinal trajectories of science identity development.

First, the varied approaches to defining science identity in the collection of studies emerged as a distinct finding. The definition of identity most frequently focused on identity-in-practice, which occurred in a setting where science identity was fostered or inhibited in conjunction with knowledge construction involving socialization and enculturation into a science community (Holland et al., 1998). Social settings such as classrooms, laboratories, science enrichment programs, and professional meetings, enable students to converse, engage in and perform science related tasks. Science identity frequently intersected with and was negotiated with other social identities, primarily racial, ethnic and gender identity, which may not have been acknowledged in the theoretical frameworks or study results.

The second finding is that researchers reported a range of factors that either support the development of a science identity or act as barriers to science identity development, which may impact learning in a manner that is unique to students from underrepresented groups. The third set of findings suggests a somewhat consistent sequence of events linked to developing science identity trajectories. As shown in Fig. 13.4, there are different spheres of science identity formation that may be accessed during the process of negotiating science identity development. A range of trajectories are apparent as students move from juncture to juncture, sometimes becoming immobilized and sometimes advancing, reaching access points, utilizing supports, and encountering barriers along the way to acceptance or marginalization in science. The trajectories that students navigate during this process may affect science learning and engagement (Nasir & Hand, 2008). The findings are discussed in more detail in the next sections.

13.12 Defining Science Identity

The first research question asks, "What is science identity and what factors impact science identity development in secondary and post-secondary students from underrepresented groups in varied learning contexts?" The study findings that address definitions of science identity are described first and the next section will discuss factors impacting science identity development. Researchers in the 18 studies included in the metasynthesis used multiple definitions of identity. While these definitions ranged from individual to group identity and were derived from theories in anthropology, psychology and sociology, the definitions did not always describe science identity even when science identity was the focus of the study. Examples include affinity-identity (Adams et al., 2014), discursive identity (Archer et al., 2015; Brown, 2004), culturally situated models of identity (Carlone & Johnson, 2007; Chapman & Feldman, 2017; Villa et al., 2016), and social identity (Sayman, 2013). In some instances, multiple theoretical frameworks were combined from a variety of disciplines and identity was not defined in terms of science but rather as a component of a sociocultural theory such as communities of practice or identity-in-practice.

The utilization of multiple definitions of science identity highlights the complexity of identity and the challenge faced when attempting to synthesize studies. There is concern by some scholars that sociocultural approaches overlook race and ethnicity and that few studies consider how multiple social identities intersect with "practice-based identities." However, the incorporation of intersectionality into the metasynthesis theoretical framework, in addition to a culturally situated definition of identity, seeks to ensure that the impact of race, ethnicity and gender was taken into consideration (Esmonde et al., 2009). The impact of the intersection of science identity and other social identities emerged in the coding of participant quotes and is discussed in the next section on factors impacting science identity development.

13.13 Factors Impacting Science Identity Development

In response to the first research question, the findings revealed multiple factors that serve as barriers to science identity development as well as a host of factors that support science identity development. The barriers will be discussed first.

Three types of barriers to identity development emerged from the qualitative metasynthesis: *Stereotypes, systems, and policies; microaggressions; and social group rejection.* The term social group rejection refers to a situation in which students are not acknowledged or invited to join a casual science-related social group such as a study group or lab team. Six studies were identified with stereotypes, systems, and policy related findings, five with microaggressions, and six that address social group rejection (Fig. 13.1). Stereotypes may be gender specific when girls are made to feel unwelcome in a physics classroom or race-based when teachers infer that African-American males are less likely to succeed in STEM professions (Gudyanga, 2016; Haun-Frank, 2011). Beyond the impact of individual teachers, students may also feel affected by systemic factors and school policies. As an example of a stereotype, system, and policy issue, a Black female graduate student made the following comment:

> But it looks like the African American community is being more adversely affected by it. There are people who are getting screwed over because some of the policies are stupid and are very subjective. There are no concrete rules, but it just seems like those concrete rules—when you don't have rules, it's easy to bend those non-rules and it looks like the minority population is the ones getting hurt most. There's a lot of people getting hurt but when you look at the number of people graduating in a certain number of years, yeah, you're bringing in all of these African American students but how many of them are leaving with the degree they came for? How many people switched to a different department? Switched to an MBA? Left with a Master's prematurely? (Malone & Barabino, 2009, p. 20).

As an example of a microaggression, a Black male graduate student expressed:

> My advisor — he's — when he tells you stuff, it's kind of — so if you approach him, and he asks you questions, and it was like some fundamentals you don't know, or you were never taught software stuff — I don't want to say he belittles you, but he'll kind of be like, "yea, you learned this as a junior in undergrad, and the sophomores here are doing it." And I'm like, "ok" (Burt et al., 2018, p. 986).

Findings also suggest that it is not unusual for students to experience microaggressions and social group rejection simultaneously, as described by high school students who note that teachers assume that they are not interested or incapable of learning and provide little encouragement (Brand et al., 2006). Another example microaggressions and social group rejection was also described by an undergraduate student who recounted how majority professors appear to be more welcoming of White students as future colleagues while simultaneously making students of color feel like they are somehow trespassing in science settings (Carlone & Johnson, 2007).

As these findings show, barriers to science identity development are frequently manifested by stereotyping students from underrepresented groups as lacking in knowledge and competence, and the creation of systems and policies designed to exclude minorities and women (Carlone & Johnson, 2007). The pursuit of science and development of science identity therefore requires active negotiation for group membership by underrepresented students, while group membership is often readily accessible for many majority students. Students from underrepresented groups, however, are doubly penalized for their racial and ethnic identity, or for gender, if they are female (Crenshaw, 1991; Tajfel, 1974; Tajfel & Turner, 2004). Microaggressions are often used as a deterrent, to discourage continuation in science studies and negatively impact science identity development. However, because many students from underrepresented groups encounter multiple barriers in one setting, it is essential to understand the impact that microaggressions may have on science identity development. Although the intersection of science identity with racial, ethnic, and gender identities, coupled with microaggressions and social group rejection may seem discouraging, that is not the full story.

While several studies reveal barriers to science identity development, there is also comparable or even greater evidence of factors that support science identity development. Findings indicated that factors such as persistence, competence, performance, recognition, and role models support science identity development, which may act singly but frequently overlap. Figure. 13.2 indicates five studies in which students provided examples of competence, six referenced performance, four addressed recognition, six addressed persistence, and six studies described the value of role models for students from underrepresented groups. A few examples of these factors will be described.

This high school student's strong science identity is apparent as he expresses how his performance in the science lab made him feel competent: "It (the project) made me feel like an actual scientist, like it made you feel good. It made me feel like you could do it and showed you how to do it and stuff and get you prepared" (Chapman & Feldman, 2017, p. 477).

The findings suggest that recognition comes from multiple sources and may include peers and authority figures. In Sayman's (2013) study of Latinas in math and science schools, a female undergraduate student describes how she struggled to gain confidence in her ability to compete, but felt appreciated when her male peers accepted her as an equal:

> I was really nervous [in high school] actually. Because I was actually the only girl in Computer Maintenance 1, so I had, I felt intimidated, you know. But I also was excited about it, and also had this feeling of like, I'm a girl, I can beat all these guys. I can be just as good as them. So, it was really exciting, and like later on in the year they were really accepting of me, so that really helped me out, you know, working on the different projects and everything (Villa et al., 2016, p. 120).

The findings across many studies suggest not only the importance of recognition as a scientist, but also the need for acknowledgement by professors (Carlone & Johnson, 2007). While the Black female student (in the example below) felt isolated and marginalized, it was still important for her to be recognized and acknowledged by her principal investigator. In the excerpt, she explains her desire to prove her worth when describing how she applied to present at a professional workshop unbeknownst to her principal investigator:

> No, he didn't know that I applied. I just saw the poster and I applied, and I figured if I don't get it, then no one will know (laughs).... But I got it, so you know. But I mean, you know, part of it is, you know, you wanna do well and…and you want make those you know, whom you work for, proud for some reason (Malone & Barabino, 2009, p. 13).

It was therefore not surprising that most of the studies lamented the lack of role models for students of color, both male and female (Adjapong et al., 2016; Archer et al., 2015; Brand et al., 2006; Burt et al., 2018), and described sometimes painful experiences with professors who were more accustomed to Asian graduate students. An African American male graduate student who participated in a summer bridge program articulated the importance of having diversity among role models:

> Well, I guess coming in through the Bridges program and having the opportunity to meet other professors that went to Ivy League schools or came from a different system. It's good to meet other African American scientists that have made it. So, it's kind of like inspiration. So those are the highlights. And then to see students that I met as a summer student matriculate and get their PhDs, these are good things too, because it kind of gives you the inspiration that I can finish too (Gibau, 2015, p. 7).

Several qualitative studies also emphasized the value of bridge and cohort programs and same-sex groups which appear to dampen the effects of isolation. Hence, students from diverse racial and ethnic groups, and women could have access to peers who were from the same background, who provide friendship and support (Gibau, 2015; Sayman, 2013; Sriram & Diaz, 2016; Villa et al., 2016).

Finally, persistence was another factor that emerged in student quotations that contributed to science identity development. Some students realized that they persisted because of their passion for science and demonstrated ability to compete in rigorous academic environments, such as a young woman who had participated in a high school enrichment program and expressed "When I was in denial about science, I thought about how much I loved it at Lang, and it kept me going" (Adams et al., 2014, p. 19). Others, however, were determined to push beyond a constellation of negative barriers and stereotypes and succeed. When this occurred, the intersection of science identity with racial, ethnic, and gender identities ultimately culminated in a robust science identity, highlighting the importance of agency in science identity development for students from underrepresented groups (Urrieta & Noblit, 2018). However, the desire to overcome negative barriers also resulted in resentment and isolation for some students (Malone & Barabino, 2009).

13.14 Trajectories of Science Identity Development and Learning

The final set of findings from the metasynthesis address the second research question, "How is science learning helped or hindered by science identity development in varied learning contexts?" The findings suggest that an important relationship exists between science identity and learning. Specifically, when students have a strong identity that is tied to a particular context, such as a classroom or laboratory, then they may be more likely to be engaged in learning (Nasir & Hand, 2008). The metasynthesis studies listed in Fig. 13.3 revealed that science identities were constantly evolving and while learning continued, it took a while for students to express that they were indeed scientists, at any stage of development. These findings infer that there are different stages associated with science identity development.

Figure 13.4 shows the various stages exhibited by students as they transitioned from novice to an intermediate stage of science identity development to a point where they expressed self-recognition and were recognized by others as being part of the science enterprise with all of the benefits of full membership (Sandelowski & Barroso, 2007). For example, a student who might be characterized as being at the novice stage, stated, "right now I'm trying to find my niche" as a rationale for participating in a STEM program after graduation and obtaining experience working in a lab (Gazley et al., 2014). However, others evolved beyond this point such as a Black male student who noted:

> You design experiments. I got really into science there. Then [there's] a competition where the city congratulates the top [high school] students of chemistry in the county. I was one of the winners, and that's what really propelled me and let me know that this is something that I'm good at maybe I should pursue (Gazley et al., 2014, p. 1032).

This student felt competent, included, and was given the opportunity to engage in scientific research.

By contrast, other students, such as this African American male, who despite being in a science graduate program and obtaining recognition as a scientist from peers and professors, was still uncertain about his science identity. He could be identified as being at the intermediate stage of science identity development. He stated:

> Is science part of my identity? You know, I think it's starting to become part of my identity. I didn't think of it as my identity before, but now that you mention it, yeah, because, I mean I've done so much, considering that I recently was published. I think that kind of bolstered that identity, you know...but do I see myself as a scientist? In other people's eyes, yes. If I were to rephrase the question, do I see myself through my eyes as a scientist? No. Not at all. I don't know (Remich et al., 2016, p.10).

At the intermediate stage, students at the secondary and postsecondary level may vacillate between certainty and uncertainty as they attempt to balance other social identities such as race and gender, with a strong science identity (Gazley et al., 2014; Sayman, 2013; Sriram & Diaz, 2016; Villa et al., 2016). As students from underrepresented groups advance from novice to expert, they are exhibiting and gaining competence as they transition to full acceptance into a scientific community as indicated by recognition by others of one's standing as a future scientist or practicing scientist (Carlone & Johnson, 2007; Remich et al., 2016).

The emerging stages of the science identity development, depicted in Fig. 13.4, show how multiple studies captured students at various points along this trajectory. By conducting this metasynthesis, the authors were able to see the longitudinal nature of the trajectory, which might not otherwise be revealed (Sandelowski & Barroso, 2007). The studies suggest that in some cases, science identity development is incremental and takes time, although not all students persist and complete science studies or pursue STEM careers. Opportunities to engage in science practices and demonstrate their abilities were sufficient for some and spurred further learning and science identity development. However, findings from the previous section suggest that this may not be sufficient to advance to full membership in various science groups when multiple barriers interfere with science identity development. The intersection of science identity with other social identities must be considered at all stages in the trajectory and the longitudinal nature of the science identity development process must be addressed concurrently.

13.15 Discussion

The framework for this study which included social identity, science identity, and intersectionality, provided an appropriate lens for the interpretation of the metasynthesis findings. While others have pointed out that a common definition for science identity is difficult to pinpoint for researchers within the same domain and across domains, this study also highlights the importance of incorporating frameworks such as intersectionality to explain that even in studies where race and ethnicity are not the primary focus, these constructs are relevant in discussions of identity. The interface between race, ethnicity, and gender and science identity and learning are obligatory for all students of color. The manner in which the other social identities influence science identity development cannot be ignored since social interactions shape and mold the identity of students of color, including African Americans, Blacks, Hispanics, Latinx, and Native Americans. The pressure associated with thwarting racism is believed by some to induce a type of psychological stress dubbed as "role strain" that is unique to African American students who must manage a barrage of stereotypes with creative coping mechanisms (McGee et al., 2019).

The findings from this qualitative metasynthesis provide evidence that the elements of Carlone and Johnson's (2007) model of science identity are present in the science identity of secondary and postsecondary students from underrepresented groups. While all three components are important, recognition seems most impactful. The findings further suggest that recognition is needed from peers (peers similar to and from different racial and ethnic groups) and faculty (also in similar and different racial and ethnic groups), as well as any persons deemed "important science others" (Remich et al., 2016). Finally, self-recognition also emerged as being critical to science identity development.

It is important for definitions and discussions of science identity and intersectionality to address race, gender, and social identity and to incorporate the salience of other social identities. The findings suggest that secondary and post-secondary students are challenged to balance an array of social identities with their science identity. Agency certainly plays a role as students may opt to embrace multiple social identities, but agency alone may not be enough. Students may also encounter social structures that intersect with their identities, representing additional impediments to science identity development (Phoenix, 2006; Urrieta & Noblit, 2018).

The results of this qualitative metasynthesis resonate with the science identity elements of recognition, performance, and competence identified by Carlone and Johnson (2007). Self-recognition as a science person, in addition to recognition by peers and "important science others" was a recurring theme (Remich et al., 2016). Performance is also important, but students from underrepresented groups first need to gain access to opportunities that are controlled by social networks (such as Asian or White professors). It appears that competence alone is not sufficient for a strong science identity. As McGee (2015) notes, the racialized system repeatedly demands proof at every juncture that students from groups that are underrepresented in STEM are qualified to be members of the STEM enterprise. The STEM environment may cause African American students at all levels to question their ability to succeed, even when they have made great progress, which ultimately may impact their trajectory of science identity development. McGee et al. (2019) purports that "role strain" is partially the cause. Additionally, the findings suggest that supports such as family, community, personal, and science cohesion among peers are important. McGee (2015) asserts that it is important for scholars to tell stories of success, and not just highlight deficits and failure. Black, Latin and other students from racially and ethnically underrepresented groups must navigate a host of obstacles in STEM academic and professional environments and many demonstrate great resilience in the face of microaggressions and often outright racism.

This qualitative metasynthesis shows that non-cognitive factors such as identity in science are important for diverse racial and ethnic groups. Also, the social identities of the authors, who are both African American females with undergraduate and graduate degrees in science and terminal degrees in science education, bring a worldview to the integration and interpretation of the findings from studies that focus on racial and ethnic minorities.

Our conclusion is that the metasynthesis demonstrates that for secondary and post-secondary students from underrepresented groups, there is a longitudinal science identity development trajectory which may include advocacy and self-authorship, juxtaposed with negative imposed identities, that collectively contribute to a developing science identity. Hence, ideally, a high school or undergraduate student from an underrepresented group may perceive himself or herself as a scientist and a person who belongs in a chosen scientific community who possesses the appropriate knowledge and skills and is able to reconcile this perception with other social identities, while discarding imposed negative identities (Nealy & Orgill, 2019).

13.16 Limitations

This qualitative metasynthesis had limitations which may impact the findings. First, as noted, while there has been a noticeable increase in studies focused on identity development since 2000, there is a paucity of qualitative studies examining science identity development in students from underrepresented groups in secondary and post-secondary settings. Second, the inclusion of studies that used STEM identity as a proxy for science identity and qualitative dominant mixed-methods studies might dilute the findings.

13.17 Future Studies

By conducting a metasynthesis on qualitative studies that address identity among students underrepresented in STEM, insights have emerged on factors that impact science identity development and the relationship between science identity development and learning. One option for future research might be to look more closely at students from underrepresented groups who have successfully navigated the barriers to science identity development.

Many students encounter science instructors with the power to deny access and recognition needed to validate a developing science identity. Additional studies could provide a better understanding of how these students successfully exercised agency and intentionally developed strategies that enabled them to be unapologetically, a Black male and a scientist or a female and a Hispanic engineer (Urrieta & Noblit, 2018). Future studies could also explore the role of intersectionality in science identity development and examine how students from underrepresented groups balance competing social identities with science identity while striving to achieve social group inclusion and recognition as a member of the science community. In order to do so, they must resolve conflicting social identities which may suggest that their race, gender, or ethnicity are not compatible with a science identity. How do they avoid falling into an intersectional trap and conforming to the practices of the majority just to survive (Sparks, 2017)?

Finally, an additional metasynthesis could include a more in-depth exploration of the longitudinal science identity development trajectory that emerged from this study. Does the longitudinal timeline for science identity development differ across racial and ethnic groups? Is this longitudinal timeline different for students from underrepresented groups who do not attend predominately White institutions? This research highlights the need for more qualitative studies addressing science identity trajectories in students from underrepresented groups studying science in order to understand how to attract, retain, and support them as they pursue science degrees and careers.

References

Adams, J., Gupta, P., & Cotumaccio, A. (2014). Long-term participants: A museum program enhances girls' STEM interest, motivation, and persistence. *Afterschool Matters, 20*, 13–20.

Adjapong, E., Levy, I., & Edmin, C. (2016). Empowering girls of color through authentic science internships. *Journal of Urban Learning, Teaching and Research, 12*, 23–24.

Archer, L., Dewitt, J., & Osborne, J. (2015). Is science for us? Black students' and parents' views of science and science careers. *Science Education, 99*(2), 199–237. https://doi.org/10.1002/sce.21146

Brand, B. R., Glasson, G. E., & Green, A. M. (2006). Sociocultural factors influencing students' learning in science and mathematics: An analysis of the perspectives of African American students. *School Science and Mathematics, 106*(5), 228–236. https://doi.org/10.1111/j.1949-8594.2006.tb18081.x

Brickhouse, N. W. (2001). Embodying science: A feminist perspective on learning. *Journal of Research in Science Teaching, 38*(3), 282–295.

Brickhouse, N. W., & Potter, J. T. (2001). Young women's scientific identity formation in an urban context. *Journal of Research in Science Teaching, 38*(8), 965–980.

Brown, B. A. (2004). Discursive identity: Assimilation into the culture of science and its implications for minority students. *Journal of Research in Science Teaching, 41*(8), 810–834. https://doi.org/10.1002/tea.20228

Burt, B. A., Williams, K. L., & Smith, W. A. (2018). Into the storm: Ecological and sociological impediments to black males' persistence in engineering graduate programs. *American Educational Research Journal, 55*(5), 965–1006. https://doi.org/10.3102/0002831218763587

Carlone, H. B., & Johnson, A. (2007). Understanding the science experiences of successful women of color: Science identity as an analytic lens. *Journal of Research in Science Teaching, 44*(8), 1187–1218. https://doi.org/10.1002/tea.20237

Chapman, A., & Feldman, A. (2017). Cultivation of science identity through authentic science in an urban high school classroom. *Cultural Studies of Science Education, 12*(2), 469–491. https://doi.org/10.1007/s11422-015-9723-3

Charmaz, K. (2006). *Constructing grounded theory: A practical guide through qualitative analysis*. Sage.

Crenshaw, K. (1991). Mapping the margins: Intersectionality, identity politics, and violence against women of color. *Stanford Law Review, 43*(6), 1241–1299. https://doi.org/10.2307/1229039

Erikson, E. H. (1968). *Identity: Youth and crisis*. W. W. Norton & Company.

Erwin, E. J., Brotherson, M. J., & Summers, J. A. (2011). Understanding qualitative metasynthesis. *Journal of Early Intervention, 33*(3), 186–200. https://doi.org/10.1177/1053815111425493

Esmonde, I., Brodie, K., Dookie, L., & Takeuchi, M. (2009). Social identities and opportunities to learn: Student perspectives on group work in an urban mathematics classroom. *Journal of Urban Mathematics Education, 2*(2), 18–45. https://doi.org/10.11575/PRISM/35652

García, H. A., Hotchkins, B. K., & McNaughtan, J. (2019). "Why not?": How STEM identity development promotes black transfer and transition. *The Journal of Negro Education, 88*(3), 343–357. https://doi.org/10.7709/jnegroeducation.88.3.0343

Gazley, J. L., Remich, R., Naffziger-Hirsch, M. E., Keller, J., Campbell, P. B., & McGee, R. (2014). Beyond preparation: Identity, cultural capital, and readiness for graduate school in the biomedical sciences. *Journal of Research in Science Teaching, 51*(8), 1021–1048. https://doi.org/10.1002/tea.21164

Gee, J. P. (2001). Identity as an academic lens for research in education. *Review of Educational Research, 25*, 99–125. https://doi.org/10.2307/1167322

Gibau, G. S. (2015). Considering student voices: Examining the experiences of underrepresented students in intervention programs. *CBE—Life Sciences Education, 14*(3), 1–12. https://doi.org/10.1187/cbe.14-06-0103

Gudyanga, A. (2016). Zimbabwean female participation in physics: Factors of identity formation considered as contributing to developing an orientation to physics by female students. *Journal of Education and Practice, 7*(26), 159–171.

Haun-Frank, J. (2011). Narratives of identity in everyday spaces: An examination of African-American students' science career trajectories. *Science Education International, 22*(4), 239–254.

Hogg, M. A. (2003). Social identity. In M. R. Leary & J. P. Tangney (Eds.), *Handbook of self and identity* (pp. 462–479). The Guilford Press.

Holland, D., Lachicotte, W., Jr., Skinner, D., & Cain, C. (1998). *Identity and agency in cultural worlds*. Harvard University Press.

Kernis, M. H., & Goldman, B. M. (2003). Stability and variability in self-concept and self-esteem. In M. R. Leary & J. P. Tangney (Eds.), *Handbook of self and identity* (pp. 106–127). The Guilford Press.

Lincoln, Y. S., & Guba, E. G. (1985). *Naturalistic inquiry*. Sage.

Malone, K., & Barabino, G. A. (2009). Narrations of race in STEM research settings: Identity formation and its discontents. *Science Education, 93*(3), 485–510.

McGee, E. O. (2015). Robust and fragile mathematical identities: A framework for exploring racialized experiences and high achievement among Black college students. *Journal for Research in Mathematics Education, 46*(5), 599–625. https://doi.org/10.5951/jresematheduc.46.5.0599

McGee, E. O., Griffith, D. M., & Houston, S. L. (2019). "I know I have to work twice as hard and hope that makes me good enough": Exploring the stress and strain of Black doctoral students in engineering and computing. *Teachers College Record, 121*(4), 1–38. https://doi.org/10.1177/016146811912100407

Merriam, S. B. (2002). *Qualitative research in practice: Examples for study and discussion*. Jossey-Bass.

Nasir, N. S., & Hand, V. (2008). From the court to the classroom: Opportunities for engagement, learning, and identity in basketball and classroom mathematics. *Journal of the Learning Sciences, 17*(2), 143–179. https://doi.org/10.1080/10508400801986108

Nasir, N. S., & Saxe, G. B. (2003). Ethnic and academic identities: A cultural practice perspective on emerging tensions and their management in the life of minority students. *Educational Researcher, 32*, 14–18. https://doi.org/10.3102/0013189X032005014

Nealy, S., & Orgill, M. (2019). Postsecondary underrepresented minority stem students' perceptions of their science identity. *The Journal of Negro Education, 88*(3), 249–268. https://doi.org/10.7709/jnegroeducation.88.3.0249

Phoenix, A. (2006). Interrogating intersectionality: Productive ways of theorizing multiple positioning. *Kvinder, Køn & Amp; Forskning, 2-3*, 21–30. https://doi.org/10.7146/kkf.v0i2-3.28082

Remich, R., Naffziger-Hirsch, M. E., Gazley, J. L., & McGee, R. (2016). Scientific growth and identity development during a postbaccalaureate program: Results from a multisite qualitative study. *CBE-Life Sciences, Education, 15*(3). https://doi.org/10.1187/cbe.16-01-0035

Sandelowski, M., & Barroso, J. (2007). *Handbook for synthesizing qualitative research*. Springer Publishing Company.

Sayman, D. M. (2013). Quinceañeras and quadratics: Experiences of Latinas in state-supported residential schools of science and math. *Journal of Latinos and Education, 12*(3), 215–230. https://doi.org/10.1080/15348431.2013.765805

Shenton, A. K. (2004). Strategies for ensuring trustworthiness in qualitative research projects. *Education for Information, 22*, 63–75.

Sparks, D. M. (2017). Navigating STEM-worlds. *Journal for Multicultural Education, 11*(3), 162–175. https://doi.org/10.1108/jme-12-2015-0049

Sriram, R., & Diaz, C. (2016). STEM as minority: A phenomenological case study of how students of color perceive their experiences in a STEM living-learning program. *Journal of Living Spaces, 5*(1), 9–18.

Stets, J. E., Brenner, P. S., Burke, P. J., & Serpe, R. T. (2017). The science identity and entering a science occupation. *Social Science Research, 64*, 1–14. https://doi.org/10.1016/j.ssresearch.2016.10.016

Tajfel, H. (1974). Social identity and intergroup behavior. *Social Science Information, 13*(2), 65–93. https://doi.org/10.1177/053901847401300204

Tajfel, H., & Turner, J. C. (2004). The social identity theory of intergroup behavior. In J. T. Jost & J. Sidanius (Eds.), *Political psychology: Key readings* (pp. 276–293). Psychology Press. https://doi.org/10.4324/9780203505984-16

Urrieta, L., & Noblit, G. W. (2018). Theorizing identity from qualitative synthesis: Implications and conclusions. In L. Urrieta & G. W. Noblit (Eds.), *Cultural constructions of identity: Meta-ethnography and theory* (pp. 246–266). Oxford University Press.

Villa, E., Wandermurem, L., Hampton, E., & Esquinca, A. (2016). Engineering identity through the Latina lens. *Journal of Education and Learning, 5*(4), 113–125.

Wellman, B. (1971). Social identities in black and white. *Sociological Inquiry, 41*, 57–66. https://doi.org/10.1111/j.1475-682X.1971.tb01200.x

Chapter 14
Representing STEM Identities as Pragmatic Configurations

Ruurd Taconis

Identities play an essential role in learning and teaching in STEM (STEM here also used to include the separate STEM domains). Identities relate importantly to (not) choosing STEM (Henriksen et al., 2015). Learner identities play a role in the accessibility of STEM education and STEM occupations (Stets et al., 2017). For all students, but for students carrying self-identities with characteristics uncommon in traditional STEM in particular. Identities also impact learning processes and learning results in STEM. Within vocational education, developing a professional identity ('becoming an engineer') is recognized as the core of education. In all of these, STEM teachers, and hence STEM teacher training, play an important role.

However, most of the research underpinning this consists of qualitative and descriptive studies (e.g., Jansen, 2016), which use a wide variety of concepts of identity. This hinders reaching overarching conclusions and limits persuasiveness. For example, "The challenge of understanding teachers' professional identity lies within determining a definition of this concept. [...] Unless these challenges are overcome, theory building in this field will remain complicated" (Canrinus, 2011, p. 101). A lack of solid identity models hinders the progress of identity research (Knez, 2016).

In this chapter, we develop a model of STEM identity and a method for mapping STEM identities accordingly. To this end, we first explore the concept of identity before zooming in on STEM identity and its relationship to expertise in particular. We then elaborate on this using schema theory.

R. Taconis (✉)
Eindhoven University of Technology, Eindhoven, Netherlands
e-mail: r.taconis@tue.nl

H. T. Holmegaard, L. Archer (eds.), *Science Identities*, Contributions from Science Education Research 12, https://doi.org/10.1007/978-3-031-17642-5_14

14.1 Views of Identities

There is no universally valid and accepted definition of identity. However, some central aspects are widely recognized. These are rooted in three fundamental existential dilemmas, to which the construction of identities is seen as a response. These are the self/other dilemma, the agency/structure dilemma, and the stability/adaptation dilemma (Bamberg, 2011).

The self/other dilemma evokes the construction of the self, which is the key to all types of identities (Burke & Stets, 2009). The agency/structure dilemma implies that identities are mediating constructs. Identities form an interface that links situated persons to situated behaviour (Schachter, 2005). This mediating function goes two ways. Identities influence how an individual perceives an environment and the self in that environment. At the same time, identities *are influenced* by that environment. In the definition of Archer et al. (2010), *"[identity is] an embodied and a performed construction that is both produced by individuals and shaped by their specific structural locations"* (p. 617).

In this agency|structure dialectics (King, 2016), creating structure (in the environment) is a collective process that transcends the individual. *"Structures are the practices of people, but it is also the practices of people that form (and reproduce) structures"* (Sewell, 1992). Behaviour (or agency) is formed or directed in a complex way by identities and structure, while behaviour and agency also contribute to a joint structure formation over time.

This dynamic constitutes a *"developmental"* dimension of identity, which corresponds to Bamberg's third dilemma, the stability/adaptation dilemma. Identities are semi-stable and are continuously (re) constructed (Ping et al., 2018). While identities guide how people see and behave, they simultaneously slowly develop through interactions with the environment. This is a partly conscious and deliberate process of interpreting, acting and giving meaning to these experiences from the current identities. Therefore identities are simultaneously a state and a process (Beauchamp & Thomas, 2009).

It is often assumed that this agency|structure dialectics occurs primarily (or exclusively) in the social arena (Scott, 2016). But there does not seem to be a compelling rationale for this assumption, and for STEM, it may even be hard to sustain. For example, the impact of technology on the identity of people is considerable (Lamb & Davidson, 2002). In STEM practice, one develops new ways of understanding the physical environment, creates artefacts, and finds new ways of manipulating the environment. All these are likely to affect the environment, one's perception of it, how one deals with it and ultimately one's identity.

14.1.1 Types of Identities

Burke and Stets (2009) distinguish three types of identities. Self-identities, like all identities, mediate between the self and the environment. Their content includes self-views, styles of action, and negotiations where a balance between the self and the environment is dynamically maintained. However, self and self-identity are fundamentally not the same. Following Bamberg, we consider the self an answer to the demarcation dilemma, while identities answer the other two dilemmas. While the self is about demarcation (what is "me", and what is "not me") and could be static in principle, self-identities "happen" in the agency|structure dialectics (Bamberg, 2011). While for the self, one could defend the idea that the self-view and the self are indistinguishable, this is not possible for self-identities that are essentially interactive and dynamic. Self-identities cannot coincide with self-perceptions alone and must include elements that refer to this interaction, such as the perception by others.

The second type of identity, role identities, are rooted in certain activities or professions. These include perceptions of in-role behaviour and the-self-in-role, how people see themselves and are seen as agentive actors. Role identities are constructed in practice, guided by (professional) standards, self-reliance and complementary behaviour (Burke & Stets, 2009). This distinguishes them from the third type of identities: social identities, which develop under the influence of prototypes, self-esteem, and 'similarity behaviour' (Burke & Stets, 2009). However, this distinction is principally analytical, and in reality, identities show a mixture of properties. Being a farmer is primarily a role identity, but with the profession comes a formal and informal community, adding elements of social identity to it.

Burke and Stets (2009) also emphasize that the self and self-identities underlie all other identities. Conflict may arise when a situation requires the construction of an identity that is partially incompatible with the self or self-identity. For example, a traditional engineering community may be characterized as a culturally and socially masculine environment. This implicitly (or explicitly) requires newcomers to construct identities that correspond to this. Identities must be constructed that are sufficiently productive and supported by a fair degree of recognition. Constructing such identities can be (very) difficult for individuals with self-characteristics that do not easily align with aspects of STEM environments, such as the masculinity mentioned above (Carlone et al., 2015). Some of these self-characteristics are (almost) non-negotiable, such as physical characteristics, personality, and cultural traits. Laws can enforce respect for such self-characteristics. Identity negotiations should be respectful and focused on participation and openness to cultural change (Roth & Tobin, 2007).

14.1.2 Identity Configurations

The developmental character of identities and the variability of identity guided behaviour produces a paradox first recognized by Erikson (Schachter, 2005). A concept of identity in the sense of a 'completely coherent' entity could not account for the variability in behaviour, experiences and developmental paths in real life. Erikson: "A 'monolithic' interpretation of the concept of identity could not explain the flexible and situation related complex ways in which identities relate to behaviour and perceptions. They should rather be considered as parts within a complex configuration of various interlocked and/or subordinate "identities" together yielding a "reasonably coherent" and developing whole," Erikson (1968, as cited in Schachter, 2005, p. 382), therefore, concludes that: "The final identity ... includes all significant identifications but it also alters them in order to make a unique and reasonably coherent whole of them ... It is a configuration gradually integrating constitutional givens, idiosyncratic libidinal needs, favoured capacities, significant identifications, effective defences, successful sublimations and consistent roles". Schachter (2004) has elaborated this idea and sees a person's various identities as situational and goal-oriented complex constructs involving various mutually conflicting and competing "mono identities" that he calls a "configuration". Other authors have also used the term identity configuration to study multiple identities and their effects (e.g., Hatmaker, 2013).

Within the configuration, identities can intersect and contradict each other (Horowitz, 2012). They can compete in generating perceptions or behaviour. For example, a person could both have the identity of a STEM teacher and the identity of an engineer in an industrial engineering team. The value of teamwork plays a role in both situations. However, moderate competition is accepted in an engineering team, while this element is usually absent - or undesirable - in a classroom situation. An identity conflict could occur as one would welcome competition as an engineer but not as a teacher. Resolving this conflict, for example, by deciding that mutual competition in school can only apply to advanced students with similar backgrounds, also means a further differentiation of the STEM teacher's identity configuration.

Identities can also have sub-identities, that is: be nested. Through specification, these can gradually arise from frequently reoccurring specific cases. Such sub-identities become independent and can then conflict with the "parent identity". An example is the identity of a sports coach as a sub-identity of being a parent. Parents can cuddle their children. As soon as it becomes an independent (sub) identity, it is apparent that hugging young athletes does not fit the role of a sports coach. Resolving internal tensions within the identity configuration is a crucial factor that makes identity configurations flexible and development-oriented.

Other authors use the word "configuration" closer to its general meaning of a structured entity comprising different mutually interconnected elements of various types. Leeferink et al. (2015) emphasise that configurations are "*structured*" totals comprising different categories of elements. They propose to analyse narratives by:

"unravelling the storied data into different categories and identifying relationships among the categories. [...], [hence] insight can be obtained into the contribution of and the relationships between the aspects, situations, and contexts on student teachers' learning experiences" (Leeferink et al., 2015, p. 9).

In this chapter, the term identity configuration encompasses both of the above perspectives. We see it primarily as a structured entity composed of various elements, such as values, perceptions/perceptual filters (e.g., ways of "seeing"), knowledge, particular ways of attribution, and specific ways of doing things ("methods"). Of which both the elements and their structural relationships contribute to the constituted (mono-) identity. For example, whether or not the value' environmental awareness' is associated with "taking a bus" as a behavioural preference. But, in addition, we see configurations as the 'imperfect but reasonably coherent whole' that includes several such mono-identities, which overlap, intersect, or are nested. Hence, each identity element plays a role in one—but possibly more than one - of the constituent mono-identities within the configuration as a whole.

A "visual" metaphor from chemistry would be a poly-crystal (e.g. pyrite – Fig. 14.1). The poly-crystal comprises interlocked and intersecting cubic mono-crystals (mono-identities). The individual particles (identity elements) can be considered part of several of these intersecting mono-crystals. The mono-crystals are well-ordered constellations of particles. In contrast, the ordering in the poly-crystal (the "ultimate identity" in Erikson's terms) is much more complex and multi-layered and to some extent ambiguous.

14.1.3 Identity and Schema Theory

One approach to identity is through the use of schema theory (Anderson, 2005). Schemas are functional structures that help organize and interpret information in the light of the self (Berzonsky, 2004; Horowitz, 2012). They shape experience and practice but are also themselves shaped by experience and exercise (King, 2016).

Fig. 14.1 Pyrite polycrystal

Schema theory in cognitive research denotes a particular theory that cognitively explains the behaviour of experts. However, the concept of schema can be understood much more broadly as a metaphor for understanding human behaviour as such, linking three elements: knowing/feeling when what, and how. It is used in clinical psychology (e.g., schema therapy). In educational research, various scholars have used it in identity research (e.g., Alexander & Rudd, 1984; Lloyd & Boyd, 2014), some concerning STEM identities (e.g., King, 2016). Berzonsky (2004) has described a social-cognitive model of identity using schema and describes a schema "as a self-theory, a conceptual structure composed of self-representational and self-regulatory constructs. [...] It contains procedural knowledge or operative schemes for solving problems and making decisions, and representational schemas or personal constructs for understanding and making sense of events and personal experiences" (pp. 303–304).

14.1.4 Elements and Structure of Schemata

In cognitive theory, schemas combine three elements: "*situational*" knowledge that function as active perceptual filters; "declarative" knowledge that, for instance, comprises concepts, models and theories; and "*procedural*" knowledge, which comprises the ability to know and perform practical procedures or actions. We will define these fitting the aim of this chapter below). Schemas as such work automatically because of the close interrelationship between the elements. When a given situation is recognized through the perceptual filter, relevant theoretical knowledge and possible actions automatically become mentally available. Schemas can be mapped in diagrams showing their structure of interrelated elements that, in fact, are small "semantic networks" (Taconis, 1995).

In expert theory, expert knowledge is characterized by extensive, correct, sophisticated, and well-integrated schemata. Moreover, experts have embedded these schemata in a meta-structure that is both functionally and theoretically correct (Taconis et al., 2001). As a whole, the knowledge base of experts is densely interconnected and has a theoretically correct practical structure. In contrast, the knowledge base of novices is characterized by missing elements, inclusion of "alternative" elements, small schemas, fragmented schemas, and a lack of an adequate meta-structure.

If we apply these ideas to identity, we must modify this model somewhat. First, we must ensure that identity schemas include all the elements that characterize identities, such as beliefs and values. An issue discussed in detail below. Second, we must move away from the normative perspective in the cognitive model above, which assumes that there is only one best meta-structure: the expert knowledge base. We need to move to a descriptive perspective that recognizes that disciplinary knowledge can take different forms depending on the situation and the individual (Stevens et al. (2008), p. 356). Even experts are only partially experts and more than just cognitively defined. Finally, in line with Erikson's argument (p.3), we must ensure that we create a structure that fits the flexible interaction with the environment that characterizes identities.

Since identities intersect and can be nested, a schema-based identity description will be a semantic network. It is complexly structured and contains elements of the different types found in identities. Within this network description, different (mono) identities can be recognized in the same way as recognizing the cubic crystals in the pyrite polycrystal. The use of such semantic network representations of knowledge in artificial intelligence research illustrates the potential of this approach. There, they are used to model expert knowledge and simulate behaviour. As a whole, such a configuration functions as a system that produces ideas, insights, and informed actions based on the situations and how they are (selectively) perceived.

Having described general theories of identity as the basis for our model under construction, we must now examine STEM identities as they are defined and used in research on STEM education.

14.2 Identity in STEM Educational Research

Literature shows various definitions of STEM identities. A frequently used description of a person with a science identity is a person that "recognizes him/herself and gets recognized by others as a science person" (Carlone & Johnson, 2007, p. 1190). This recognition results from the dialectical interplay between the self and the environment in identity construction. Carlone and Johnson (2007) also have proposed a much-used (a priori) model for STEM identity comprising three interrelated dimensions of a science identity: competence, performance, recognition. Elaborating on this, Kim et al. (2018, p. 612) added to it: perception of scientists and interest in science careers. For professional identities (e.g. STEM teachers, engineers), additional elements are 'how one perceives the profession, and perceptions to "the self as a professional" (Avraamidou, 2014). There are (at least) two lines of research in STEM identity research which will be discussed next.

14.2.1 STEM in General Education

The first line of STEM identity research focuses on self-identity as a factor in choosing and successfully participating in STEM education (e.g., Holmegaard et al., 2014; Taconis & Kessels, 2009). Upon entry into a STEM environment, the negotiation to form a self and self-based STEM identity that fits the situational demands of the STEM (learning) environment begins (Beijaard et al., 2004; Pillen et al., 2013). In principle, this applies to all learners (Aikenhead, 1996). The process, however, is sensitive if one or more aspects of the "self" do not readily align with the implicit demands of (traditional) STEM education. Recognition (by the self and by others) as a STEM person is essential. However, it can be challenging to achieve, especially if such recognition is not always received, even when the competencies are present (Carlone & Johnson, 2007).

On the side of the learning environment, STEM tasks and classroom culture are critical factors (Kock et al., 2013). If STEM only comprises "nerdy" tasks, it can easily repel students, e.g. if they will not identify as "nerds". STEM tasks should enable and encourage various approaches and behaviours, not just those consistent with 'traditional STEM'. A STEM student identity that promotes sharing ideas can only contribute to solving STEM problems if the learning environment includes group work on contextual, fairly open-ended problems. In that case, new kinds of STEM identities can be constructed, strengthened, and recognized in a welcoming atmosphere.

Critical theorists have argued that STEM in secondary education should promote student agency to help innovate STEM practice to improve the accessibility of STEM education. And to renew the professional STEM practices themselves to make these more socially aware by "creating identities among citizens that would create new forms of science culture" (Roth & Tobin, 2007, p. 340). However, modern STEM education seeks to connect with students' interests and align classroom practice with professional science and technology practices through using open, contextually rich and authentic tasks. Professional practices themselves are also evolving to become broader and more connected to society (Graham, 2018).

14.2.2 Higher and Vocational Education

The second line of research focuses on vocational or higher education that prepares students to "become" STEM professionals, engineering and STEM teachers in particular (e.g., Dehing et al., 2013; Nylen et al., 2018; Perrenet & Taconis, 2009). Here, "STEM professional identity" not only is a scientific concept defined and used by educational researchers to explain and facilitate educational participation learning; however, on top of that, it has very concrete importance for society and the students (Stets et al., 2017). For example, STEM professional identity can be useful for attracting new students to STEM education and occupations (Dehing, 2012, p. 13), and in particular for the students themselves. Acquiring a 'STEM profession identity' through education and getting recognition as a STEM professional is of significant practical importance for their lives. This recognition is the key to their employability and admission to a professional community.

What counts as a STEM professional identity is determined first and foremost by the professional community and the field. These can explicitly describe what knowledge, skills and attitudes they want. In addition, professional practice implicitly shapes roles and identities through practice. Important aspects may also be explicitly prescribed through professional codes, a codes of conduct, handbooks, accreditation of programs, certification committees, and district assemblies (Dehing, 2012). In these processes, the educational researcher can play only a modest role as one of many stakeholders.

STEM identity research in higher and vocational education, as in general education, focuses on analysing and facilitating the process of developing STEM identities. Specifically, this research focuses on those identities that qualify as professional STEM identities, whose definition is primarily external. Various perspectives on identity development are used (Dehing, 2012). Some authors see it as a more or less "focused process" aimed at meeting professional standards. For example, "... to serve the public with specialized knowledge and skills through a commitment to the public goals and ethical standards of the field" (Sheppard et al., 2008, p. 170). Others see identity development as an open individual construction process within the context of the professional work environment. A process mainly guided by social embedding and (cognitive) apprenticeship (Collins et al., 1989).

In both approaches, the outcome of identity development contains room for 'personal colouring' of the STEM professional identity. Ibarra (1999) defines professional (role) identity as "... the relatively stable and enduring constellation of attributes, beliefs, values, motives and experiences in terms of which people define themselves in a professional role" (p. 2). Hence it is possible that two individuals both identify as an engineer, but do not share the same values, motives, etc. There may be various professional identities that would count as valid "professional STEM identity". Van Driel et al. (2007) show that there are different types of identities of STEM teachers, each with different values, norms, and beliefs.

Combining the two, we see that STEM identity research focuses on identity development from some 'initial' student identities to 'STEM identities' and facilitating this process. In both lines of research, individual differences between acquired STEM identities are recognized and welcomed. However, professional STEM identities seem to act as a gauge: either directly as a goal of identity development in higher and professional education, or indirectly as one of the anchors for defining STEM educational practices and requirements in secondary education, which guides identity development at that level.

14.2.3 Elements of STEM (Professional) Identities

Various groups of researchers have described elements of professional and/or STEM-related identities. Research taking a general perspective yields: (Farrell, 2011; Fishbein & Ajzen, 2011; Ibarra, 1999; Kaplan & Garner, 2017): the self-in-role, self-esteem, self-views; personal role-related norms, values, beliefs and attitudes; perceptions of role-related norms, values, beliefs and attitudes; perceptions concerning the work such as jobs, tasks, responsibilities; and perceptions of the professional future. Some of these primarily refer to the self, while others appear anchored in professional practice and the community. In this list, there is no reference to the competence concerning the profession.

Of researchers focussing on STEM identity, only a part includes competence (or knowledge) in their measurements or models of STEM identities (Blank et al., 2016; Carlone et al., 2015; Herrera et al., 2012; Kim et al., 2018; Steinke, 2017).

However, an analysis of papers shows that most studies do not measure STEM competence as such. Moreover, if STEM competence is measured at all, it is usually in the form of students' overall perception or self-assessment of "being competent in STEM". This, for example, is the case in the vast majority of studies selected for the review by Kim et al. (2018).

Concerning engineering education, there is broad recognition of identity d as a crucial aspect of becoming an engineer. Though this type of identity is mostly seen as separate from 'learning' considering identity formation taking place "on top of" and after acquiring knowledge and skills, particularly during internships (e.g., Dehing et al., 2013; Hatmaker, 2013).

Studies of identity in engineering education that measure STEM competency usually is as self-perceived competency or by grades. For example, Stevens et al. (2008) recognize the acquisition of disciplinary knowledge and skills as one of the three components to becoming an engineer, but he does not address its content. Instead, he focuses on students' beliefs about what counts as engineering knowledge. Tonso (2007) shows how engineering students used definitions of expertise as a criterion to accept or refute identity claims. She concludes that this generally served to support a conservative and closed culture. Hatmaker (2013) shows how the monopolization of "the concept of expertise" is used as a tool to preserve privileges as well.

We conclude that only STEM identity research generally sees competence (or expertise) connected to STEM identities. However, in these studies, competence is usually measured as 'self-perceived competence' or "belief in what counts as STEM expertise".

14.2.4 Mapping (STEM) Identities

Identities and STEM identities are usually mapped by reconstructing these from narratives (Jansen, 2016). Narratives reveal pictures of students' developing STEM identities within changing contexts (Connelly & Clandinin, 1999, in Anspal et al., 2012, p. 200). A minority uses questionnaires (e.g., Burke & Reitzes, 1981; Hazari et al., 2010; Taconis et al., 2010).

Narratives are constructed based on various assignment types, such as open writing and interviews. Generally, narratives comprise particular elements like Who, What, When, Where, Why, How (Freitag et al., 2021). In principle, narratives already include argumentative and explanatory elements (the Why). However, narratives are structured and employ particular components such as plots, heroes, resistance/opponents, resolution or victory (Thorndyke, 1977). Facts presented in narratives are not pure facts as they also serve the coherence and purpose of the story. The reconstruction of descriptive data in the form of a narrative should be understood as a process of "emplotment" (Polkinghorne, 2005). "This process involves the development of a narrative configuration (the plot) in which the events and activities in the data are linked and given meaning. Activities in the data are linked and given meaning in the light of a particular outcome" (Leeferink, 2016, p. 36).

So, narratives are not objective disclosures of the self, self-identities or other identities. They are "emplotted" accounts built upon self-perception and self-perceived identities. Hence, narratives do not necessarily reflect the "self" or identities others interpret and see demonstrated in behaviour (Polkinghorne, 2007). For example, a self-report about being competent may be a relatively accurate account of how one perceives one's STEM competence, but is not automatically an accurate actual account of the STEM competencies displayed, showed or perceived by others in actual STEM situations.

Bamberg (2011) distinguishes between two types of narratives approaches used in identity research. The first, the "big story" approach, emphasizes the overarching story of "the self". These are ".... stories that we tell in an attempt to make sense of how we experience ourselves and how we want to be understood in order to bring structure to our personal lives" (Søreide, 2006, p. 529). Undoubtedly, such reflection on one's life path, current position, and future steps contributes to and help guides identity development. Bamberg fully acknowledges the contributions that the analysis of "big story" narratives has made to educational research. However, he also points to some opportunities to strengthen the methodology and further improve identity research.

First, it is unclear what criteria people use to select things they include in such a "big story". Thus, there is no guarantee that all experiences relevant to identity construction are included, remembered correctly and placed in the proper perspective. Second, it seems that there must be more behind the dynamics of identity than a reflection on life paths, especially for STEM. Bamberg (2011) emphasizes that "the struggle of everyday practice" is the main arena where role identities are constructed and that the narrative structure of "big story" downplays the role of such interactive experiences. It causes "life as reflected" to be credited and "life as lived" to be discredited (p. 14).

An example (from experience): When teachers are asked in retrospect to explain why they started using an interactive whiteboard, they are likely to tell about their pedagogical motives for doing so. Suppose they were only asked to write down *when* they started using an interactive whiteboard (and thus not construct a narrative). In that case, it may happen that all teachers in the school name the year in which the school principal proactively purchased a set of interactive whiteboards that were placed in the classrooms during the summer break. By telling the story, events and motives may get highlighted, which at that time may not have been "that important". Unintentionally and subconsciously, the process of becoming ICT-teachers becomes "emplotted", making it appear in the form of a "life as reflected".

The identity dynamic that big stories show may (in part) result from constructing the "big story" narrative itself. This dynamic is not necessarily identical to the dynamic of the identity construction actually taking place at the time. Finally, Bamberg (2011) observes that such narratives often lack referential evidence. Hence, the validity of "big story" analysis entirely depends on "emplotted" facts and self-perceptions. Specifically, this obscures the view of how identities are formed in the interplay of *actual* facts and *actual* actions.

As a solution, Bamberg suggests using "small story" narratives. These focus on one's actions and agency within the role, not life stories. When "placing emphasis on small stories [this] allows for the study of how people as agentive actors, position themselves—and in doing so become positioned" (Bamberg, 2011, pp. 15–16). Small stories are personal and narrative accounts of the agency/structure dialectic that takes place during practice. They are rich sources of information that provide clues about what happened, why, and reveal underlying ideas. Small stories have a narrative structure, but this structure is not supposed to reflect (or be) the structure of the identity (construction) itself. Like all narratives, information is "emplotted". They are analyzed as data sources and need to be supplemented by other sources of information. If so, they offer a rich and credible account of the dialectic between agency and structure that takes place in practice and from which identity and its development are reconstructed.

14.2.5 A Pragmatic Perspective on STEM

We must now examine the nature of STEM and STEM subjects to elaborate the conceptualization of STEM identities. At the heart of STEM lies a productive engagement with the physical world. It may be tempting to call the latter "reality", but that quickly suggests that there would be only one such reality and only one 'true' way to understand it. It seems valuable to keep the concept of the physical world alive while avoiding such philosophical pitfalls.

One way out is the constructivist approach of Von Glasersfeld (2001). Rejecting the principles of "one reality" and science as a unique model that fits this reality in a superior way, one can assume that all one needs regarding science is a "viable" way of understanding. A way of understanding that - for now - is not refuted by physical experience or experimentation. There may be different ways of understanding the physical world, and they may not necessarily be consistent with canonical (or "traditional Western") science. However, STEM is not all about science and math alone. It is about engineering as well. Here, engineering is understood as "the complex practice of technology development" (Nia et al., 2019). With respect to STEM and specifically engineering, an additional principle that complements viability emerges: pragmatic value (Worren et al., 2002). Pragmatic value means a practical fruitfulness of ideas, not limited to supporting further understanding, but especially for (professional) practice itself, either directly or indirectly through technology development. A "rule of thumb" is an example of this.

On learning STEM, a corresponding view retains the idea of a physical world as the area of operation for STEM but avoiding the above pitfalls. The pragmatist, Charles Sanders Peirce, emphasised the role of inquiry as the primary semiotic activity and source of learning about 'the real' (Rosiek & Atkinson, 2005). Peirce states that "knowledge of reality arises from the mediation of the experiences of members of the community with reality" (Pierce, quoted in Rosiek & Atkinson, 2005, p.434). Pierce grants experiential learning the leading role within STEM and STEM learning. However, experiential learning itself is a social endeavour. But sharing and discussing must be ab*out experiences with the real.* It is not sharing hypothetical ideas or discussing theories; these are secondary activities. This identification of this primary semiotic activity of discussing authentic STEM experiences reflects the core of STEM. In a way, it defines STEM. Hence, it also is the royal way of learning STEM. Even though in education much time may be spent on learning STEM theory and doing exercises, the latter must be seen as often necessary supporting, but not a defining element.

Following these basic principles of viability and pragmatic value, and through research, STEM activities have accumulated an enormous amount of knowledge. Constructed, shared and expressed in and through different languages - mathematics in particular. This knowledge base is highly formalised, theoretical and abstract (Perrenet et al., 2000). Therefore, the study of theory has become an intrinsic element within STEM practice, not only in education but also in professional life (Kock et al., 2013). However, it remains a complementary activity, not one that defines the nature of STEM.

Modern professional STEM practice, particularly in engineering, involves a wide range of substantive and social activities. In industry, people work on broadly defined complex problems in multidisciplinary and internationally composed teams. However, STEM education is sometimes (implicitly) understood in terms of "mastering theory". Reducing STEM education to the reproduction of theory is an inadequate representation of STEM, especially if it does not include inquiry or the active application of the viability and pragmatic value principles.

When studying theory or sharing ideas without clear reference to actual experience, there is a danger of becoming "encapsulated" in theory and its language forms. This leads to the construction of "shared meaning" and knowledge, which may not correspond to "reality" and is most likely a copy of "textbook knowledge". Moreover, this narrow practice leads to the construction of "scholastic" STEM identities and leaves little room for the invention of new kinds of STEM identity.

14.3 STEM Identities as Essentially Including STEM Expertise

Before we can describe our model of STEM identities, we first argue why it follows from the above that STEM competence (or personal STEM expertise) is an intrinsic part of STEM identity. We provide three main arguments: the nature of identities, the sources of gaining recognition in STEM identities, and the dynamics of developing STEM identities.

14.3.1 First Argument: The Mediating Nature of Identities

Identities are inherently mediating and connect individuals dialectically to situations. This mediation is often assumed to be a mainly (or sometimes an exclusively) social process. STEM identities are about playing STEM roles which are not just social roles but pivot around "doing STEM", which is an activity that requires considerable mastery over abstract and theoretical knowledge. Therefore, understanding and skilled operation in the STEM environment take a pivoting role in STEM identity development. Although interaction with the physical environment is central to STEM behaviour, it is by no means an entirely individual process. The social and the physical are probably intertwined. Professional interaction with the physical STEM environment and social activities may alternate or combine, influencing the other and vice versa.

For example, social processes will influence the development of STEM expertise, such as through competition among learners. Communicative styles, such as being "shy" for personal or cultural reasons, will influence whether or not certain learning opportunities are obtained. On the other hand, social processes may favour certain ways of gaining expertise. Student framing "what counts as STEM" can discourage other students from pursuing their own lines of thought (Hatmaker, 2013). Specific cognitive strengths, such as programming skills, may help in a particular group to be recognized as STEM persons (Ping et al., 2021).

Within this dynamic interaction, success in "doing STEM" plays an additional role. Mainly success in STEM practice, but theoretical insight as well, provide opportunities to reinforce identity claims. For example, personal mastery of a particular piece of mathematical theory will influence the discussion in a student group working on a problem for which this piece of expertise is essential. This can contribute significantly to one's recognition as a "mathematical person".

Successes in STEM do not come from a perception of competence or self-perceived expertise. They come from what we actually achieve and only partly from what oneself and others think we have achieved or are capable of. Not from our self-perception of what we think we are achieving or could achieve. Actual STEM competence has a direct effect on performance. Forming and maintaining STEM identities requires a viable and pragmatic understanding of the physical world. However, this does not necessarily coincide with traditional concepts of "canonical" STEM. It can be, and probably will be in part, alternative and personal, as long as it is viable and pragmatic. Fundamentally, STEM expertise can take many forms. However, "canonical" STEM provides one path to achieving an advanced level of expertise that meets these standards.

As the practical activity of "doing STEM" is pivoting in developing a STEM identity (p. 10), the "big story" approach (Bamberg, 2011) seems less appropriate for studying STEM identities. However, the "small story" approach, which focuses on everyday practices, fits STEM very well.

14.3.2 Second Argument: Success in STEM and Gaining Recognized in STEM Identities

Gaining recognition in an identity is a complex process that depends on many variables. It is a general characteristic of role identities that there is a relation between being competent for a role and getting and remaining recognised in that identity. How would a musician who cannot really "play" legitimately claim that professional identity of "being a musician"?

This relationship is fundamental but complicated in practice. One can get recognition from only part of the community, for example. A painful situation in STEM education is that people can be competent yet not receive recognition (Carlone & Johnson, 2007). This inequitable situation is critical to potential STEM careers must be resolved. However, in the opposite case, in which someone gets recognised in STEM without sufficient competence, it may prove that sustaining such recognition may be problematic, and recognition may erode over time. Discrepancies between being competent and getting recognised may be smaller when the criteria for showing competence are clearer. Professional or curricular standards can help delineate the area within which competent performance can be contested for professional identities. For STEM fields, it is usually relatively straightforward what the quality of work is and whether a project has been successful or not, based on more or less irrefutable material evidence. This is due to the precise criteria of viable and has pragmatic value (p. 10).

Nevertheless, these norms leave room for various types of STEM identities. Hence, one can be a "STEM-person" in various ways. However, if one uses methods that do not work (have any pragmatic value) or produces theories that are not "viable", one undermines one's STEM identity. For reasons of principle, canonical STEM is only one of the possible ways to meet these standards. Still, on the other hand, it is probably the richest source of viable and pragmatic knowledge available. Moreover, criteria for successful performance in STEM may be relatively objective compared to the arts of the social sciences, but they are not entirely objective. What exactly gets recognised as competent behaviour is also in part constructed within the community, and this construction can affect factual opportunities to learn in the classroom (Gresalfi et al., 2009).

Gaining recognition in a STEM identity also takes place outside of education. Acquiring recognition in a STEM professional role by society is critically dependent on expertise. Dehing (2012) points out that acquiring a professional identity is of great importance to the learner, particularly in vocational education. "Acquiring a scientific identity through the acquisition of that professionalism promises a satisfying sense of identity and personal achievement by contributing to values we share" (Sullivan, 2004, p. 6). It increases self-esteem and provides professional and social recognition and reward (Van der Sanden, 2004). All of this is critically dependent on STEM professional knowledge rather than on self-perceptions, beliefs or even the internalisation of professional standards. In short: society recognises STEM expertise as an essential part of the STEM professional identity.

14.3.3 Third Argument: Learning STEM and the Development of STEM Identities

It is a fundamental quality of identities that these develop. Various mechanisms contribute to identity development, such as socialization, identification, modelling and resolution of identity conflicts, situational change and deliberate alignment with identity norms. These mechanisms have social as well as cognitive aspects and are experience-based (Burke & Stets, 2009).

In experiential learning, identity development takes place through behaviour. The results it produces and the feedback it invokes are mentally processed and may lead to changes in identity, a process described by the well-known TRA and RAA models by Fishbein and Ajzen (2011). In these models, identity beliefs, norms and attitudes, produce intentions that lead to the deployed behaviour under the moderation of 'actual control'. This actual control comprises skills, actual activities and environmental factors. The outcomes of that behaviour are then mentally processed, leading to possible adaptations of behavioural beliefs and control beliefs. How would this apply to the development of STEM identities? The critical point here is that STEM fields are complex, partly abstract and theoretical (p.10). Doing STEM requires complex and comprehensive theories that cannot be learned quickly when a concrete task requires it (Perrenet et al., 2000). Brushing up on STEM knowledge "just in time" is often impossible. Therefore, in the case of STEM, "actual control" in Fishbein and Ajzen's (2011) models is often critically dependent on STEM expertise built up in advance. More specifically, "actual control" will often require theoretical understanding and complex skills, and this is likely to be a critical success factor.

A first consequence is a bootstrapping problem: identity development as in the model cannot begin without sufficient expertise. The model assumes that norms, beliefs, attitudes, and akin factors effectively guide the approach to STEM tasks. However, for STEM, this is not very likely. Successful ideas for solving STEM problems in most cases come from knowledge, insight, and strategy use, not from beliefs or norms. Moreover, putting plans into practice adequately also critically depend on expertise. Lack of expertise will stand in the way of successful action.

Expertise is also essential for further improvements. For example, someone who has never heard of differential equations probably has no idea how to solve a problem that requires that particular mathematical approach. STEM values or beliefs give no clue as to which mathematical tool is needed. Without sufficient expertise to at least effectively search for solutions, failure is likely, with negative consequences for one's self-esteem and beliefs. At least without further help, hints or explanations of the theory needed. So one fails, and identity development comes to a halt. At least in a positive sense.

Without sufficient expertise, the success of new promising and potentially fruitful approaches may be blocked. It cannot be reinforced as would happen in the case of success, according to Fishbein and Ajzen's models. For example, the strategy of searching the Web for new ideas to solve a complex socially embedded STEM problem will not lead to success if the expertise is insufficient to understand and apply the theories found. Without success, the development of STEM identities based on such novel but fruitful approaches is easily discouraged rather than encouraged.

On the other hand, developing expertise can lead to a change of identity. A STEM professional may come to believe in the approaches in which he/she is most proficient by education or training. These are well-known, are applied with greater ease and probably produce relatively quick and good results. Hence, one is most likely to achieve successes, and all these factors underpin a developing preference. For example: Not so long ago, most STEM teachers may have believed that systematic explanations and memorizing definitions and lemma's by students were important methods in their teaching profession. The modern insight that being cognitively active in (re)construction knowledge is much more valuable to learning has altered the beliefs and behaviour of most STEM teachers. It ultimately changed the professional standard. The new insights on effective professional behaviour (expertise) have changed the professional identity of STEM teachers.

The intertwined nature of acquiring STEM identities and STEM expertise has been addressed before. Yackel and Cobb (1996) essentially drew the same conclusion: "that mathematical learning is both a process of active individual knowledge construction and a process of acculturation into the mathematical practices as performed by professionals". Pure knowledge acquisition is (professionally) meaningless if one cannot make professional choices on what to apply, when and how. Their argument seems to run in the opposite direction but actually is supplementary. On engineering education, Stevens et al. (2008) write that: "learning involves more than the acquisition of skills and knowledge, but also involves changes in what types of people we become and in how someone understands him or herself in relation to a particular disciplinary practice (e.g., engineering)" (p. 357).

14.3.4 Summary

In summary, we conclude that the mediating role of STEM identities is both "social" and "physical" and that STEM expertise plays a crucial role in both. Moreover, STEM expertise plays a vital role in the development of STEM identities. It also seems impossible to accept that someone could be a "STEM person" without the expertise to perform STEM tasks—whether in an educational setting or in a professional environment. Most people probably consider "being able to perform in a STEM role" to be the most typical characteristic of a "STEM person". Yet expertise or competence is absent from various lists of identity components in the literature. STEM identity research that takes expertise into account mainly employs the reduced form of self-perception of general STEM competence. The next section will explore how we can build a model that integrates expertise as a central component in STEM identities.

14.4 A Pragmatic Model of STEM Identities

As we conclude that STEM expertise is an intrinsic part of STEM identity, we define developing a STEM identity pragmatically as "*developing increasingly effective and authentic professional behaviour based on personal expertise, self-awareness and (professional) recognition.*" STEM identity development is about becoming both proficient and recognized as a STEM person while doing STEM and participating in STEM practices and community.

To reflect this, we propose to distinguish two distinct but partially overlapping components within STEM identities: an emerging "professional self" (PS) consisting of self-perceived values, internalized norms, beliefs, assumptions, emotions, etc. And a developing "personal professional expertise" (PPE) consisting of personal STEM knowledge, understanding, skill, etc. (Fig. 14.2). The term professional here indicates that STEM identities directly or indirectly refer to professional STEM practices (not implying a level of competence). In higher and vocational education, the reference to professional STEM practices is relatively direct. In a general educational context, STEM identity development is aligned only indirectly with possible STEM professional identities through the images of STEM practices that emerge in that educational context.

14.4.1 A Schema View on Elements and Structure of PPE

To elaborate on this model, we propose to use schema theory to model the PPE and its connections to PS as a configuration of schemata. We now discuss the different components of such identity schemata.

Situational Knowledge An essential aspect of PPE is how it shapes our perceptions. For example, perceptions of the self as a professional and perceptions of the self-in-action (Canrinus, 2011). These are not 'objective' but shaped by professional cognition and other factors. How we perceive, recognise and interpret a professional situation is based on situational knowledge forming an active cognitive filter (Taconis et al., 2001). Recognising a situation in-depth means associating it with certain theoretical concepts and 'seeing' possible professional actions. For example, expert and novice STEM teachers see different things happening in the

Fig. 14.2 Pragmatic model of STEM role identity

classroom. Other aspects become salient to them, and experts' observations are connected more often and more directly to relevant knowledge and intelligible actions (Van den Bogert, 2016).

Values, norms and beliefs all influence situational knowledge. These can be professional values and norms defined in standards by a professional community. However, values and beliefs can also be personal. For example, a person keen on gender equality in STEM education is probably better able to detect covert signs of gender inequality (e.g., an un-balanced division of labour during chemistry experiments) without deliberately focussing on that. That person has built situational knowledge that automatically implies sensitivity for gender inequality.

Declarative Knowledge Schemata in PPE comprise particular concepts, models and theories that provide ways to interpret and understand professional tasks such as physics problems for physicists and classroom management issues for a STEM teacher. In practice, an abundance of canonical and informal schemata provides competing ways to interpret, understand, and complete professional tasks, and professionals must make choices.

Trained physicists understand problems in physics using their advanced knowledge and sophisticated understanding of the domain. Still, some physicists will perceive and solve various types of problems differently than others. For example, many problems in thermodynamics can be solved using statistical theory as an alternative to thermodynamic theory, which some may prefer. Generally, in STEM, one can solve quantitative problems using analytical calculations, experimentation, or numerical methods. Yet, a particular scientist may have a clear preference for one of these approaches, expressing a fragment of his professional identity, for instance, as an "empirically oriented physicist".

The corresponding knowledge and mental models will probably be (come) more developed and sophisticated than that of colleagues with a different style. Hence, PPE is formed based on personal preferences, limitations and circumstances, such as current and previous education. Within the limits of the standards of the professional community, professionals interpret and master their profession in a personal way. For example, Van Driel et al. (2007) found that STEM teachers had different views on the profession and the central classroom learning objectives and usually favoured learning theories corresponding to these views.

Procedural Knowledge PPE also includes procedural knowledge. This is the mastery of professional techniques considered applicable in a specific type of situation (Berzonsky, 2004). A trained STEM teacher may see an opportunity for an exciting field trip to a nearby factory to reinforce learning. One that is not recognized/seen as meaningful by another teacher with a different professional identity. One that implies that 'field trips are clearly outside the curriculum as defined in the textbook'. Design engineers also have different approaches to solving design problems and can be, for example, "seekers" or "treasure hunters" (Kim & Kim, 2015). Values and beliefs about what can or will work are crucial to developing ideas about possible ways to solve a particular professional problem and, more importantly, also for developing their ideas about how they want to tackle that problem. Over time, professionals are likely to become more proficient in the professional techniques they prefer and perform more often. Such preferences influence the skills gained and help shape the development of PPE in the future.

Professional Beliefs The knowledge types listed above are all classic components of cognitive schemas. We believe professional beliefs are also an intrinsic part of PPE.

We define professional beliefs as beliefs that influence professional behaviour. Professional practices require a professional way of understanding situations. The theory is the backbone of this professional understanding, but the complex reality is never fully covered by the theory. A purely theoretical understanding of reality is never complete, and not all insights and approaches required in professional practice can be "evidence-based". In addition, even professionals will have an imperfect understanding of theories. So, professional beliefs are an inevitable and essential element in professionalism. Beliefs make professional knowledge more complete, flexible and personal. Sometimes they can even be the seeds from which new ideas and approaches emerge.

Professional definitions of STEM identities inherently leave room for personal, professional beliefs and styles. Some beliefs are accepted (or tolerated) within the professional community. Different paradigms and 'schools' coexist within the professional community within many disciplines, each school with its own way of seeing, interpreting, and understanding. Moreover, with its own way of acting, all connected to certain beliefs and underlying values. Van Driel et al. (2007) found that different and independent belief systems coexisted among chemistry teachers. One was subject-oriented, the other student-oriented. Their difference concerned both perceived learning outcomes and beliefs about the best teaching method. Moreover, "researchers have repeatedly found strong links between teachers' beliefs and their practice (planning, instructional decisions, and classroom practices)" (Van Driel et al., p. 157).

The genus of "beliefs" includes various types. Some beliefs are "convictions" that a person consciously cherishes. Others are "tacit" and remain hidden from the person and the outside world. Patrick and Pintrich (2001) state about STEM teachers: *"these beliefs are assumed to be implicit or tacit because teachers may not be aware of how they influence their behaviour" (p. 118).*

Both tacit and explicit beliefs can contradict canonical theory or even facts. In STEM education theory, idiosyncratic student beliefs are well known as "misconception" (Gilbert & Watts, 1983). Although they lack formal logical coherence and may conflict with "canonical theories", to students, these appear intuitively coherent and viable and are practically indistinguishable from knowledge.

Whether they are true or false, implicit or conscious, people hold on to beliefs as if they 'know' they are "true". People act on them and define themselves by them. Or, put another way, perhaps knowledge should be seen as "beliefs that happen to be true" (Sartwell, 1992). In practice, it is almost impossible to distinguish between one's knowledge and beliefs (Van Driel et al., 2007). Hence, we treat beliefs as "alternative knowledge", an inherent part of the PPE that colours it and makes it more complete.

Finally, there is a variation in the scale of beliefs. Above, we have focused primarily on belief systems. Belief systems are relatively comprehensive entities rooted in deeper cognitive and normative structures (Pajares, 1992). Beliefs can also be small-scale, such as a single idea, a self-conceived, or systematically misused concept. Presumably of the declarative type. However, if a belief is consistently associated with a particular situation or implies a specific alternative action to be taken, it may also be situational or procedural, respectively. Belief systems are represented by "alternative" schemata, presumably containing one or more small beliefs, and with a structure reflecting unusual relationships between elements.

Values and Norms of the Profession We see norms as more or less codified lines of conduct. Professional standards define the boundaries of acceptable professional behaviour. Professional values translate into positive guidelines for desired professional behaviour. In either case, professionals must internalize them, meaning they accept the value/standard, understand the situations in which it applies, and use it to generate appropriate professional behaviour. Many - if not all - professional norms and values are, in addition, the generators of very general "prototypical" behavioural schemas. These not only tell the individual how and when to act according to these norms but also are templates for building specific normative (sub) schemata that guide behaviour for specific situations in more detail.

Such (sub-) schemata help to produce the desired behaviour more sophisticatedly and in a more situation-specific and personal way with appropriate flexibility. For example, the professional value "teachers must take care of the student as well as they can" (Søreide, 2006) may imply to a novice STEM teacher "to explain particular subject matter over and over again to students that did not yet get it". However, once this novice STEM teacher has learned about misconceptions in STEM and activation pedagogies, the schema associated with this professional value may differentiate to form two (sub-) schemata. A first schema that implies the procedure: "check that the student has made enough effort and, if not, put him to work". And a second schema that suggests "finding out about and discuss the specific assumptions and reasoning of the student".

14.4.2 Descriptional Format

Using the various components defined above, we can define a format for describing STEM identities and PBMs in semantic diagrams. Conventions can be used for this purpose, for example those described by Taconis (1995). There is a pictogram for each knowledge type: trapezoids for situational knowledge, blocks for declarative knowledge, curved block arrows for procedural knowledge. Small beliefs of different types are represented by circling the knowledge symbol with a cloud-like outline. Schemas and configurations are formed by the various connections between different elements indicating their relationship, such as "being a subtype of" and "being an attribute of".

Now that we have completed our analysis of the concept of professional identity in STEM and a possible format for mapping it, it is time to explore how this approach can be used in practice.

14.5 Mapping STEM Identities: Pragmatic Identity Analysis

A "pragmatic identity analysis" for the study of STEM identities is proposed based on the above. It is a "small story" approach (p. 12) that focuses on how STEM identities unfold and develop as people engage with STEM. Hence, personal accounts should be supported by referential evidence, such as written materials, questionnaires, or recordings. It builds a schema representation (p. 15) of STEM identities as pragmatic configurations (p. 3) centred on PPE. These structured diagrams explain how people act, think, make trade-offs, and express themselves as they engage in STEM tasks.

We provide two examples to demonstrate the model and its use. The examples are drawn from a practice that preceded the development of the approach described here, and therefore contain some methodological imperfections, and their results cannot be considered empirical evidence for the cases described. Moreover, within the scope of this chapter, we can only describe the analytical approach very briefly and must essentially limit ourselves to a description of the circumstances and the result of the analysis.

14.5.1 An Example from Modern Engineering Education

In modern engineering education, students learn domain/specific and general engineering skills by working on broadly defined challenges (Graham, 2018). This is in contrast to traditional engineering education, which focuses on applying previously learned content, science and mathematics. This so-called Challenge Based Learning (CBL) directly impacts the formation of STEM identities, PPE in particular.

The example concerns a course within the 3rd-year undergraduate engineering program. Students work in small teams on a complex real-life challenge drawn from business practice in healthcare, renewable energy, or agriculture. Before the CBL-project, the students viewed that problems in applied physics have exact solutions, which must be represented mathematically in an analytical way (i.e., using formulas) or, if not mathematically impossible, by numerical approximation (i.e., using an algorithm). They generally saw their role as building a suitable math model and doing the calculations. During the projects, it became clear that this was usually not what the companies were after. And that there were several limitations in terms of time and available resources. Students were initially confused about this and could not accept as 'real solutions' the "quick and dirty" approximation that some companies wanted in response.

After the challenge, the students had changed their mind. They understood that the company was not asking for an "exact solution" but a workable and practical solution. The companies did not care much about exactness, while timely results were paramount. The students' conceptualization of problem-solving was differentiated. They now recognized that problems have practical characteristics, such as (minimum) specifications of the solution and limitations (time frame, available resources). At the same time, they learned that exploring these limitations with the companies is part of their professional role as engineers. Moreover, their professional values shifted from "engineers deliver mathematically exact solutions" to "engineers deliver feasible and pragmatic solutions within constraints". In addition, their view of communicative aspects of their role as an engineer changed.

14.5.2 An Example from STEM Teacher Training

A second example builds on data collected in earlier research on "identity workshops" conducted within the initial STEM teacher training at University (Mommers et al., 2016; Schelings et al., 2017). The two years' fulltime teacher training program includes courses on educational theory and the pedagogy of STEM teaching, projects and a series of "group supervision meetings". It also includes 300 h of school internship. Students are supervised by an experienced STEM teacher and a supervisor from the university.

Participants are university students studying to become secondary school teachers in a STEM subject (e.g., chemistry, mathematics, general STEM). They enter their initial teacher training after (or as a component within) their academic engineering studies. All have at least an academic bachelor degree in a STEM subject. At the start, students usually are 22–25 years old. Participants are newcomers to education and may have "naïve" views about education and their teaching role at the beginning of their training. These views are rooted in their own experiences in high school or college. A standard view is that teachers' main task is to show the beauty of their STEM subject to high school students, especially by explaining it well and coming up with clear and engaging examples. Most students do not initially have a clear idea of their role in classroom management. For example, they may not recognize early signs in the classroom that might indicate growing unrest and what actions they could (or should) take. The most common career perspective is a mixed one, alternating between working as an engineer and teaching STEM or combining both.

Identity Workshops Group supervision consisted of a series of "identity workshops" with various assignments that support the identity development of STEM teachers. Students explored personal and situational factors that influence their STEM identity development and were challenged to connect elements of their selves, STEM education theory, and their school experiences and encourage reflection.

In the workshops, the students got several assignments. The rationale behind these is to bring forward and then help resolve tensions between various identity aspects to stimulate STEM teacher identity development (Pillen et al., 2012). One assignment, for example, demands the students to make (and explain) a graphical poster showing their view of their relative position in the school towards the pupils, their colleagues, and the school management (Mommers et al., 2017). The assignments were completed as homework and turned in digitally. In the meetings (after consent), the students discuss the main features as a group. Students can request a one-to-one talk with the supervisor from the teacher training institute.

14.6 Assignment

This chapter chose an assignment that provided students with ample opportunities to express their professional identity in terms of both their professional self (PS) and their personal professional expertise (PPE). This assignment asked the student teachers to describe a recent experience of their interaction with their students in their STEM classroom (Fig. 14.3).

The assignments existed before the writing of this chapter and were not designed to serve for a pragmatic identity analysis. They focus on the professional self and do not explicitly ask about all the different PPE elements. Moreover, they do not include a request for referential evidence. Nevertheless, the chosen task is sufficiently triggering to serve as an illustrative example in this chapter, especially since we are convinced that the PPE will come to the fore - at least in part - because it is an intrinsic part of identity.

Assignment: Interactions as a STEM teacher

This assignment aims to make you more aware of the various aspects of student-teacher interaction. How do you relate to your student? We ask you to think of an imagined or actual moment in school that made you either anxious or particularly enthusiastic.

1. Please describe your expectations, your intentions regarding your relationship with the students, aspects in the school environment that you consider relevant, your personal feelings when contact with the students would evolve as you would have liked, and your personal feelings when this would not happen.
2. Please describe the situation more specifically (who, what, when, where, how). Also, reflect on yourself 'as a person who is becoming a teacher'. This can be about intentions, (factors in) the environment, the perspective you have taken (e.g., as a teacher or an authority), your feeling of friction or flow, and anything that you consider relevant for the communication with your students.
3. As a (developing) teacher, what behaviour do you consider appropriate in the communication with your students, and what does that mean for the relationship you currently maintain with your students?

Fig. 14.3 Assignment by Schellings et al. (2017)

Analysis The analysis of the data took place in two rounds. First, the text was carefully read. Possible PPE elements were marked. The elements were examined and sorted by type in the second round, and we drew up a provisional schema. We used the relationships explicitly established by the students in their text to decide to include that connection in the schema. Relationships implied in the text, e.g. by multiple proximate occurrences of elements. We then checked the text for possible components not yet included and if the current diagram was complete. Where needed, we completed the schema.

A Description Using Schemata Figure 14.4 shows a pragmatic description of (a fragment of) the professional identity of one of the beginning STEM student teachers. The student selected (student A) is a communicative female student that has recently completed her master as a chemical engineer. When she did the assignment, she had approximately eight weeks of in-school experience and less than 20 h of teaching experience. The diagram depicts a reconstructed PPE using the schema model developed above. The diagram shows the expertise system underlying the dynamics of the professional behaviour rooted in her STEM identity as reconstructed from her completion of the assignment.

Student A's emerging professional STEM teacher identity comprises two main role-activities (i.e., procedural knowledge) she has to accomplish as a STEM teacher: explaining to students and motivating students. She named two sub-activities concerning motivating students: "have a positive interaction with students" and "reward students". She also named the role-activity of "being a policeman", but she distances herself from it and states that "she is not going to do

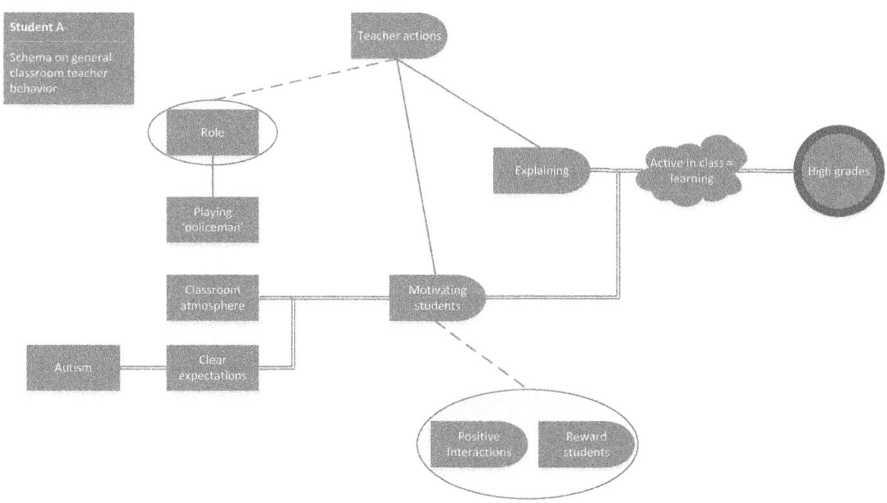

Fig. 14.4 Example of the PPE of Student A

that". Thus, it is not represented as procedural knowledge, but as an attribute of the general role of STEM teacher.

From various passages in the text, it becomes clear that Student A strongly believes that "explaining the subject to sufficiently motivated students" will make her reach her goal: students with high grades. For example, there is no mention of actively involving students in learning activities or its possible impact on student learning. Neither does she show to have explored what "effective learning" is (e.g., students asking questions, students explaining to each other, empowering students to develop agency concerning their learning …).

"Declarative knowledge" occurs in a few places in the text and primarily relates to motivating students. The declarative knowledge elements do not show any interrelationships. They do not constitute an explanatory model that could underpin professional action. Nor are they connected to the pedagogical theory discussed in the Institute's courses. The isolated topic of autism is, it was a topic in one of the courses two weeks prior.

Figure 14.4 shows no "situational" elements, suggesting that the diagram as a whole is probably an omnivalent view in the eyes of the student. However, the assignment asked for specifics about circumstances and particulars. Various aspects in Fig. 14.4 show that this student is a beginning academic STEM teacher. Firstly, due to the lack of situation-specific schemata. The diagram is poorly integrated, contains no situational elements, and includes no declarative knowledge related to educational theory. Secondly, it demonstrates the specific belief that explaining is enough for motivated students.

The student is enthusiastic and motivated to become a teacher who takes care of her students as much as possible, a common element in many teacher identities (Søreide, 2006), which she currently interprets as 'high grades'. However, at this stage, she has no expertise to realize this in practice. She maybe will not succeed, and disappointment and discouragement will likely result (Fishbein & Ajzen, 2011). Without further growth in expertise, this student is at risk of becoming stuck in her development as a professional STEM teacher and in her professional identity.

In conclusion, despite using an assignment not tailored to the requirements of a pragmatic identity analysis, the above provides a compact and differentiated picture of (a part of) student A's STEM teacher identity that connects PPE and PS. In addition, Fig. 14.4 provides cues for the further development of this STEM teacher's identity. In the case of student A, expertise development seems needed concerning motivation theories and active student learning. This knowledge can then replace the "naïve beliefs" that currently limit the students PPE. Also, the student will have to learn what to apply and when – "situational knowledge". Only then will successes be achieved that lead to enhanced identity development. With the growth of her expertise, a certain margin develops within which her behaviour is sufficiently professional and successful. Within this range, she can experiment with prosperous and less effective behaviour to shape her professional identity and develop her particular style. Implying also that she can transform her beliefs and preferences into successful practice, making her the kind of STEM teacher she wants to be.

14.7 Discussion

In our exploration into STEM identities, we argued that identities should be viewed as configurations. Next, we made a case for the inclusion of Personal Professional Expertise (PPE) within the concept of STEM identity for both professional and educational STEM identities, which we modelled using an elaboration of schema theory. Finally, we presented examples to illustrate both this model and the corresponding method to analyse STEM identities: "pragmatic identity analysis".

14.7.1 Research Contribution

Our plea for including STEM expertise in the concept of STEM identity, professional and educational does not detract anything from the merits of STEM identity research that does not include expertise. A majority of the STEM identity researchers may not come from a science or engineering tradition. Their work has indeed added valuably to STEM education research that before may have been overly focussing on knowledge, understanding and skills; and at times present a bias to just one tradition: that of canonical western science and technology.

We aimed to demonstrate that including personal STEM expertise in the concept of STEM identity can contribute to the study of STEM identity. In addition, we sought to further clarify the concept of STEM identity by exploring the idea of identities as structural configurations, by using schema theory to open avenues for more detailed identity descriptions, and finally, by showing pragmatic identity analysis as a method for mapping STEM identities. This led to identity descriptions that are more or less standardized, which can facilitate the comparison of identities; for example, between students, but especially when measuring the same student repeatedly at different times. Comparing different 'identity maps' constructed using the standard described can help to gain a better understanding of identity development and its dynamics. It could also support efforts to provide more and more detailed evidence about the relationship of STEM identities to, for example, learning outcomes.

14.7.2 Reflection on Pragmatic Identity Analysis

The second example illustrates how a pragmatic identity analysis arrives at a compact and structured description of PPE and connects it with PS to form a STEM identity description. However, there are some practical shortcomings in the assignment used to write this chapter. To improve the pragmatic identity analysis, ensuring completeness and referential validity are essential. The procedure needs elaboration. After analysing the narrative as shown in the example, a second step must follow, in which students are asked to complete the schema and add referential evidence. The results can be verified in a dialogue with the learner. Then an additional third step is needed, in which the student is invited to reflect on the whole of the STEM identity map, emphasizing further integration of the PPE-based schema with the self and to explore options for further identity development.

Apart from providing a map of PPE, pragmatic identity analysis provides concrete, specific clues on how a particular STEM identity could be further developed. In the second example, it appears that substituting knowledge for beliefs and challenging undifferentiated ideas about the applicability of views seem the first steps to take. This may signify a general implication of the presented view on STEM identity: expertise development may precede and underlie identity development.

14.7.3 Implications for STEM Education

The model developed also has implications for the practice of STEM education. The main message is that we need forms of education that address and connect both STEM expertise development and STEM identity development. Modern engineering education (Graham, 2018), in particular Challenge Based Learning (CBL), comes to mind here. According to Van den Beemt et al. (2021), CBL triggers student team learning by using open-ended challenges that are relevant to real life and authentic in the sense that they are derived from professional engineering practice. These challenges align with global issues (e.g., climate change), while typically engaging students with real stakeholders. Generally, at the beginning of the project, students do not have all the expertise needed to complete the project, but are asked to supplement their knowledge "on the job."

While the obvious merits of such authentic and rich learning environments for the formation of STEM identities have been argued previously, our model can contribute to a better understanding of how the construction of STEM identity occurs in relation to the learning process, and what would be particular concerns. For example, it is clear that on the job learning in such complex projects will be particularly demanding of students, especially when learning complex theoretical STEM concepts. Overloading students could easily happen in CBL, but it follows from our explorations in this chapter that this would limit the scope for STEM identity construction. After all, workload implies that only the most immediate and familiar ways of approaching things still seem legitimate.

On the other hand, because the construction of STEM identity and the development of STEM expertise are intertwined, teachers might consider choosing certain contents that contribute to the development of certain (alternative) types of STEM identities. For example, by emphasizing certain aspects of STEM, such as social aspects inherent in assuming the role of expert advisor, or by addressing a topic for which there are multiple alternative scientific models, each of which is incomplete, approximate, and presuppositional. In this way, a situation is created in which students can demonstrate the value to STEM of certain beliefs, attitudes, and behaviours that are not included in traditional science and thus put them forward as components of legitimate STEM identities.

References

Aikenhead, G. S. (1996). Science education: Border crossing into the subculture of science. *Studies in Science Education, 27*(1), 1–52. https://doi.org/10.1080/03057269608560077

Alexander, C. N., & Rudd, J. (1984). Predicting behaviors from situated identities. *Social Psychology Quarterly, 47*(2), 172–177. https://doi.org/10.2307/3033945

Anderson, J. R. (2005). *Cognitive psychology and its implications*. Macmillan.

Anspal, T., Eisenschmidt, E., & Löfström, E. (2012). Finding myself as a teacher: Exploring the shaping of teacher identities through student teachers' narratives. *Teachers and Teaching, 18*(2), 197–216. https://doi.org/10.1080/13540602.2012.632268

Archer, L., DeWitt, J., Osborne, J., Dillon, J., Willis, B., & Wong, B. (2010). "Doing" science versus "being" a scientist: Examining 10/11-year-old schoolchildren's constructions of science through the lens of identity. *Science Education, 94*(4), 617–639. https://doi.org/10.1002/sce.20399

Avraamidou, L. (2014). Studying science teacher identity: Current insights and future research directions. *Studies in Science Education, 50*(2), 145–179. https://doi.org/10.1080/03057267.2014.937171

Bamberg, M. (2011). Who am I? Narration and its contribution to self and identity. *Theory & Psychology, 21*(1), 3–24. https://doi.org/10.1177/0959354309355852

Beauchamp, C., & Thomas, L. (2009). Understanding teacher identity: An overview of issues in the literature and implications for teacher education. *Cambridge Journal of Education, 39*(2), 175–189. https://doi.org/10.1080/03057640902902252

Beijaard, D., Meijer, P. C., & Verloop, N. (2004). Reconsidering research on teachers' professional identity. *Teaching and Teacher Education, 20*(2), 107–128. https://doi.org/10.1016/j.tate.2003.07.001

Berzonsky, M. D. (2004). Identity processing style, self-construction, and personal epistemic assumptions: A social-cognitive perspective. *European Journal of Developmental Psychology, 1*(4), 303–315. https://doi.org/10.1080/17405620444000120

Blank, L. M., Almquist, H., Estrada, J., & Crews, J. (2016). Factors affecting student success with a Google earth-based earth science curriculum. *Journal of Science Education and Technology, 25*(1), 77–90. https://doi.org/10.1007/s10956-015-9578-0

Burke, P. J., & Reitzes, D. C. (1981). The link between identity and role performance. *Social Psychology Quarterly*, 83–92. https://doi.org/10.2307/3033704

Burke, P. J., & Stets, J. E. (2009). *Identity theory*. Oxford University Press.

Canrinus, E. T. (2011). *Teachers' sense of their professional identity*. University of Groningen.

Carlone, H. B., & Johnson, A. (2007). Understanding the science experiences of successful women of color: Science identity as an analytic lens. *Journal of Research in Science Teaching, 44*(8), 1187–1218. https://doi.org/10.1002/tea.20237

Carlone, H. B., Johnson, A., & Scott, C. M. (2015). Agency amidst formidable structures: How girls perform gender in science class. *Journal of Research in Science Teaching, 52*(4), 474–488. https://doi.org/10.1002/tea.21224

Collins, A., Brown, J. S., & Newman, S. E. (1989). Cognitive apprenticeship: Teaching the crafts of reading, writing, and mathematics. *Knowing, Learning, and Instruction: Essays in Honor of Robert Glaser, 18*, 32–42.

Dehing, A. J. M. (2012). *Preparing students for workplace learning in higher engineering education*. Eindhoven University of Technology. https://doi.org/10.6100/IR733802.

Dehing, A. J. M., Jochems, W. M. G., & Baartman, L. K. G. (2013). Development of an engineering identity in the engineering curriculum in Dutch higher education: An exploratory study from the teaching staff perspective. *European Journal of Engineering Education, 38*(1), 1–10. https://doi.org/10.1080/03043797.2012.742866

Farrell, T. S. (2011). Exploring the professional role identities of experienced ESL teachers through reflective practice. *System, 39*(1), 54–62. https://doi.org/10.1016/j.system.2011.01.012

Fishbein, M., & Ajzen, I. (2011). *Predicting and changing behavior: The reasoned action approach*. Taylor & Francis.

Freitag, D., Stautz, S., Melton, R., & Benson, H. S. (2021). *Elements of narrative - A structural approach to writing narrative*. https://www.rcboe.org/cms/lib010/GA01903614/Centricity/Domain/4395/Elements%20of%20a%20Story.pdf

Gilbert, J. K., & Watts, D. M. (1983). Concepts, misconceptions and alternative conceptions: Changing perspectives in science education. *Studies in Science Education, 10*(1), 61–98. https://doi.org/10.1080/03057268308559905

Graham, R. (2018). *The global state of the art in engineering education*. Massachusetts Institute of Technology (MIT) Report.

Gresalfi, M., Martin, T., Hand, V., & Greeno, J. (2009). Constructing competence: An analysis of student participation in the activity systems of mathematics classrooms. *Educational Studies in Mathematics, 70*(1), 49–70. https://doi.org/10.1007/s10649-008-9141-5

Hatmaker, D. M. (2013). Engineering identity: Gender and professional identity negotiation among women engineers. *Gender, Work and Organization, 20*(4), 382–396. https://doi.org/10.1111/j.1468-0432.2012.00589.x

Hazari, Z., Sonnert, G., Sadler, P. M., & Shanahan, M.-C. (2010). Connecting high school physics experiences, outcome expectations, physics identity, and physics career choice: A gender study. *Journal of Research in Science Teaching, 47*(8), 978–1003. https://doi.org/10.1002/tea.20363

Henriksen, E. K., Dillon, J., & Ryder, J. (Eds.). (2015). *Understanding student participation and choice in science and technology education*. Springer. https://doi.org/10.1007/978-94-007-7793-4

Herrera, F. A., Hurtado, S., Garcia, G. A., & Gasiewski, J. (2012). *A model for redefining STEM identity for talented STEM graduate students* [Conference presentation]. American Educational Research Association Annual Conference.

Holmegaard, H. T., Ulriksen, L. M., & Madsen, L. M. (2014). The process of choosing what to study: A longitudinal study of upper secondary students' identity work when choosing higher education. *Scandinavian Journal of Educational Research, 58*(1), 21–40. https://doi.org/10.1080/00313831.2012.696212

Horowitz, M. J. (2012). Self-identity theory and research methods. *Journal of Research Practice, 8*(2), 14.

Ibarra, H. (1999). Provisional selves: Experimenting with image and identity in professional adaptation. *Administrative Science Quarterly, 44*(4), 764–791. https://doi.org/10.2307/2667055

Jansen, M. (2016). *Professional identity of first-year students at the academic teacher training college: A narrative approach.*

Kaplan, A., & Garner, J. K. (2017). A complex dynamic systems perspective on identity and its development: The dynamic systems model of role identity. *Developmental Psychology, 53*(11), 2036–2051. https://psycnet.apa.org/doi/10.1037/dev0000339

Kim, A. Y., Sinatra, G. M., & Seyranian, V. (2018). Developing a stem identity among young women: A social identity perspective. *Review of Educational Research, 88*(4), 589–625. https://doi.org/10.3102/0034654318779957

Kim, E., & Kim, K. (2015). Cognitive styles in design problem solving: Insights from network-based cognitive maps. *Design Studies, 40*, 1–38. https://doi.org/10.1016/j.destud.2015.05.002

King, D. (2016). Teaching and learning in context-based science classes. In I. R. Taconis, P. J. den Brok, & A. Pilot (Eds.), *Teachers creating context-based learning environments in science* (pp. 71–85). SensePublishers.

Knez, I. (2016). Toward a model of work related self: A narrative review. *Frontiers in Psychology, 7*(331), 1–14. https://doi.org/10.3389/fpsyg.2016.00331

Kock, Z.-J., Taconis, R., Bolhuis, S., & Gravemeijer, K. (2013). Some key issues in creating inquiry-based instructional practices that aim at the understanding of simple electric circuits. *Research in Science Education, 43*(2), 579–597. https://doi.org/10.1007/s11165-011-9278-6

Lamb, R., & Davidson, E. (2002). Social scientists: Managing identity in socio-technical networks. In *Proceedings of the 35th annual Hawaii international conference on system sciences*, pp. 1132–1141. https://doi.org/10.1109/HICSS.2002.994034.

Leeferink, H. (2016). *Leren van aanstaande leraren op en van de werkplek*. Eindhoven University of Technology.

Leeferink, H., Koopman, M., Beijaard, D., & Ketelaar, E. (2015). Unraveling the complexity of student teachers' learning in and from the workplace. *Journal of Teacher Education, 66*(4), 334–348. https://doi.org/10.1177/0022487115592163

Lloyd, D., & Boyd, W. E. (2014). An exploration of the role of schema theory and the (non-indigenous) construction of indigenous identity. *Environmental Education Research, 20*(6), 795–813. https://doi.org/10.1080/13504622.2013.833592

Mommers, J., Schellings, G. L. M., & Beijaard, D. (2016, May 25). *Identity workshops voor beginnende docenten: Herkennen en omgaan met identiteitsspanningen.* Onderwijs Research Dagen (ORD 2016), Rotterdam, The Netherlands.

Mommers, J., Schellings, G. L. M., & Beijaard, D. (2017). Het opbouwen van een professioneel zelfbeeld: In een workshop of digitale omgeving. 17.

Nia, M. G., Harandi, M. F., & de Vries, M. J. (2019). Technology development as a normative practice: A meaning-based approach to learning about values in engineering-damming as a case study. *Science and Engineering Ethics, 25*(1), 55–82. https://doi.org/10.1007/s11948-017-9999-7

Nylen, A., Daniels, M., Pears, A., Cajander, A., McDermott, R., & Isomottonen, V. (2018). Why are we here? The educational value model (EVM) as a framework to investigate the role of students' professional identity development. *IEEE Frontiers in Education Conference (FIE), 2018*, 1–8. https://doi.org/10.1109/FIE.2018.8659055

Pajares, M. F. (1992). Teachers' beliefs and educational research: Cleaning up a messy construct. *Review of Educational Research, 62*(3), 307–332. https://doi.org/10.3102/2F00346543062003307

Patrick, H., & Pintrich, P. R. (2001). Conceptual change in teachers' intuitive conceptions of learning, motivation, and instruction: The role of motivational and epistemological beliefs. Understanding and teaching the intuitive mind. *Student and Teacher Learning*, 117–143.

Perrenet, J. C., Bouhuijs, P. A. J., & Smits, J. G. M. M. (2000). The suitability of problem-based learning for engineering education: Theory and practice. *Teaching in Higher Education, 5*(3), 345–358. https://doi.org/10.1080/713699144

Perrenet, J. C., & Taconis, R. (2009). Mathematical enculturation from the students' perspective: Shifts in problem-solving beliefs and behaviour during the bachelor programme. *Educational Studies in Mathematics, 71*(2), 181–198. https://doi.org/10.1007/s10649-008-9166-9

Pillen, M. T., Beijaard, D., & den Brok, P. J. (2012). Tensions in beginning teachers' professional identity development, accompanying feelings and coping strategies. *European Journal of Teacher Education, 36*(3), 240–260. https://doi.org/10.1080/02619768.2012.696192

Pillen, M. T., den Brok, P. J., & Beijaard, D. (2013). Profiles and change in beginning teachers' professional identity tensions. *Teaching and Teacher Education, 34*, 86–97. https://doi.org/10.1016/j.tate.2013.04.003

Ping, C., Kleingeld, A., Rispens, S., & Taconis, R. (2021). *Perceptions of multicultural group work in the international classroom.* The 17th CDIO International Conference, Chulalongkorn University, Bangkok, Thailand.

Ping, C., Schellings, G. L. M., & Beijaard, D. (2018). Teacher educators' professional learning: A literature review. *Teaching and Teacher Education, 75*, 93–104. https://doi.org/10.1016/j.tate.2018.06.003

Polkinghorne, D. E. (2005). Language and meaning: Data collection in qualitative research. *Journal of Counseling Psychology, 52*(2), 137–145. https://doi.org/10.1037/0022-0167.52.2.137
Polkinghorne, D. E. (2007). Validity issues in narrative research. *Qualitative Inquiry, 13*(4), 471–486. https://doi.org/10.1177/1077800406297670
Rosiek, J., & Atkinson, B. (2005). Bridging the divides: The need for a pragmatic semiotics of teacher knowledge research: Bridging the divides. *Educational Theory, 55*(4), 421–442. https://doi.org/10.1111/j.1741-5446.2005.00003.x-i1
Roth, W.-M., & Tobin, K. (2007). *Science, learning, identity: Sociocultural and cultural-historical perspectives* (Vol. 7). SensePublishers.
Sartwell, C. (1992). Why knowledge is merely true belief. *The Journal of Philosophy, 89*(4), 167. https://doi.org/10.2307/2026639
Schachter, E. P. (2004). Identity configurations: A new perspective on identity formation in *Contemporary society*. *Journal of Personality, 72*(1), 167–200. https://doi.org/10.1111/j.0022-3506.2004.00260.x
Schachter, E. P. (2005). Context and identity formation: A theoretical analysis and a case study. *Journal of Adolescent Research, 20*(3), 375–395. https://doi.org/10.1177/0743558405275172
Schellings, G. L. M., Mommers, J., Koopman, M., & Beijaard, D. (2017). *Exploring configurations of personal and situational factors in teachers' identity development.* 17th Biennial EARLI conference for research on learning and instruction (EARLI 2017).
Scott, S. (2016). *Negotiating identity: Symbolic interactionist approaches to social identity*. Wiley.
Sewell, W. H. (1992). A theory of structure: Duality, agency, and transformation. *American Journal of Sociology, 98*(1), 1–29. https://doi.org/10.1086/229967
Sheppard, S. D., Macatangay, K., Colby, A., & Sullivan, W. M. (2008). *Educating engineers: Designing for the future of the field.* Book Highlights. Carnegie Foundation for the Advancement of Teaching.
Søreide, G. E. (2006). Narrative construction of teacher identity: Positioning and negotiation. *Teachers and Teaching, 12*(5), 527–547. https://doi.org/10.1080/13540600600832247
Steinke, J. (2017). Adolescent girls' STEM identity formation and media images of STEM professionals: Considering the influence of contextual cues. *Frontiers in Psychology, 8*(716), 1–15. https://doi.org/10.3389/fpsyg.2017.00716
Stets, J. E., Brenner, P. S., Burke, P. J., & Serpe, R. T. (2017). The science identity and entering a science occupation. *Social Science Research, 64*, 1–14. https://doi.org/10.1016/j.ssresearch.2016.10.016
Stevens, R., O'Connor, K., Garrison, L., Jocuns, A., & Amos, D. M. (2008). Becoming an engineer: Toward a three dimensional view of engineering learning. *Journal of Engineering Education, 97*(3), 355–368. https://doi.org/10.1002/j.2168-9830.2008.tb00984.x
Sullivan, W. M. (2004). *Vocation: Where liberal and professional educations meet*. The Fourth Annual Conversation on the Liberal Arts.
Taconis, R. (1995). *Understanding based problem solving: Towards qualification-oriented teaching and learning in physics education*. Eindhoven University of Technology.
Taconis, R., Ferguson-Hessler, M. G. M., & Broekkamp, H. (2001). Teaching science problem solving: An overview of experimental work. *Journal of Research in Science Teaching, 38*(4), 442–468. https://doi.org/10.1002/tea.1013
Taconis, R., & Kessels, U. (2009). How choosing science depends on students' individual fit to 'science culture.' International Journal of Science Education, 31(8), 1115–1132.
Taconis, R., de Putter-Smits, L. G. M., Henry, S., den Brok, P. J., & Beijaard, D. (2010). The construction of a questionnaire to evaluate the science orientedness of students' identities as learners from a cognitive perspective. *Educational Research and Evaluation, 16*(4), 383–400. https://doi.org/10.1080/13803611.2010.528206
Thorndyke, P. W. (1977). Cognitive structures in comprehension and memory of narrative discourse. *Cognitive Psychology, 9*(1), 77–110. https://doi.org/10.1016/0010-0285(77)90005-6
Tonso, K. (2007). Learning to be engineers: Welding together expertise, gender, and power. *Science, Learning, Identity: Sociocultural and Cultural-Historical Perspectives, 7*, 103–120.

Van den Beemt, A., van de Watering, G., & Bots, M. (2021). *Challenge-based learning in a higher education context*. Eindhoven University of Technology.

Van den Bogert, N. J. (2016). *On teachers' visual perception and interpretation of classroom events using eye tracking and collaborative tagging methodologies*. Eindhoven University of Technology.

Van der Sanden, J. (2004, December 3). *"Ergens goed in worden"* [becoming a professional] [Inaugural speech]. https://pure.tue.nl/ws/portalfiles/portal/1114054029/Sanden2004.pdf

Van Driel, J. H., Bulte, A. M., & Verloop, N. (2007). The relationships between teachers' general beliefs about teaching and learning and their domain specific curricular beliefs. *Learning and Instruction, 17*(2), 156–171. https://doi.org/10.1016/j.learninstruc.2007.01.010

Von Glasersfeld, E. (2001). The radical constructivist view of science. *Foundations of Science, 6*(1–3), 31–43. https://doi.org/10.1023/A:1011345023932

Worren, N., Moore, K., & Elliott, R. (2002). When theories become tools: Toward a framework for pragmatic validity. *Human Relations, 55*, 1227–1250. https://doi.org/10.1177/0018726702055010082

Yackel, E., & Cobb, P. (1996). Sociomathematical norms, argumentation, and autonomy in mathematics. *Journal for Research in Mathematics Education, 27*(4), 458–477. https://doi.org/10.2307/749877

Chapter 15
How Activity Frames Shape Situated Identity Negotiation: Theoretical and Practical Insights from an Informal Engineering Education Program

Smirla Ramos-Montañez and Scott Pattison

A central challenge for research on science identities is developing tools and theories to understand how individuals exercise agency in defining and negotiating their identities and how social and cultural structures constrain and afford this process. Furthermore, researchers must address these dynamics at a scale that is meaningful to practitioners, such as program design or educator facilitation strategies. Currently most of the research related to the development of science identities has been conducted in the context of formal science education with a primary focus on understanding how youth develop their science identities and how educators can shape this process and even their own professional identities (Archer et al., 2010; Calabrese Barton et al., 2013; El Nagdi et al., 2018). While this focus on science identity has generated valuable insights, the conceptualization of science identity as situated, and context dependent demands the study of this construct across a variety of domains and contexts.

For example, if we consider that the development of science identities can be influenced by educators, culture, family, and peers, then the work to construct a science identity takes place both inside and outside the classroom (Archer et al., 2010; Tan et al., 2013). In fact, much of the work that youth do to develop and negotiate their science identities happens in social contexts among groups of children and adults (Esmonde, 2009; Falk, 2009; Gee, 2000; Norris, 2011; Penuel & Wertsch, 1995). Informal learning environments, like science centers and afterschool programs, have been championed as important opportunities for youth identity development, especially since youth often have more influence in defining what happens in informal learning activities than in the classroom (Falk & Dierking, 2013; National Research Council, National Academy of Engineering and National

S. Ramos-Montañez (✉) · S. Pattison
TERC, Cambridge, MA, USA
e-mail: smirla_ramos@terc.edu; scott_pattison@terc.edu

H. T. Holmegaard, L. Archer (eds.), *Science Identities*, Contributions from Science Education Research 12, https://doi.org/10.1007/978-3-031-17642-5_15

Research Council National Academy of Engineering and National Research Council, 2009).

Despite the perceived value of informal learning activities in the development of science identities, few studies have specifically focused on describing how identity negotiation happens in these contexts. Even fewer studies have considered identity negotiation in informal learning environments with topics other than science. Engineering, for example, is a STEM domain that has become a prominent focus across education curricula, with many educators looking at engineering education as a way to unify STEM domains and improve student learning and achievement in science and math (Cunningham & Carlsen, 2014; National Academies of Sciences, Engineering, and Medicine, 2020; Truesdell, 2014). In informal education spaces, engineering has been seen as a way to engage families in hands-on, interest-driven challenges with opportunities for creative problem-solving that can shape the way individuals engage with STEM in general (Ata-Aktürk & Demircan, 2020; Caspe et al., 2018; Svarovsky et al., 2018). While there are clearly many connections between science and engineering, we suspect that there are key differences in the dynamics of identity negotiation relative to each discipline and that the study of identity negotiation within the context of each is important to develop a broader understanding of STEM identity in general (see more discussion of these disciplinary distinctions later in the chapter).

In this chapter, we draw from 4 years of identity research through the NSF-funded *Designing Our World* (DOW) project to expand on science identity work conducted in formal classrooms and contribute to a growing body of engineering identity research in informal engineering education contexts (Cardella, 2020; Dou et al., 2019; Pattison et al., 2018a, b, 2020; Svarovsky et al., 2018). The DOW project provided a unique perspective on engineering-related identity negotiation by combining two theoretical perspectives, *situated identity* and *activity frames*, to describe the ways that youth negotiate their engineering-related identities with peers and adults. We believe this perspective is of value for both research and practice. From a research perspective, we introduce a framework and approach that accounts for the complexity of moment-by-moment interactions in informal STEM learning contexts, acknowledging both agency and structure dynamics relevant to identity development (Gutiérrez & Calabrese Barton, 2015). From a more practical perspective, we contribute to educators' understanding and awareness of activity frames, especially those that are at play in social and collaborative learning environments. Educators can incorporate this perspective in the development and delivery of programs, as well as the fostering of environments that support the STEM-related identity development of youth.

We begin the chapter by introducing the *Identity-Frame Model*, which describes engineering-related identity negotiation in the context of informal engineering learning experiences (Pattison et al. 2018a, b, 2020). The Identity-Frame Model is based on a situated perspective on identity and posits that identity negotiation involves both performance and recognition work by peers and adults. Through this performance and recognition work, youth develop context-specific identities and understandings about the nature and goals of the interaction, which in turn can further afford or constrain identity negotiation. It is this emergent understanding of the program and activities, which we refer to as activity frames, that we primary focus on in this chapter. Through a series of descriptive examples, we explore the concept of activity frames and the influence of different engineering-related frames on identity negotiation, how youth can impact the identity negotiation of their peers by shaping activity frames, and the tensions that can arise between educator- and youth-negotiated frames.

Finally, we discuss the relationship between science and engineering as it relates to STEM identity and present implications for research and practice of the Identity-Frame Model. In particular, we draw attention to the in-the-moment interactional dynamics that can influence the identity development of youth, specifically highlighting the ways that educators shape activity frames that can either afford or constrain youth's identity development.

15.1 Study Overview and Context

The work presented in the chapter emerged from two research studies implemented between 2014 and 2018 as part of the NSF-funded *Designing our World* (DOW) project, which was led by the Oregon Museum of Science and Industry (OMSI) in partnership with two local community-based organizations (Boys and Girls Club and Adelante Mujeres). The DOW project focused on engaging adolescent girls (ages 10–14) with engineering through science center exhibits, existing partnerships with local girl-serving organizations, educator and parent professional development, and afterschool programming designed to spark engineering-related interests for participants from communities where representation in the engineering workforce is scarce (National Science Board, 2018). DOW programming was developed based on research by the National Academy of Engineering (NAE) suggesting that framing engineering as an altruistic, collaborative, and personally relevant endeavor can motivate girls to pursue engineering careers in the future (National Academy of Engineering, 2008).

The DOW engineering programming lasted 3 months on average and involved a series of large-group discussions, presentations from guest engineers, and small-group engineering activities. OMSI educators facilitated the program and focused on highlighting altruistic and collaborative aspects of engineering while providing participants strategies to collaborate with their peers, persist through failure, and overcome frustration. For example, small group activities included engineering design challenges focused on designing surgical tools that could remove obstructions located in the internal organs of a dog or a human, building ziplines to carry an injured person safely to get treatment, and creating an assembly line to prepare supplies for people in areas that have suffered from natural disasters.

15.2 Theoretical Framework: Identity and Activity Frames

In our work, we take a *situated perspective on identity*, focusing on how the STEM-related identities of youth and adults emerge and are actively negotiated within specific social interactions and contexts (Gee, 2000; Jones, 2005; Penuel & Wertsch, 1995). To better conceptualize how these identities are constrained and afforded in these moments, and the complex social, cultural, and historical contexts underlying the interactions, we also draw from the theoretical notion of *activity frames* (Goffman, 1974; Tannen, 1993). Both perspectives are described below. In Table 15.1, we provide definitions of key concepts underlying these perspectives.

15.2.1 Situated Identity

Rather than a static trait of an individual, we conceptualize identity as situated and contextual—actively negotiated between particular individuals in specific situations and contexts (Carlone & Johnson, 2007; De Fina & Georgakopoulou, 2012; Gee, 2005; Penuel & Wertsch, 1995). This *situated identity* perspective has become pervasive across STEM education and learning research, although it can be challenging to operationalize and study (Pattison et al., 2020). From this perspective, identity can best be thought of as a rhetorical or discursive process—the ongoing act of trying to convince others (and oneself) through talk and action that you are a certain type of person while at the same time those around you respond to this performance and, through their talk and behavior, make their own claims about their identities (De Fina & Georgakopoulou, 2012; Gee, 2000). It is not enough to ask whether a

Table 15.1 Theoretical framework definitions

Term	Definition
Situated identity	Identities (who we are and how we want others to perceive us) that we actively negotiate with others in specific situations and contexts. These emerge from an ongoing process of performance and recognition or positioning by others during social interactions within the situation
Activity frame	Context- specific, emergent understandings or expectations, either implicit or explicit, about the nature and goals of an interaction within a specific situation and context. These emerge from ongoing participant talk and behavior that references or implies particular types of situations shared within a culture or society
Identity-frame model	A descriptive model combining the theoretical concepts of situated identity and activity frames to understand youth identity negotiation during informal STEM education programs (Fig. 15.1). The model posits that situated identity negotiation is an ongoing process of performance and definition work by an individual and recognition and positioning work by other adults and peers that creates emergent, context-specific, and interrelated identities and activity frames
Critical identity moment	A type of situation within a social interaction that can lead to more intense and explicit identity and activity frame negation among participants

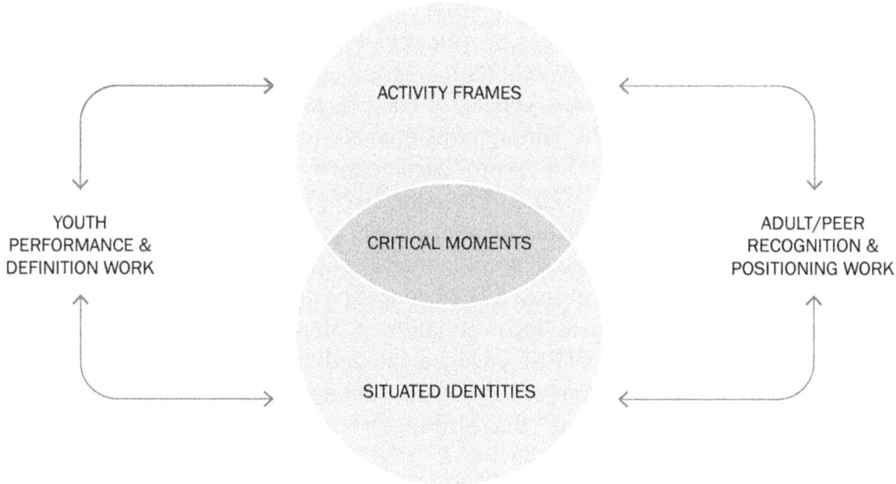

Fig. 15.1 Identity-frame model of youth situated identity negotiation. The Identity-Frame Model emerged through our work and is an organizing framework for thinking about and analyzing the intersections between activity frames and situated identities (Pattison et al., 2018a, b, 2020)

person thinks of themselves as having a "STEM identity" but rather how that person performs their STEM-related identity and whether or not other social participants recognize and support this performance (Carlone & Johnson, 2007; Kane, 2012; Ryu, 2015; Tan & Barton, 2008). An individual's STEM-related identity can, therefore, shift across interactions and contexts (e.g., inside and outside of school, with family members or peers) as that person is able to negotiate an identity as someone who is connected with STEM in some situations and with some groups but not others (Calabrese Barton et al., 2013; Calabrese Barton & Tan, 2019; Jones, 2005).

Recognition is a critical part of this negotiation process. Communicating and establishing an individual's identity within specific interactions is influenced and constrained by the ways others recognize and position that individual (Rodriguez et al., 2017; Tan & Barton, 2008). In the rhetorical and discursive process of situated identity negotiation, one's identity performance is only meaningful to the extent that it is recognized by others (Carlone et al., 2015; Carlone & Johnson, 2007; Fields & Enyedy, 2013; Hughes et al., 2021). Furthermore, the identity performance of an individual often makes claims about the identities of others. At the same time as individuals are negotiating their own identities, they are also being explicitly or implicitly positioned by others in the interaction relative to specific identities or types of people (Ryu, 2015; Tan et al., 2013).

Research has highlighted how individuals in positions of authority or expertise, such as teachers within a classroom, have particular power to recognize and position individuals with respect to their situated identities (Carlone & Johnson, 2007; Hazari et al., 2015; Tan et al., 2013). Peers, and especially those with higher perceived status in the classroom or other learning contexts, can also play a powerful role in positioning the situated identities of others (Pattison et al. 2018a, b; Ryu,

2015; Shim & Kim, 2018). This positioning can be directly relevant to an individual's relationship to STEM, or it can be salient to other aspects of identity that open up or constrain situated identity negotiation—such as positioning relative to social stereotypes connected with race, gender, or class (Carlone et al., 2015; Rodriguez et al., 2017; Tan et al., 2013). Through this process, teachers and peers can act as supports or "gate closers" for youth participation and identity development (Aschbacher et al., 2009; Ryu, 2015; Takeuchi, 2016; Tan & Barton, 2008).

Although the notion of situated identity clearly privileges the detailed study of social interaction and discourse in specific moments (Norris, 2011), and especially moments such as formal or informal STEM education program when STEM-related identities are made more salient, the negotiation of situated identities is also critical for long-term patterns of STEM engagement and participation. For example, Carlone and Johnson (2007) argued that these routines of identity negotiation can become "patterned and habitual," thus shaping how youth come to perceive themselves and be perceived by others beyond a specific interaction (Kane, 2015). As we highlight throughout this chapter, situated identity negotiation patterns can also shape the expectations and understandings that individuals have for specific types of contexts or interactions (e.g., how to act in a classroom or engage with a science activity), which in turn constrains and affords the types of identities that are available and supported in those contexts (Carlone et al., 2015; Hutchison & Hammer, 2009; Jimenez-Aleixandre & Rodriguez, 2000).

In the research described in this chapter, conducted in the context of an after-school informal engineering education program for adolescent youth, we focused on directly observing how participants made bids for identity recognition during the programs, how other participants and adults recognized or responded to these bids, and how other participants and adults positioned youth relative to being involved, taking active roles, and proficiently and competently using age-appropriate engineering practices. We focused on situated identities related to participation, engagement, and competence with the engineering activities and practices, rather than engineering careers or the field of engineering more broadly, because most of these youth were encountering this topic for the first time and were just beginning to develop ideas about engineering as a subject area or career. The engineering practices introduced by the DOW program, while similar to science practices, ultimately have different goals (National Academies of Sciences, Engineering, and Medicine, 2020; National Academy of Engineering & National Research Council, 2009). For example, the goal of the engineering activities was not guided by inquiry or the exploration of scientific phenomena but by the creation of a design or object as a solution to a specific problem with defined constraints and success criteria. Engagement in this process can foster values, attitudes, and skills that shape participants' outlook, experiences, and identities related to engineering. Ultimately, we believe the negotiation of the situated identities related to these engineering activities and practices likely has implications for the participants' long-term engagement with engineering, inside and outside of school (Calabrese Barton et al., 2013; Calabrese Barton & Tan, 2018).

15.2.2 Activity Frames

Through our research as part of the DOW project, we came to realize that the situated identity negotiation of participants was intimately linked to a second emergent characteristic of the interactions: the ways that participants seemed to understand the nature, goals, and expectations of the engineering activities. To conceptualize this relationship, we turned to the theoretical concept of "activity frames" (Goffman, 1974; Hand et al., 2012; Tannen, 1993), also known as situation definitions (Rowe, 2005; Scollon, 2001), from sociolinguistics and mediated discourse. In our work, we define activity frames as the context-specific, emergent understandings or expectations, either implicit or explicit, about the nature and goals of the interaction and the engineering activity (Pattison et al. 2018a, b, 2020).

Activity frames provide meaning to our daily interactions and a background context against which all action and talk is interpreted (Goffman, 1974; Hutchison & Hammer, 2009). For example, a business meeting is a commonly shared understanding of a specific type of situation that many individuals in our society have experienced. Within this activity frame, the acts of discussing the agenda, prioritizing topics, of raising one's hand to provide input all make sense, whereas they might seem ridiculous in the context of a social gathering of friends or family. Like situated identities, activity frames are not static but instead are actively negotiated. Many situations that we encounter in our daily life may be so routine that we are hardly aware of the subtle cues that prompt particular activity frames. In the example above, the combination of context and participant behavior can help to quickly establish a shared understanding that participants are part of a business meeting. However, in other cases, such as the engineering activities that we describe below, the process of negotiating a shared understanding about the nature and goals of the interaction may be more complex, the activity frame may shift throughout the interaction, and no single activity frame may come to be shared across all participants at a given time (Hutchison & Hammer, 2009; Jones, 2005; Shim & Kim, 2018).

Activity frames are important because, although they are context specific, they draw their meaning from broader cultural understandings and patterns (Jones, 2005; Rowe, 2005; Wertsch, 1998). Based on the cues used to invoke them and the social structures from which these cues draw their meaning, activity frames create constraints and affordances around what is possible in a given situation and the roles and identities that individuals are able to negotiate and perform (Jones, 2005; Rowe, 2005). By establishing the context for making meaning, activity frames privilege certain meanings, roles, and identities over others, and thus are associated with dynamics of power and authority (Rowe, 2005), including expectations about "right" or "wrong" ways of behaving (Gee, 2000; Hutchison & Hammer, 2009; Jimenez-Aleixandre & Rodriguez, 2000). For example, raising one's hand can invoke not only an activity frame related to school but also associated ideas about a teacher or expert being the source of knowledge and authority. As described above, just as certain individuals have more power and authority to recognize and position others relative to their situated identities, these individuals may also have more influence over the activity frames that are salient within a specific experience and ultimately shape how situated identities are negotiated (Pattison et al. 2018a, b; Ryu, 2015; Shim & Kim, 2018).

Prior research has described the types of frames that are common in STEM classrooms and the ways these frames shape classroom discourse, knowledge transfer, and argumentation patters. Hammer and others (Berland & Hammer, 2012; Elby & Hammer, 2010; Hammer et al., 2005; Hutchison & Hammer, 2009; Jimenez-Aleixandre & Rodriguez, 2000; Shim & Kim, 2018) have explored how different understandings of classroom activities or discussions, such as playing the "classroom game" (Lemke, 1990), making sense of the science, doing the lesson, or understanding the phenomenon, shape how students approach science learning, relate to each other, recognize the authority of teachers and students, and engage in scientific discourse and practices. Similarly, Berland and Hammer (2012) described how student expectations about whether science argumentation was about getting the right answer or persuading others influenced the depth and nature of the scientific discourse. Collectively, this classroom research has highlighted the importance of frames for shaping learning, the dynamic and shifting nature of these frames even within a particular interaction, the importance of certain "sticky" frames (Hammer et al., 2005) that are more likely to take hold and persist, and the ways that teachers can both passively and actively work to shape classroom frames through their words and actions, their design of the lessons, and their explicit discussion of learning goals with students.

15.2.3 Intersection of Frames and Identity

Through the DOW project, we extended the research described above to explore how activity frames shape, interact with, and emerge from the process of STEM-related identity negotiation for youth. We also extend this research to an informal learning context—where scholars have argued there is the potential for participants to more flexibly define what is going on and who can be what in a given situation (Calabrese Barton & Tan, 2019; Falk & Dierking, 2013; National Research Council, National Research Council, 2000). However, there may also be more uncertainty about how these dynamics will emerge in these informal learning environments, with either positive or negative consequences for the STEM-related identities of participants (Pattison et al. 2018a, b).

Through our study of adolescent youth participating in the DOW program, we developed the *Identity-Frame Model* as an organizing framework for thinking about and analyzing the intersections between activity frames and situated identities (Pattison et al. 2018a, b, 2020). The model posits that situated identity negotiation is an ongoing process of performance and definition work by an individual and recognition and positioning work by other adults and peers that creates emergent, context-specific, and interrelated identities and activity frames (Fig. 15.1). The identities that emerge through negotiation among participants help to define the understandings of what the situation is about (i.e., framing) and the understandings of the situation constrain and afford what participants are able to be positioned and recognized as (i.e., situated identities). These identities and frames are made particularly salient during critical identity moments, when the understanding of the situation and the roles and identities of participants within that situation are "at stake" (Penuel & Wertsch, 1995) or "being called into question" (Hammer et al., 2005).

Other researchers have described the two-way relationship between epistemological framing and positional framing (Greeno, 2009; Shim & Kim, 2018). However, our notion of situated identity goes beyond "students' perceptions of themselves and other group members in the knowledge building process" (Shim & Kim, 2018, p. 132) to encompass how participants position themselves and are positioned by others relative to the each other, the STEM activity, and the STEM domain more broadly. Aligned with Gee's notion of situated identity (Gee, 2000, 2014), our conceptualization extends to how an individual sees themselves and is seen by others as a STEM learner and integral participant within and STEM learning community. Other theories have also looked at group interaction dynamics (Rabbie et al., 1989; Terry et al., 1999). The Identity-Frame model builds on and contributes to this work (Pattison et al. 2018a, b) while focusing specific attention on interaction dynamics related identity negotiation that potentially contribute to long-term patterns of identity development.

We recognize the inherent challenge of distinguishing and studying the connections between situated identity negotiation and activity frames, since both are inextricably linked. What an individual communicates about himself or herself in a particular situation often carries implications for the implied understandings of that situation (e.g., positioning yourself as a teacher implies a set of expectations about learning goals, the learning environment, and the learners). Similarly, making bids for a particular activity frame often accompanies claims for the type of identities and roles the individual would like to assume within that frame (e.g., a classroom frame is evoked by someone claiming to be the knowledgeable teacher). In our qualitative analyses, we often found that the same youth action or utterance provided evidence for both situated identities and activity frames.

In the remainder of the chapter, we draw from our research to illustrate the way we applied the Identity-Frame model to explore the types of activity frames that emerged during the DOW program, how these appeared to interact with participants' identity negotiation, and the factors that may influence this process, such as the role of leadership-oriented youth and the tension between youth- and educator-negotiated activity frames.

15.3 Methods for Exploring the Connection Between Framing and Identity

During the DOW project, we conducted two qualitative studies that included in-depth video analysis, case study development, interviews, and embedded assessment data collected during DOW afterschool engineering programs. Throughout the process, we followed best practices in culturally responsive research, such as engaging a multicultural research team with similar lived experiences to study participants, collecting and analyzing data in two languages (English and Spanish), conducting member checks with community partners, and disseminating findings in broad and relevant ways to communities and stakeholders (Falk et al., 2007; Gutiérrez & Rogoff, 2003; Okazaki & Sue, 1995; Thomas & Parsons, 2017).

Data collection included videotaping youth during their participation in the engineering activities, researcher observations and reflections, reflective activities with youth, and interviews with program staff and the participants' caregivers. Data analysis drew from techniques in grounded theory (Charmaz, 2014; Glaser & Strauss, 2009). Using this approach during in-depth video analysis, we developed a coding framework to describe the critical aspects of identity negotiation. The coding framework included *micro* codes that were applied to specific actions and behaviors for individual youth and *macro* codes, which were applied to whole video segments for each individual youth. Overall, we collected and analyzed approximately 25 h of video divided across 21 program sessions. From this we analyzed data for a subset of participants (a total of 17 focal youth between the two studies). Criteria for focal youth selection included level of participation and attendance to the program sessions. All video data were coded per focal youth and detailed video coding summaries were produced for each program session.

Following video coding, the research team implemented a case study approach (Calabrese Barton et al., 2013; Yin, 2018) to develop narrative descriptions ("youth profiles") for select program participants. A total of 12 youth profiles were developed across the two studies from the 17 focal youth initially analyzed. The narratives included the detailed video coding summaries, researcher observations and reflections, secondary data collected from participants, and information from parents and program staff. They also included a general youth description, a summary of their involvement during the program, and a description of their patterns and routines for identity negotiation. In conjunction with the video analysis, we utilized the youth profiles to explore emergent activity frames during the interactions, as well as other factors that could influence identity negotiation (see Pattison et al. 2018a, b, 2020 for additional details about the data collection and analysis methods).

15.4 Activity Frames in STEM-Related Identity Negotiation

The activity frames that emerged through our analysis ultimately reflected both the program's intended goals as well as emergent understandings influenced by the interactions among participants and facilitations. The engineering activities that were part of the programming highlighted altruistic and collaborative aspects of engineering, both because these are considered important engineering habits of mind (Cunningham & Kelly, 2017) but also because it has been shown to be a promising framing of engineering for our target audience (National Academy of Engineering, 2008). Program facilitators presented the engineering activities as collaborative design challenges focused on helping people or animals. Aligning with engineering practice, they discussed how collaborative work is essential in the engineering design process and framed failure as a normative and even expected occurrence in engineering that can lead to learning and improvement.

Regardless of the way that the programs were presented by facilitators, however, we observed participants expressing a variety of understandings about the nature and goals of the activities. This is not unexpected, since activity frames are not static but actively negotiated by youth participants and adult facilitators. Within the context of the DOW program, we observed negotiation of participant expectations related to both the activity goals and peer collaboration. The activity frames described here appeared to be closely linked with situated identity, since different understandings around the activity can create conflict among participants and salient moments of identity negotiation. Additionally, the way youth were positioned also held significance relative to the meaning and purpose of the engineering activities. To begin, we present examples of two sets of activity frames at play during the engineering activities that appeared to influence STEM-related identity negotiation.

15.4.1 Engineering Activities as Competitive or Collaborative

The first set of activity frames we describe involves negotiated expectations around the engineering activities as "individual and competitive" or "collaborative and supportive". In the following transcript, we observed three youth from the Boys & Girls Club, Jaime, Britany, and Ariel, working on the "Surgical Solutions" activity (please note that all participant names used in this chapter are pseudonyms).

This activity is similar to the game Operation, where participants had to design "surgical tools" that could be used to remove a blockage from the internal organs of a dog. Jamie, Britany and Ariel were all active participants throughout the duration of the engineering program. Ariel and Jamie were brother and sister (11- and 12-years old) and Britany (12 years old) also participated in the program with her sibling. In the "Surgical Solutions" challenge, they first built separate tools to target different design challenges. Once they had completed this task, they were encouraged to build a single tool that would incorporate the function of individual tools into a “master tool”. The three participants worked on the tools and alternated between working individually or in pairs. An adult facilitator checked-in periodically. The transcript starts during one of these check-in moments.

Initially, we saw a “collaborative activity frame” emerge when the adult facilitator asked the group what they had been working on. Britany and Jaime, who had been working together on one tool, started describing their progress (lines 4–5). They referred to the group as “we” and implied a collaborative activity frame, even coordinating to show the facilitator what they had done. As Jaime started describing and demonstrating one of the designs (line 7), Britany positioned him negatively by pointing out, according to her, the failure of the tool (line 7). At this moment the activity frame changed and more of a “competitive frame” emerged, with Britany focusing on the success and failure of individual group members and showing the adult facilitator how “successful” she was with a paper clip at a different challenge (line 8–9). The facilitator acknowledged her success and made a few comments to shift the focus towards collaboration (lines 8 and 19). However, the facilitator inadvertently ended up supporting the individual and competitive activity frame by complementing Britany as she took over from Jaime (lines 9–10), emphasizing individual ownership of the design ideas (line 13) and positioning Britany’s knowledge and skills in contrast to her peers– even declaring Britany the lead surgeon. (line 17). Towards the end of the interaction, we saw Britany positively positioning Ariel and her design (line 18) but also take the opportunity to highlight her own success (lines 14–16), which was supported by the facilitator (line 17).

In this transcript, Britany’s talk and behavior suggested a more individual and competitive activity frame. Through the interaction, Britany positioned herself as competent and knowledgeable while intentionally or unintentionally minimizing Ariel and Jaime’s contributions to the group. Throughout the program, Britany’s actions often implied an individual and competitive frame that contrasted from the activity frame communicated by the program educators. On the other hand, Jaime and Ariel communicated a more collaborative activity frame, often choosing to work together in the small-group engineering activities.

15.4.2 Failure as Either Negative or Positive

A second set of engineering-related activity frames that we observed influencing identity negotiation is related to the interpretation of failure as either negative or positive. Program facilitators presented persistence through failure as an integral part of engaging with engineering. In large-group settings, engineering role models discussed their approach towards failure and program facilitators helped participants brainstorm strategies to overcome frustration. In our studies, we identified a number of indicators related to seeing failure as a positive aspect during the engineering activities, such as persisting through failure, trying multiple iterations, recognizing failure and continuing to engage with the process, and using humor in response to challenges with the activity. In contrast, participants that seemed to recognize failure as a negative often expressed frustration or helplessness, blamed others or materials for failure, and sometimes stopped working on the design all together. In the prior transcript with Britany, Ariel, and Jamie (Table 15.1), Britany communicated both an individual and competitive activity frame as well as a negative view towards failure. She highlighted Jaime's failures (line 7) and seemed to focus on individual success and failure (lines 7–9).

Similar to Britany, Rosey, a 10 year old participant from the Boys and Girls Club, also tended to shape the dominant activity frames during the engineering activities. She often seemed to communicate a more individual and competitive frame and view failure as a negative. When she encountered failure, Rosey became visibly frustrated, blaming the materials or other participants for the perceived failure. In the following transcript, Rosey worked in a small group that was designing a zip line that could take an injured person safely from the top of a mountain to the ground. One participant was testing her initial design while Rosey observed.

At the beginning of the interaction, we observed a participant claim success at the activity after she tested her design (line 1). After the participant declared the design was working, Rosey quickly pointed out that in fact, the design did not work because it did not reach the end of the line (lines 1–2). Rosey communicated a negative view around failure. Her focus on enforcing the "rules" for the activity eventually led to a small argument between Rosey and the other participant (lines 6 and 7). She then proceeded to build and test an individual design, which also did not meet the group's expectations. At this point the participant that "failed" at the beginning of the interaction pointed out Rosey's failure as well (line 14). Rosey continued to work on this activity by herself, while the other participants eventually left the group and proceeded to find other youth to work with. In general, Rosey was very vocal about highlighting and celebrating her individual success throughout the engineering activities while highlighting other youths' perceived failures.

15.5 Role of Activity Frames in Shaping STEM-Related Identity Negotiation

In the previous section, we described the concept of activity frames and provided examples of possible activity frames at play during informal engineering education activities. A closer review of the transcripts also elucidates the intersection between activity frames and situated identity. For example, in Table 15.2, Britany's actions communicated more individual and competitive activity frames that dominated and were implicitly supported by others. This seemed to afford the negotiation of her role as a successful and confident participant while constraining the identity development of Ariel and Jamie, who through their actions communicated more of a collaborative activity frame. Britany's behavior throughout the program often supported an individual and competitive view of the engineering activities. She compared herself to others, blamed others for failure, and took credit for the group's success.

Table 15.2 Jaime, Britany and Ariel at the surgery solutions activity

Line no.	Conversation	Behavior
1	Facilitator: Look at this mess on top of the body! What is going on here?	Approaching table and remarking on the messy appearance of the table
2	Britany: We needed to get the tools that were in the bottom	Referring to the group sorting through the materials bin and making a mess at the work station
3	Facilitator: If this was an operating table, I don't know… So, what have you guys figured out so far?	Other participants laughing. Brittany tries to respond right away. Ariel and Jaime keep working
4	Britany: Umm, we used chopsticks for the ice cube. We sucked the penny out with a straw and poked the Play-Doh	Quickly moving through the stations and showing the different tools and items they used
5	Jaime: And we are going to go fishing.	Showing a tool that he built
6	Facilitator: Let's see!	Ariel, Brittany, and the facilitator move to watch Jaime try to work with a tool he and Brittany worked on prior to the facilitator arriving
7	Britany: Can't touch the tissue! Tissue! Yeah, that thing is not working!	Telling Jaime as he accidentally touches the border representing the tissue
8	Facilitator: That's okay	
9	Britany: Paper clip!	Grabbing a paper clip and trying to remove obstacle with it. Ariel and Jaime step back
10	Facilitator: Nice!	
11	Facilitator: I think you guys are going to get far with this, but you are going to have to design one tool that can do this thing. Right now it's taking two of you to do the same thing. Can you work together to make something?	
12	Jaime: This is falling apart	Referring to one of the challenge setup as he steps back from station
13	Facilitator: So, who came up with each idea?	
14	Britany: I came up with the pokey thing.	Responding immediately as Ariel tries to talk at the same time
15	Ariel: Mmm, I…	Interrupted by Brittany
16	Britany: I came up with the straw thingy and I came up with the chopstick thingy	
17	Facilitator: Sounds like you are the lead surgeon here	
18	Britany: She started doing it with the popsicle thing	Pointing to Ariel and mentioning a design Ariel used previously
19	Facilitator: Cool! You want to show me?	Asking Ariel to show her what she had tried. Ariel proceeds to show her

Similar to Britany, Rosey had a similar influence in shaping the activity frames at play during the engineering activities. In Table 15.3, Rosey communicated a negative view of failure to the members of the group. This allowed Rosey to position herself as knowledgeable and successful in the activity but limited the participation from other youth in the design process and ultimately undermined their identity negotiation.

In the next section, we provide two examples of group dynamics and factors that we observed influencing situated identity negotiation through activity frames. The first example shows how leadership-oriented youth that are part of the engineering activities can influence the dynamics in a small group, shaping dominant activity frames that afford or constrain the identity negotiation of their peers. The second example demonstrates how tensions that arise during the facilitation of educational activities can implicitly or explicitly undermine the way that the activities are framed.

Table 15.3 Rosey at the zip line rescue activity

Line no.	Conversation	Behavior
1	Participant 1: It's working!	Watching her design slide down the zip line
2	Rosey: No, it didn't go all the way. It has to go all the way	Participant 1 walks away, looking frustrated
3	Participant 2: Can I use that cup?	Asking for a plastic cup that Rosey is using for her design
4	Rosey: No!	Moving the plastic cup away from participant 2, who sits back and looks frustrated
5	Rosey: Let me see the person. Let me see, let me see, let me see!	Trying to take the "test" person from participant 3, who is still testing the design that participant 1 created
6	Participant 3: No, it doesn't have to go all the way	Arguing with Rosey
7	Rosey: Yes it does	
8	Participant 3: Does it have to go all the way?	Asking an adult program facilitator approaching the table
9	Adult: What do you think? If this was a real zip line and it went to here and this was the land...	Indicating a stop point before the bottom of the zip line
10	Participant 3: Yeah	Smiling and agreeing with the educator that the cart has to go to the end of the zip line to safely deliver the person
11	Adult: Hmm, it might be dangerous	
12	Rosey: Ready?	Preparing to test her design again
13	Participant 2: I don't want to be in this group anymore	Sitting back and looking frustrated
14	Participant 3: Hah, yours stopped!	Playfully wagging her finger at Rosey's design, which stopped before the end of the zip line

15.5.1 Shaping of Activity Frames by Leadership-Oriented Youth

Throughout their participation in the DOW engineering activities, all youth were observed negotiating situated identities and activity frames. They also responded to the identity work of others, sometimes recognizing and supporting identity work and other times undermining it. In this collaborative learning environment, some youth like Britany and Rosey repeatedly worked to position themselves and be recognized by others as leaders. In the context of classroom peer groups, Mercier et al. (2014) defined leadership as "attempts to move the group forward, either by addressing issues of organization of the group, such as turn management, or addressing the intellectual aspects of the activity, such as idea management and development" (p. 401). Similarly, we defined a group leader as a youth whose bids for leadership or "leadership moves" are acknowledged and taken up by the group (Pattison et al. 2018a, b). Leadership-oriented youth seemed to exert specific influence on the identity negotiation of others by shaping the activity frames that dominated throughout the engineering activities and by either supporting or undermining the identity negotiation of their peers.

In our studies, Rosey and Britany were two of these leadership-oriented youth. During the engineering activities, Rosey and Britany regularly positioned themselves as knowledgeable and competent in relation to the activities and leaders within the small groups. As observed in the previous transcripts, these youth seemed to have a central role in shaping the experience of other participants, which in turn could afford or constraint their peers' identity negotiation. In the transcripts, both Rosey and Britany undermined the activity frames presented by the program facilitators by highlighting failure as negative and by promoting the engineering activities more as individual and competitive rather than collaborative. These dominant activity frames did not support the identity negotiation of other participants and in some cases, seemed to alienate and even limit the space for other participants to engage in their own identity work.

In contrast to Britany and Rosey, other leadership-oriented youth created space for and supported the identity negotiation of other participants. Amparo, a 10 year old participant from the Chicas Youth Development Program (Adelante Mujeres), continuously modeled and reinforced the messages included in the engineering activities and often recognized and positioned other youth as knowledgeable and competent. In the following transcript, Amparo supported an activity frame of failure and iteration as a positive aspect of the engineering design process. She worked in a small group with other youth to create surgical tools designed to remove a blockage from a dog's internal organ (Table 15.4).

Amparo initially recognized the activity as challenging (line 1). However, she immediately provided suggestions to other youth in her group (line 3). She joined another participant to try a series of different strategies that failed (lines 5–8 and 12). Regardless, the group continued to work on the challenge and after a couple of

Table 15.4 Amparo at the surgical solutions activity

Line no.	Conversation	Behavior
1	Amparo: That one is going to be really hard	Approaching the activity station
2	Adult: So what are some ideas of how you would approach it?	
3	Amparo: Mira! (Look!) toothpicks, we could go like this!	Exclaiming her idea to other youth and beginning to work on the design. Other participants begin working on other challenges at the table
4	Adult: Nice, so, what is that, what is that for?	
5	Amparo: It's like chopsticks to get the marble. Let's try this out. OK help me out here, do we need more rubber bands?	Trying her design as the adult walks away. She is unsuccessful on her first attempt and passes on the tool to another youth at the table while she tries to improve her design by adding more rubber bands at the top
6	Participant 4: This won't get it, we need something that will hold it tight	Trying her design without success
7	Amparo: Right? I was trying to tie rubber bands to this. This is the right size to get it	Pointing to another design she had been working on that involved a clothes pin
8	Amparo: We can suck it out	Amparo grabs a straw. Participant gives her a disgusted look, grabs the clothes pin, and tries to tie the end open
9	Participant 4: Oh! Gross! Would a real surgeon actually do that?	Looking at Amparo in disbelief
10	Amparo: If they have to, yea! Let me try it	Amparo uses suction with straw to try to remove marble from tube
11	Amparo: Yea, it's too heavy. Oh, it has other straws in it!	She is unsuccessful but continues to work, realizing the straw she had been using had other straws inside, impeding the airflow. She removes the straws and tries again without success
12	Adult: I like that technique. I haven't seen anyone else do that	Amparo and other youth abandon suction and try another design

tries Amparo stated that doctors might need to try different and sometimes strange approaches when faced with a difficult challenge (line 10).

Through her actions and behavior, Amparo communicated an activity frame of failure as a positive aspect of the engineering design process. She shared ideas with other (line 1), used inclusive language, and invited others to brainstorm ideas and try different strategies (line 5–7). Amparo reinforced the messages from the program staff and contributed to a space where other youth could engage in identity negotiation work. Compared to activity frames associated with failure as a negative, like those in Table 15.1 and Table 15.2, this collaborative activity frame appeared to contribute to a decrease in competition among participants, reduce instances of internal group conflict, and avoid marginalizing or constraining the identity negotiation of other participants.

15.5.2 Tensions Between Educator and Youth-Negotiated Frames

Facilitation by adults of educational activities can also influence situated identity negotiation through the shaping of activity frames involved in the small group interactions. Similar to youth participants, educators communicate information that contributes to the ongoing negotiation of activity frames during the interactions. Educators generally have activity goals, outcomes, and messages they want to convey when they engage in the development and facilitation of learning activities. For example, with the DOW programming, facilitators wanted to highlight altruistic and collaborative aspects of the activities as well as persistence through failure. The activities were purposely framed this way, the design challenges reflected situations that were intended to be relevant to participants, and the group often discussed strategies to overcome frustration and foster iteration. However, despite the intentionality behind the framing for the activities, the emerging activity frames sometimes differed from those set forth by educators.

In Table 15.2 for example, we saw the facilitator explicitly encouraging collaboration when she asked youth to work together to make one surgical tool (line 13). However, later in the interaction she implied a competitive frame when she asked which youth had come up with each idea (line 13). The adult complemented Britany's design when she took over from Jaime (lines 9–10), emphasized individual ownership of the design ideas (line 13), positioned Britany's knowledge and skills in contrast to her peers (line 17), calling her the "lead surgeon." The competitive frame implied by the facilitator and the positioning of Britany could contribute to the marginalization of other youth or constraint their identity negotiation process.

A challenge for educators during program sessions like these is that they often have to divide their time between small groups as they also manage the logistics of the program. Therefore, when they approach a group, they might not be aware of what activity frames are emerging or what identity negotiation patterns are at play. In Table 15.3, the educator approached after Rosey had been communicating an activity frame related to the negative aspect of failure. The educator answered a question from a participant (line 9) but was not aware of the frustration the participant was experiencing. Eventually the participant left, most likely due to the conflict with Rosey. The educator inadvertently sent a subtle message that contradicted the framing of failure as an opportunity to revise and learn.

15.6 Implications for Supporting STEM Identity Development

This chapter provides a perspective of identity negotiation that accounts for the complexity of moment-by-moment interactions in informal engineering learning contexts and acknowledges both agency and structure dynamics relevant to identity development. We believe this perspective has implications for both educators and researchers interested in broadening their understanding of science and engineering identities.

15.6.1 Implications for Practitioners and Educators

Educators in both formal and informal learning spaces play important roles in creating and delivering programs and curricula that shape and support productive learning environments. Due to the nature of this work, educators have the opportunity of creating environments that support STEM-related identity negotiations for learners. One way that educators can do this is by understanding and being aware of activity frames at play in social and collaborative learning environments and the ways these activity frames influence STEM-related identity development. The dynamics in social and collaborative learning environments are complex, constantly evolving, and influenced by many factors, such as competition, power and control, friendship, and perceived expertise (Emmer & Stough, 2001; Takeuchi, 2016). Activity frames are an important part of the dynamics of learning environments and can be especially influenced by educators, as well as youth that take on leadership roles relatives to their peers, in ways that influence identity development. As seen in the examples in this chapter, in some cases educators and learners can promote stereotypes or dynamics that could negatively impact the learning experience. In other cases, educators and learners can promote environments where learners feel safe to perform STEM-related identities and be recognized for their contributions.

The Identity-Frame Model developed through the DOW project draws attention to the in-the-moment interactional dynamics that directly influence the identity development of youth and provides insights into the ways that teachers and educators can modify their own approaches to support STEM-related identity development. Researchers broadly agree that proximal processes, or the interactions of individuals with their social and physical environment, are the drivers of learning and development (Bronfenbrenner, 1979; NRC, 2000). In an education setting, evidence suggests that these ongoing interactions create normative expectations, patterns, and routines of behavior and discourse, including what counts as allowable and legitimate disciplinary practices, which in turn shape how individuals are able to negotiate their identities and how youth come to perceive themselves and be perceived by others beyond a specific interaction (Carlone, 2017; Carlone & Johnson, 2007; Hegedus et al., 2014; Polman & Miller, 2010; Ryu, 2015).

To leverage the importance of these moments and encourage deeper understanding of the complex dynamics of learning environments, including activity frames, the DOW project team developed a tool to engage educators in reflective practice (Pattison et al. 2018a, b). This tool encourages educators to reflect on their programs or curriculum-related goals, define possible indicators related to these goals, and identify programmatic aspects that can influence STEM-related identity negotiation, such as activity frames. The facilitation reflection tool also encourages educators to identify strategies for guiding or framing activities to provide opportunities for learners to continue negotiating their STEM-related identity. In our experience, we have observed educators in informal learning spaces such as museums and after-school programs benefit from this type of reflective practice, especially when multiple educators engage in observing and reflecting on the same program session. Having multiple perspectives on a program session fosters further discussion about facilitation and the development of programs, including possibly high-impact practices that support identity negotiation.

It is important to understand that every program will have unique needs and different dynamics influenced by participants, so the team encouraged educators to use the facilitation reflection tool in an exploratory rather than prescriptive way. This tool, along with other professional development strategies, can help identify educational practices that support opportunities for youth to negotiate their STEM identities for themselves. Additional research is needed to further understand situated identities and frames. Therefore, we encouraged educators to not jump to conclusions about specific activity frames but rather use the tool to increase their awareness of frames that emerge in their programs and their role in shaping STEM identity negotiation. Educators should also carefully consider and be aware of how they approach program design and facilitation to best align not only with their educational goals but also the activity frames most relevant to the learning area or discipline that their programs focus on. For example, since our research took place within the context of an engineering education program and the goal of the programs were to highlight the altruistic and collaborative aspects of engineering, some of the activity frames that educators focused on related to collaboration and viewing failure as a positive aspect of the engineering design process.

15.6.2 Implications for Research

The concepts of situated identity and activity frames presented here also have implications for researchers interested in understanding and describing the process of STEM-related identity negotiation, especially in informal learning environments. Our theoretical approach suggests that engineering identity is negotiated and constructed in particular moments, through interactions with peers, adults, and in our case, engineering activities. The negotiation of these situated identities is closely linked to the negotiation of emergent activity frames, which influence the understanding of educational activities for youth. As observed in the examples above, these activity frames may afford the identity development for some youth while constraining it for others.

As more research is conducted around science identity, it is important to continue considering the relationship between science and other STEM disciplines, while being careful not to use these terms interchangeably. Most of the work describing STEM identity has been constructed by studying engagement with science in the classroom. The work presented in this chapter contributes to broadening knowledge around STEM identity, specifically by studying how youth negotiate their identifies in the context of informal engineering education programs. Additional work is needed to explore the extent to which the situated identities and activity frames identified in our work are unique to engineering or are shared across other STEM domains and learning contexts. For example, as we discussed earlier, individuals participating in science and engineering activities can engage in common practices (e.g., planning, analyzing, evaluating, using math and technology). However, ultimately science and engineering have different goals (National Academies of Sciences, Engineering, and Medicine, 2020; National Academy of Engineering & National Research Council, 2009). Since engineering education experiences often involve the creation of a design or object to achieve a specific purpose, rather than the more open-ended exploration and understanding of a phenomenon or topic (NGSS Lead States, 2013), we suspect that critical moments, activity frames, and identity negotiation patterns related to failure will be particularly salient in an engineering education context. Martin (2015) identified "failure-positive" perspectives (e.g., celebrating failure as part of the creative process and recognizing the importance of "getting stuck and unstuck"), also referred to as "productive failure" (Litts et al., 2016), as critical to making, design, and engineering (see also Weiner et al., 2018). In contrast, science-related learning experiences may motivate more identity negotiation related to knowledge and expertise (e.g., Fields & Enyedy, 2013; Kane, 2015; Shim & Kim, 2018). Across both domains, other aspects of identity negotiation dynamics may be similar, since adults and youth participants often draw from the broader lexicon of school-related identities and activity frames (Hutchison & Hammer, 2009; Jimenez-Aleixandre & Rodriguez, 2000).

Regardless of the STEM domain, we believe it is important for researchers interested in using the concepts of situated identity and activity frames as tools for exploring STEM-related identity negotiation to take into account the broader social and cultural structures that intersect and influence situated identity within these STEM learning spaces. While we currently do not know the extent to which activity frames represent recognizable understandings of different situations for different participants and how this recognition influences the identity negotiation process, we suspect that these dynamics can shape long-term identity development of youth during their participation in the programs and possibly beyond. Researchers should continue to expand the understanding of identity negotiation dynamics that are unique and similar across different STEM learning contexts and should work alongside practitioners to create learning experiences that support positive and inclusive STEM identity negotiation and ultimately set the stage for life-long STEM learning and engagement.

Acknowledgements We are grateful to the many individuals that have contributed to this work as part of the Designing Our World project, including staff from the Oregon Museum of Science and Industry (OMSI), The Boys and Girls Club and Adelante Mujeres. This work was supported by the National Science Foundation under Grant No. 1322306.

References

Archer, L., DeWitt, J., Osborne, J., Dillon, J., Willis, B., & Wong, B. (2010). "Doing" science versus "being" a scientist: Examining 10/11-year-old schoolchildren's constructions of science through the lens of identity. *Science Education, 94*(4), 617–639. https://doi.org/10.1002/sce.20399

Aschbacher, P. R., Li, E., & Roth, E. J. (2009). Is science me? High school students' identities, participation and aspirations in science, engineering, and medicine. *Journal of Research in Science Teaching, 47*(5), 564–582. https://doi.org/10.1002/tea.20353

Ata-Aktürk, A., & Demircan, H. Ö. (2020). Supporting preschool children's stem learning with parent-involved early engineering education. *Early Childhood Education Journal, 49*(4), 607–621. https://doi.org/10.1007/s10643-020-01100-1

Berland, L. K., & Hammer, D. (2012). Framing for scientific argumentation. *Journal of Research in Science Teaching, 49*(1), 68–94. https://doi.org/10.1002/tea.20446

Bronfenbrenner, U. (1979). *Ecology of human development*. Harvard University Press.

Calabrese Barton, A., & Tan, E. (2018). A longitudinal study of equity-oriented stem-rich making among youth from historically marginalized communities. *American Educational Research Journal, 55*(4), 761–800. https://doi.org/10.3102/0002831218758668

Calabrese Barton, A., & Tan, E. (2019). Designing for rightful presence in STEM: The role of making present practices. *Journal of the Learning Sciences, 28*(4–5), 616–658. https://doi.org/10.1080/10508406.2019.1591411

Calabrese Barton, A., Kang, H., Tan, E., O'Neill, T. B., Bautista-Guerra, J., & Brecklin, C. (2013). Crafting a future in science: Tracing middle school girls' identity work over time and space. *American Educational Research Journal, 50*(1), 37–75. https://doi.org/10.3102/0002831212458142

Cardella, M. E. (2020, March). *Early childhood engineering: Supporting engineering design practices with young children and their families.* Conference presentation. NARST 2020 Annual international Conference, Portland, United States. https://www.researchgate.net/publication/340234317_Early_Childhood_Engineering_Supporting_Engineering_Design_Practices_with_Young_Children_and_Their_Families.

Carlone, H. B. (2017). Disciplinary identity as analytic construct and design goal: Making learning sciences matter. *Journal of the Learning Sciences, 26*(3), 525–531. https://doi.org/10.1080/10508406.2017.1336026

Carlone, H. B., & Johnson, A. (2007). Understanding the science experiences of successful women of color: Science identity as an analytic lens. *Journal of Research in Science Teaching, 44*(8), 1187–1218. https://doi.org/10.1002/tea.20237

Carlone, H. B., Johnson, A., & Scott, C. M. (2015). Agency amidst formidable structures: How girls perform gender in science class: Agency, structure, and girls in science. *Journal of Research in Science Teaching, 52*(4), 474–488. https://doi.org/10.1002/tea.21224

Caspe, M., Woods, T. A., & Kennedy, J. L. (Eds.). (2018). *Promising practices for engaging families in STEM learning*. IAP, Information Age Publishing, Inc.

Charmaz, K. (2014). *Constructing grounded theory* (2nd ed.). Sage.

Cunningham, C. M., & Carlsen, W. S. (2014). Teaching engineering practices. *Journal of Science Teacher Education, 25*(2), 197–210. https://doi.org/10.1007/s10972-014-9380-5

Cunningham, C. M., & Kelly, G. J. (2017). Epistemic practices of engineering for education: Epistemic practices of engineering for education. *Science Education, 101*(3), 486–505. https://doi.org/10.1002/sce.21271

De Fina, A., & Georgakopoulou, A. (2012). *Analyzing narrative: Discourse and sociolinguistic perspectives*. Cambridge University Press.

Dou, R., Hazari, Z., Dabney, K., Sonnert, G., & Sadler, P. (2019). Early informal STEM experiences and STEM identity: The importance of talking science. *Science Education, 103*(3), 623–637. https://doi.org/10.1002/sce.21499

Elby, A., & Hammer, D. (2010). Epistemological resources and framing: A cognitive framework for helping teachers interpret and respond to their students' epistemologies. In L. D. Bendixen & F. C. Feucht (Eds.), *Personal epistemology in the classroom: Theory, research, and implications for practice* (pp. 409–434). Cambridge University Press. https://doi.org/10.1017/CBO9780511691904.013

El Nagdi, M., Leammukda, F., & Roehrig, G. (2018). Developing identities of STEM teachers at emerging STEM schools. *International Journal of STEM Education, 5*(1), 36. https://doi.org/10.1186/s40594-018-0136-1

Emmer, E. T., & Stough, L. M. (2001). Classroom management: A critical part of Educational Psychology with implications for teacher education. *Educational Psychologist, 36*(2), 103–112. https://doi.org/10.1207/S15326985EP3602_5

Esmonde, I. (2009). Ideas and identities: Supporting equity in cooperative mathematics learning. *Review of Educational Research, 79*(2), 1008–1043. https://doi.org/10.3102/0034654309332562

Falk, J. H. (2009). *Identity and the museum visitor experience*. Left Coast Press.

Falk, J. H., & Dierking, L. D. (2013). *The museum experience revisited*. Left Coast Press.

Falk, J. H., Dierking, L. D., & Foutz, S. (Eds.). (2007). *In principle, in practice: Museums as learning institutions*. Alta Mira Press.

Fields, D., & Enyedy, N. (2013). Picking up the mantle of "expert": Assigned roles, assertion of identity, and peer recognition within a programming class. *Mind, Culture, and Activity, 20*(2), 113–131. https://doi.org/10.1080/10749039.2012.691199

Gee, J. P. (2000). Identity as an analytic lens for research in education. *Review of Research in Education, 25*(1), 99–125. https://doi.org/10.3102/0091732X025001099

Gee, J. P. (2005). *An introduction to discourse analysis: Theory and method* (2nd ed.). Routledge.

Gee, J. P. (2014). *An introduction to discourse analysis: Theory and method*. Routledge.

Glaser, B. G., & Strauss, A. L. (2009). *The discovery of grounded theory: Strategies for qualitative research*. Aldine.

Goffman, E. (1974). *Frame analysis: An essay on the organizationss of experience*. Harper & Row.

Greeno, J. G. (2009). A theory bite on contextualizing, framing, and positioning: A companion to son and goldstone. *Cognition and Instruction, 27*(3), 269–275. https://doi.org/10.1080/07370000903014386

Gutiérrez, K. D., & Calabrese Barton, A. (2015). The possibilities and limits of the structure-agency dialectic in advancing science for all: Limits of the structure-agency dialectic. *Journal of Research in Science Teaching, 52*(4), 574–583. https://doi.org/10.1002/tea.21229

Gutiérrez, K. D., & Rogoff, B. (2003). Cultural ways of learning: Individual traits or repertoires of practice. *Educational Researcher, 32*(5), 19–25. https://doi.org/10.3102/0013189X032005019

Hammer, D., Elby, A., Scherr, R. E., & Redish, E. F. (2005). Resources, framing, and transfer. In Mestre, J. P. (Ed.). (2006). *Transfer of learning from a modern multidisciplinary perspective* (p. 89). IAP.

Hand, V., Penuel, W. R., & Gutiérrez, K. D. (2012). (Re)framing educational possibility: Attending to power and equity in shaping access to and within learning opportunities. *Human Development, 55*(5–6), 250–268. https://doi.org/10.1159/000345313

Hazari, Z., Cass, C., & Beattie, C. (2015). Obscuring power structures in the physics classroom: Linking teacher positioning, student engagement, and physics identity development. *Journal of Research in Science Teaching, 52*(6), 735–762. https://doi.org/10.1002/tea.21214

Hegedus, T. A., Carlone, H. B., & Carter, A. D. (2014, June 15). *Shifts in the cultural production of "smartness" through engineering in elementary classrooms*. Conference presentation. 121st ASEE Annual Conference & Exposition, Indianapolis, United States.

Hughes, R., Schellinger, J., & Roberts, K. (2021). The role of recognition in disciplinary identity for girls. *Journal of Research in Science Teaching, 58*(3), 420–455. https://doi.org/10.1002/tea.21665

Hutchison, P., & Hammer, D. (2009). Attending to student epistemological framing in a science classroom. *Science Education, 94*(3), 506–524. https://doi.org/10.1002/sce.20373

Identifying and supporting productive STEM programs in out-of-school settings. (2015). National Academies Press. https://doi.org/10.17226/21740

Jimenez-Aleixandre, M., & Rodriguez, A. B. (2000). "Doing the lesson" or "doing science": Argument in high school genetics. *Science Education, 84*(6), 757–792. https://doi.org/10.1002/1098-237X(200011)84:6<757::AID-SCE5>3.0.CO;2-F

Jones, R. H. (2005). *Discourse in action: Introducing mediated discourse analysis* (1st ed.). Routledge. https://doi.org/10.4324/9780203018767

Kane, J. M. (2012). Young African American children constructing academic and disciplinary identities in an urban science classroom. *Science Education, 96*(3), 457–487. https://doi.org/10.1002/sce.20483

Kane, J. M. (2015). The structure-agency dialectic in contested science spaces, "Do earthworms eat apples?": Contested science spaces. *Journal of Research in Science Teaching, 52*(4), 461–473. https://doi.org/10.1002/tea.21206

Learning science in informal environments: People, places, and pursuits. (2009). National Academies Press. https://doi.org/10.17226/12190

Lemke, J. L. (1990). *Talking science: Language, learning, and values*. Ablex Publishing Corporation.

Litts, B. K., Kafai, Y. B., Searle, K. A., & Dieckmeyer, E. (2016). Perceptions of productive failure in design projects: High school students' challenges in making electronic textiles. *ICLS Proceedings*, 498–505. https://repository.isls.org//handle/1/154

Martin, L. (2015). The promise of the maker movement for education. *Journal of Pre-College Engineering Education Research, 5*(1), 30–39. https://doi.org/10.7771/2157-9288.1099

Mercier, E. M., Higgins, S. E., & da Costa, L. (2014). Different leaders: Emergent organizational and intellectual leadership in children's collaborative learning groups. *International Journal of Computer-Supported Collaborative Learning, 9*(4), 397–432. https://doi.org/10.1007/s11412-014-9201-z

National Academy of Engineering. (2008). *Changing the conversation: Messages for improving public understanding of engineering*. National Academies Press.

National Academy of Engineering & National Research Council. (2009). *Engineering in K-12 education: Understanding the status and improving the prospects*. National Academies Press.

National Academies of Sciences, Engineering, and Medicine. (2020). *Building capacity for teaching engineering in K-12 education*. National Academies Press. https://doi.org/10.17226/25612

National Research Council. (2000). *From neurons to neighborhoods: The science of early child development*. National Academy Press.

National Science Board. (2018). *Science and engineering indicators 2018* (NSB-2018-2). National Science Foundation https://www.nsf.gov/statistics/2018/nsb20181/.

NGSS Lead States. (2013). *Next generation science standards: For states, by states*. National Academies Press.

Norris, S. (2011). *Identity in (inter)action: Introducing multimodal interaction analysis*. De Gruyter Mouton.

Okazaki, S., & Sue, S. (1995). Methodological issues in assessment research with ethnic minorities. *Psychological Assessment, 7*(3), 367–375. https://doi.org/10.1037/1040-3590.7.3.367

Pattison, S. A., Gontan, I., Ramos-Montañez, S., & Moreno, L. (2018a). Identity negotiation within peer groups during an informal engineering education program: The central role of leadership-oriented youth. *Science Education, 102*(5), 978–1006. https://doi.org/10.1002/sce.21459

Pattison, S. A., Ramos-Montañez, S., & Gontan, I. (2018b). *STEM identity: A reflective tool for educators*. https://www.informalscience.org/stem-identity-reflective-tool-educators

Pattison, S., Gontan, I., Ramos-Montañez, S., Shagott, T., Francisco, M., & Dierking, L. (2020). The identity-frame model: A framework to describe situated identity negotiation for adolescent youth participating in an informal engineering education program. *Journal of the Learning Sciences, 29*(4–5), 550–597. https://doi.org/10.1080/10508406.2020.1770762

Penuel, W. R., & Wertsch, J. V. (1995). Vygotsky and identity formation: A sociocultural approach. *Educational Psychologist, 30*(2), 83–92. https://doi.org/10.1207/s15326985ep3002_5

Polman, J. L., & Miller, D. (2010). Changing stories: Trajectories of identification among African American youth in a science outreach apprenticeship. *American Educational Research Journal, 47*(4), 879–918. https://doi.org/10.3102/0002831210367513

Rabbie, J. M., Schot, J. C., & Visser, L. (1989). Social identity theory: A conceptual and empirical critique from the perspective of a behavioural interaction model. *European Journal of Social Psychology, 19*(3), 171–202. https://doi.org/10.1002/ejsp.2420190302

Rodriguez, S., Cunningham, K., & Jordan, A. (2017). STEM identity development for Latinas: The role of self- and outside recognition. *Journal of Hispanic Higher Education, 18*. https://doi.org/10.1177/1538192717739958

Rowe, S.. (2005). Using multiple situation definitions to create hybrid activity space. In *Discourse in action: Introducing mediated discourse analaysis* (pp. 123–134). Routledge.

Ryu, M. (2015). Positionings of racial, ethnic, and linguistic minority students in high school biology class: Implications for science education in diverse classrooms: Positionings in diverse science class rooms. *Journal of Research in Science Teaching, 52*(3), 347–370. https://doi.org/10.1002/tea.21194

Scollon, R. (2001). *Mediated discourse: The nexus of practice*. Routledge.

Shim, S.-Y., & Kim, H.-B. (2018). Framing negotiation: Dynamics of epistemological and positional framing in small groups during scientific modeling. *Science Education, 102*(1), 128–152. https://doi.org/10.1002/sce.21306

Svarovsky, G. N., Wagner, C., & Monica, C. (2018). Exploring moments of agency for girls during an engineering activity. *International Journal of Education in Mathematics, Science and Technology, 6*(3), 302–319.

Takeuchi, M. A. (2016). Friendships and group work in linguistically diverse mathematics classrooms: Opportunities to learn for English language learners. *Journal of the Learning Sciences, 25*(3), 411–437. https://doi.org/10.1080/10508406.2016.1169422

Tan, E., & Barton, A. C. (2008). From peripheral to central, the story of Melanie's metamorphosis in an urban middle school science class. *Science Education, 92*(4), 567–590. https://doi.org/10.1002/sce.20253

Tan, E., Barton, A. C., Kang, H., & O'Neill, T. (2013). Desiring a career in STEM-related fields: How middle school girls articulate and negotiate identities-in-practice in science. *Journal of Research in Science Teaching, 50*(10), 1143–1179. https://doi.org/10.1002/tea.21123

Tannen, D. (Ed.). (1993). *Framing in discourse*. Oxford University Press.

Terry, D. J., Hogg, M. A., & White, K. M. (1999). The theory of planned behaviour: Self-identity, social identity and group norms. *British Journal of Social Psychology, 38*(3), 225–244. https://doi.org/10.1348/014466699164149

Thomas, V. G., & Parsons, B. A. (2017). Culturally responsive evaluation meets systems-oriented evaluation. *American Journal of Evaluation, 38*(1), 7–28. https://doi.org/10.1177/1098214016644069

Truesdell, P. (2014). *Engineering essentials for STEM instruction: How do I infuse real-world problem solving into science, technology, and math?* NSTA Press.

Weiner, S., Lande, M., & Jordan, S. (2018). The engineer of 2020, in the making: Understanding how young adults develop maker identities and the implications for education reform. *International Journal of Engineering Education, 34*(2), 833–842.

Wertsch, J. V. (1998). *Mind as action*. Oxford University Press.

Yin, R. K. (2018). *Case study research and applications: Design and methods* (6th ed.). Sage.

Part VI
Conclusion

Chapter 16
Working Towards Justice: Critical Next Steps in Identity Studies in Science Education

Angela Calabrese Barton and Edna Tan

In this concluding Chapter, we build upon the key insights presented across this text, along with our own work, to make sense of the possibilities the field of identity studies in science education holds towards promoting a more just world. Indeed, inequities in opportunities to learn and become in science, especially for children and youth from nondominant communities – children and youth of Color, from low-income communities, who are undocumented, and who speak languages that institutional power-holders do not – are enduring and systemic. Such inequities are a result of structural, cultural, and relational inequities that shape their experiences in school science. The field has known, for some time, that how children and youth experience science, through curriculum, pedagogy, materials, classroom and out-of-classroom interactions, all shape opportunities for being and becoming in science.

Who one is and wishes to become is always in development, always shaped through social interaction with and against powered relationalities, always political, and always on the move. Keeping in mind the fluidity and contentiousness of authoring identities is central in efforts towards justice. The work of identity occurs in relation to how power operates in any given place – through discourses, practices, materials and structures. Understanding how people seek to engage in identity work that disrupts such powered relationalities towards more just ends is critical to the field. Indeed, across the chapters in this text, it is clear that an identity framing, by itself, is agnostic. There is nothing inherently justice-oriented in it unless one explicitly takes a justice-oriented lens.

In this concluding chapter we argue for why one must take a justice-oriented lens in identity work. Many of the chapters in this text do so either explicitly or

---

A. Calabrese Barton (✉)
University of Michigan, Ann Arbor, MI, USA
e-mail: angiecb@umich.edu

E. Tan
The University of North Carolina at Greensboro, Greensboro, NC, USA

H. T. Holmegaard, L. Archer (eds.), *Science Identities*, Contributions from Science Education Research 12, https://doi.org/10.1007/978-3-031-17642-5_16

implicitly, and how they do so is elemental to the field. Drawing upon this body of work, and our own, we argue that the imperative to the present and future of identity studies is a focus on the ethical and political dimensions of identity work. First, we describe how and why identity studies are, or can be, justice projects. Using this as a backdrop and in dialogue with the texts in this work and our own research, we problematize and consider how identity studies need to consider: (1) Whose justice, vis-à-vis intersectional oppressions borne by particular bodies are centered? (2) The relationalities between trajectories of identities and trajectories of (in)justice; and (3) The relationalities between disrupting dominant science discourses and narratives and identity work. Then, using a rightful presence framework, which calls attention to the ethical and political struggles of identity work, we outline new directions for the field to consider in efforts to expand the study of identities as a justice-oriented project. We show how the chapters in this text offer clear guidance on these next steps, as well as offer questions for further consideration.

16.1 Identities Studies in Science Education as Justice-Oriented Project

16.1.1 Identities and Identity Work as Power-Mediated

The focus on identity in science teaching and learning ought to be framed as a justice-oriented project *because it is always about power* (Urrieta & Noblit, 2018). Several themes about power arise across the texts which help to flesh out important directions of identity research in support of justice. First, several chapters call attention to how who counts as somebody, how so, and why, are shaped by normative discourses and practices as they take place in social context.

One example of the power-mediated ways in which identity work happens is presented in Chap. 5. Drawing upon sociological theories of consumption and social reproduction, Wong develops the idea of *symbolic exchange* to show student decisions to study advanced physics is related to its exchange-value and forms of symbolic identity. That is, such decisions are about more than learning or connecting to the domain, but also about the "recognition" and "gradual embodiment of the cultures of science and physics" (p. 11). This chapter speaks directly to power because it raises questions on *who* has access to such forms of symbolic exchange, and the *risks* one takes in making such an exchange, given the White and heteropatriarchal underpinnings of physics and science.

This Chapter and others remind us that access and opportunity are not benign and thus neither is the opportunities for identity work young people may have. As we see in Chap. 6, such issues are deeply but differentially related to risks and rewards depending on who one is – in relation to race, gender, sexual orientation and socioeconomic status. For example, in this Chapter, Holmegaard and Johannsen use theories of cultural production to illustrate how power is exercised in how both teachers and students make meaning of what "talent" is and how it is performed in university classrooms. Starting with the stance that higher education actively and purposefully reproduces inequities in everyday practices by "pretending to reward students who are talented", these authors show the complex unfolding of risks and rewards for students as they engage in talent negotiations as part of their university experiences. These authors explored how constructions of talent are culturally embedded, and take shape in performances, negotiations and experiences. By carefully showing how talent is produced and by whom, and how it is legitimated in higher education science learning contexts, these authors show the systemic nature of inequity as identities are enacted in everyday practice in relation to such talent negotiations.

16.1.2 Whose Justice? The Role of Intersectional Oppressions

As exemplified by these examples, studies of identities as a justice-oriented project begs the question, Whose Justice? And as Johansson and Larsson (Chap. 8) aptly ask, "Whose Knowledge? Whose Interests?" are considered when designing for learning opportunities. Such questions speak to opportunities for identity work. Consider the following two quotes from our own participatory ethnographic work with youth (e.g., Tan & Calabrese Barton, 2020):

> The racist stereotype is that Black people are not listening to science. But that is not true. Maybe it's the other way around, like science is not listening to us. I just wish that people could see what I could do, like what I am doing at home. Making home-made hand sanitizer, making masks, caring for my elders. I don't want to act White. I don't want people to tell me I'm not white enough. I want you to know how I feel as a young, Black girl in America and in STEM. I want to feel like I can be me in STEM and have that celebrated. (Jazmyn, 15-years old).

> Hi Mr. B, this is your student Quentin, the first one in the second row. I'm going to tell you about things that we should do in Science… I do things out of school and out of [my after-school STEM club] that involve science. I went to door to door and ask adults if they use CFL lights. The majority of the adults did NOT use CFL lights, I try to decrease the amount of people who use incandescent lights. I did it on Wain Wright Ave. and I did it because people's bills are up because they use just incandescent lights … But, it's not so much for energy that I get attention in school, but for being a smart alek. (Quentin, 11-years old).

Jazmyn and Quentin's words are powerful. They remind us that youths' opportunities to learn and become in science are shaped by how schools and science legitimate participation and learning. Yet, it is well documented that schooling and science operate as White, heteropatriarchal dominant spaces, discounting youths' embodied presence of youth of Color and girls (Morales-Doyle, 2017). For example, these two youth tell us that:

- They are invisible in school, made missing despite their embodied presence – this is Quentin the first one in the second row.
- They are recognized for their deficits and not their strengths

- Their community science expertise is not recognized or valued in school science
- They have to act White and fight stereotypes on their own.

Indeed, the culture and structures of science teaching and learning routinely advantage those who fit dominant culture while producing cumulative and chronic adverse outcomes for those who do not. How people talk, act, use their bodies and move – all identity performances – shape how people are recognized for who they are and what they bring to science (Calabrese Barton et al., 2012; Tan et al., 2013). As we've seen in the chapters of this text, the discourses and practices of a learning community directly impact whether or not particular people are recognized as good students, science experts or good classroom community citizens.

Several of the chapters either directly or indirectly raise hard questions for the field of identity studies with respect to its role in either reproducing or transforming social inequality. Johansson and Larsson's (Chap. 8) review of the literature in physics identities provides insights into the consequences of the various ways in which identity has been approached. Without attention, as they write, to "whose knowledge" and "whose interests" are given precedence in this work in order to challenge the epistemological basis of the role of identities in learning, White heteropatriarchy – that which undergirds science - remains the norm by which all are measured and Otherized. That identity studies can claim that "Minority students do not develop enough of a physics identity" (Chap. 8) is a clear example of such forms of oppressive reproduction in this very field of work in hyper-racialized ways.

In a very helpful move, Seiler and Kwamboka (Chap. 10) ask how race has been incorporated into science teacher identity research as one way to advance understanding of how identity work always takes place across scales of activity in ways that increase risk among those already oppressed by dominant culture. Schools are racialized spaces, and thus how these spaces mediate identity work looks and feels different given one's positioning in the space. Using Hall's (2015) construct of "diaspora", Seiler and Kwamboka show how teachers from non-dominant communities are placed in untenable positions as they have to navigate whiteness and Westernness on a daily basis through how they are constantly reminded of how they have been "displaced" and "taken away" from what they know and value. This leaves teachers from non-dominant communities with profound choices as they seek to become good science teachers. As these authors write, "Constructing one's identity under the interplay of history, culture, and power that permeates our schools is challenging for all teachers, but it has particular fractures and conflicts for those who have experienced life in figured worlds often distinct from those of the educational system and science" (Chap. 10).

16.1.3 Trajectories of Identities/Trajectories of (In)Justice

Judgements of whose knowledge, experiences, interests, discourses and thus identities, matter, also have trajectories, just as the development of identities do. They seed actions that produce and reproduce injustices. In other words, who is teaching or learning, what is taught and why, how outcomes are defined and measured, all work to position some youth as marginal or central in becoming in science just because of who they are and what they bring to the science learning community. The opportunities one has to engage in valued activity, to draw upon their wisdom, knowledge and practices of their lived lives in ways as they do so and to have that welcomed by others, and to be recognized for their contributions can be traced back to the judgements of authority figures who wield power over youth. Keeping trajectories and their critical transition points, such as from school to careers and further education is central to justice-focused work (e.g., Rabe and Kressdorf's work presented in Chap. 4).

Trajectories call attention to the need for longitudinal studies and what they make visible. For example, consider Archer et al.'s study of a Black working-class young woman, Vanessa and her parents (Chap. 2). These authors, using a Bourdieusian conceptual lens and an in-depth case study, show how schooling enacts intersectional injustices, building across settings and time, often in mundane and systemic ways, thus impacting young Black women's social futures. The authors show how Vanessa is pushed out of STEM, despite, still having a love for STEM. They further illustrate how such trajectories are set in motion through "the rules of the game" regarding *what* and *who* gets valued within a given space (Chap. 2). This point is crucial for it illustrates the importance of focusing on the system as the point of intervention rather than the individual.

A longitudinal focus on the study of identities – and thus also on the design work for programs in support of identity work – also powerfully shifts the foci from identity as a static and essentialized trait, to one that is dynamic and socially mediated. Gonsalves and Rahm et al. (Chap. 3), for example, show, for example, how girls' engagement in activity develops over time. When programs are designed in ways which give increasing access and legitimacy to non-dominant forms of course, such as the attention to the value of the currency of popular femininity, girls identification towards science increases. This helps to make visible the power of identity resources such as what the authors refer as non-dominant forms of science capital, in shifting when, where and how the girls author identities in science.

As these Chapters and our own work illustrate, youth's words, actions and performances all call into question whose justice is centered in the work of identity studies. What the field needs to come to terms with is that science teaching and learning contexts for youth who have been oppressed by dominant discourses practices and relationalities, have historically been places of both real and symbolic trauma. Identity work in these contexts is mired in these challenges (Urrieta, 2019).

Indeed, who one is, in the past, present and possible futures, is informed by the encounters one has with others while enacting new practices and activity (Lave &

Wenger, 1991). Who one is and who one is becoming, at any given moment, is always under negotiation and contingent upon the resources to which one has access (Wortham, 2006). The qualitative metasynthesis of identity studies offered by Butterfield and Marshall (Chap. 14) punctuates this point by showing not only how the complexity of identity work matters for youth of Color – that is, how it is composed of powerful non-cognitive factors, but also how, over trajectories, systemic factors such as negative imposed identities from external sources, shape these opportunities to become.

16.1.4 Disrupting Dominant Discourses/Narratives

Studies of identities in science education have been presented with a daunting tension. Important studies of STEM identity development have framed identity in terms of competence in the discipline and being recognized for it (e.g., Carlone & Johnson, 2007). As these authors state, "One cannot pull off being a particular kind of person (enacting a particular identity) unless one makes visible to (performs for) others one's competence in relevant practices, and, in response, others recognize one's performance as credible" (Carlone & Johnson, 2007, p. 1190). As these and other studies (e.g., Hazari et al., 2010) indicate, being recognized as a science person is related to what one knows, how they use what they know, and how that is valued by powerful others in the field. While it is important to make sense of how people perform identities by drawing upon shared and valued knowledge matters in one's enactments of identity work, it is also problematic to frame a science identity in terms of a settled view of the field (Bang & Medin, 2010). Being recognized as a "science person" could be contingent on a narrowly defined criteria that hews to the White patriarchal norms that characterize the culture of science. Competence – and its underlying ideologies – need to be contested. Indeed, as Madsen and Malm (Chap. 7) ask, what is at stake when becoming and negotiating a geo-science identity?

Dominant discourses of schooling are powerful mediators shaping both students' and teachers' identity work. Wade-Jaimes in Chap. 9, for example, shows how teachers engage in identity work is related to the dominant discourses of schooling as well as the relational work of teaching and learning. This chapter, in some ways, serves as a cautionary tale on the possibilities for disrupting the kinds of deficit discourses which position urban youth of Color in marginal ways. A focal point on teachers and teaching practice itself, while important, cannot be separated from the broader sociopolitical contexts of schooling and education in society.

However, such insight can provide powerful points of intervention. Consider how Richmond and Wray, in Chap. 11, link teachers' identity work not only with who they are and are becoming, but what they can do and perhaps desire to do. Echoing other work in this text, these authors show how teachers' pedagogical decisions are shaped profoundly by how teachers' identities mediate their understandings of and agency within school communities and school culture. They offer insights into interventions which support teachers' identity development in ways that provide points of access for more critically interrogating school culture, for example, by focusing on "what it is that they value and to explore how it is these values developed" (p. 14). More studies that examine the intersection of everyday pedagogical decision making and practice and how it is informed by context and culture would be productive.

Thus, acts of identity work, especially for youth and teachers from nondominant communities, are often tension-filled. When or why one bids for recognition, or how one is recognized by their teacher, peers and others, in classroom life, is an artifact of the powered relationalities that shape life in those spaces (Nasir, 2011).

Dixon et al. in Chap. 12 call attention to these powered relationalities as they examine how science itself is positioned in learning environments in ways that open up or foreclose students' identity and agency development. They show how when science is put in the service of action, in the context of community-engaged activity, new modes of interactivity arise – creating new norms, roles and practices that youth and community can take up, sometimes in alignment with but sometimes challenge those of professional science. These new activity systems produce new narratives of who is and can be successful in school science in support of the on-going formation of identities linked to youths' desired social futures.

These authors, like many in this text, draw upon social practice theory to help make sense of the ongoing struggle between personal and historical narratives influencing participation with science by integrating "the study of persons, local practice and long term historically institutionalized struggles" (Holland & Lave, 2009, p. 1). The struggle between the two forms of history influences individuals' actions in local practices. The institutional and personal narratives experienced or brought into the classroom hold specific meanings for the actors in this space. Thus, the relations between history-in-person and history-in-institutionalized struggles, erupt primarily because local practice(s) comes about in the encounters between "people as they address and respond to each other while enacting cultural activities under conditions of political-economic and cultural-historical conjecture" (Holland & Lave, 2009, p. 3). Local contentious practice can be thought of as the *critical stuff* of becoming.

Ramos-Montañez and Pattison in Chap. 15 provide a close analysis of these very kinds of negotiations, and account for the complexity of moment-by-moment interactions in informal STEM learning contexts. Using situated identity and activity frames Ramos-Montañez and Pattison describe the ways that youth negotiated their engineering-related identities with peers and adults. They show how, identity negotiation is an active process, shaped by youths' interactions with others and by their changing understanding or expectations of the nature and goals of the learning activities. What is really powerful in this chapter is the attention given to the tensions that arise between educator and youth-negotiated frames, another form of contention. The implication here is that in order to create empowering spaces for young people to develop identities in science, educators need to be aware of the activity frames they put into play in social and collaborative learning environments, and how these activity frames shape their students STEM-related identity work.

As we have argued elsewhere (Calabrese Barton et al., 2020), examining how contentious practice takes shape over time and across spaces allows for better understanding of how people negotiate new future-oriented identities in the tensions and relations between history-in-person and history-in-institutionalized struggles. This view foregrounds the importance of disrupting historicized narratives in how individuals participate in cultural activity towards new social futures. This space of becoming functions as a space of refusal, resistance, and radical possibility, as much as it does a bid for recognition. The question to ask here is this: What do people want to be recognized for, and how do such acts demand for what kinds of disruptions and transformations of science and schooling?

16.2 How Rightful Presence Extends the Identity Research Agenda

The collection of chapters in this text offers powerful directions in terms of justice-oriented identities studies. And they leave the field with the key question: How do we, as a field, take up the tension of unsettling the field as a part of identity studies? In the remainder of this chapter we focus on how a focus on rightful presence can offer one way forward.

Rightful presence emerges from critical justice studies of guest-host relationships in sanctuary cities, troubling assumptions of what it means to be inclusive in-the-present rather than in some abstract future (Calabrese Barton & Tan, 2020). It calls attention to the importance of the political struggle to re-author the rights for legitimized belonging, and how these rights are enacted in practice in ways that disrupt injustices and unequal power, rather than simply extending rights which may, in effect, reproduce unequal power.

We can think of a rightful presence in science education as a way of being and becoming in a place where one is legitimately welcomed, undoing the order that currently exists in learning environments and from which dominant groups abundantly benefit. This perspective shifts attention from an inclusionary model where people who were once outsiders are invited to conform to science, to a justice model – where the terrain of science is actively contested through re-seeing and re-making science, including its own fraught histories and possibilities, through the lives of those the field itself has made missing.

We can also think of rightful presence as a critical mode for 'making present' those who have been made "missing" by the forms of racialization and colonization manifest in schooling (Calabrese Barton & Tan, 2019). A rightful presence is not established firmly in one event, but builds over time, through many events.

If we return to the quotes of Jazmyn and Quentin, and to many of the examples in this text that focus on youth, we can see them both making bids for legitimacy in science. For many youths, engaging in science in school and/or out-of-school settings is constraining and limited (e.g., Chap. 3). Not all youth are encouraged or supported in leveraging *their* powerful cultural expertise towards meaningful identity work. We can think about this as denying youth a Rightful Presence in science.

Youth's opportunities to be and to become would look different in classrooms if students were more than guests – if students had a rightful presence in their classrooms. Youth we've worked have stated, for example, that an important form of expertise that should be valued in schools is that of "community science expert". As one 12-year old youth, Jana, explained, instead of keeping school science and every life in silos, being and becoming a community science expert through re-mixing science into one's lived life and community wisdom to make a difference, should be the goal of science education. As she explains:

> Being a community science expert is doing things that are good for the community because of what we know. We know a lot of science and we also know a lot about our community. Who else can put these ideas together?

In authoring identities as community science experts, youth are not framing community knowledge as a resource towards fitting into existing power dynamics already at play, but rather about up-ending what really counts to them as "science that matters". In other words, the youth are demanding for more than equity-as-inclusion into existing structures and norms. They are asking for a reexamination of what means to do science, why, how, when and for whom (thus echoing fundamental onto-epistemological questions raised by Johansson and Larsson in Chap. 8). This stance puts forward by the youth is not trivial. Embedded within it is an onto-epistemological shift accounting for, as Tuck (2009) has written, "the loss and despair, but also the hope, the visions, the wisdom of lived lives and communities," (p. 417).

How youth embody and bring to bear such hope, visions and wisdom of lived lives and community wisdom in their science engagement is through identity work. For such identity work to become consequential, structural changes are necessary in order for such as yet unsanctioned identity work to be recognized, supported, and built upon. This point ties to the cross-cutting theme of focusing on the system as critical points of intervention rather than the individual. This is impossible with an equity-as-inclusion model since inclusion is contingent on maintaining current norms and structures related to schooling and to science. Thus, to support identity work that is both anchored in, and that seek to amplify community wisdom historically shunned by schooling and science, is to engage in political struggle.

Rightful presence calls attention to the importance of political struggle in identity work. By political struggle in identity work, we refer to the contentious processes of re-organizing power in the science learning environment. Political struggle has both ethical and political dimensions. That is, it involves contentious processes meant to disrupt power through critical awareness and contestation in ways that affect people's lives, social relations and possibilities.

As we noted earlier, in science learning environments – whether it be classrooms or informal learning spaces, power is organized through the routines of the dominant culture – in the expectations for and enactments of discourses, practices and relationalities (e.g., see discussion of talent negotiations, Chap. 6, Holmegaard and

Johannsen). Thus, being and becoming is always filled with tensions and contradictions as resources, practices and relationships are negotiated (Gutiérrez & Penuel, 2014). Whose lives – histories, identities, communities and ideologies – are legitimately welcomed in these spaces and how, are contingent on who has power in these spaces.

How young people's lived lives and community wisdom are thus made present in the discourses, practices and relationships of becoming in, with and against science, is the work of justice. Becoming, in this sense, takes shape through embodied experiences of fraught histories and concrete injustices, and through youth's already-present brilliant, rebellious, and agentic acts of everyday practice and its transformative potential (see Selier and Kwamboka's discussion of race, Archer et al's foregrounding of Black science capital or Gonsalves and Rahm's discussion of non-dominant forms of science capital). Youth, through interactions with peers and educators, bid to have their lived lives and community wisdom made visible and integral to becoming. In these bids they are seeking to reorganize how power is distributed. They are engaged in political struggle.

One major implication is how the field may make sense of youths' *identity work* as forms of justice-oriented disruption/remediation. The political struggle to be rightfully present invokes contestation. These acts have social, material and spatial significance in how they "refuse to cooperate with and reproduce the norms, representational practices and spatial meanings" of science and schooling in orienting towards new social futures (Elwood & Mitchell, 2013, p. 276). Such acts, which bear witness to the struggles/injustices of peers and community members in relation to dominant power structures, leverage upon a wide range of knowledge and practices to imagine and participate in social transformation towards establishing a more rightful and humanizing presence (see Ramos-Montañez and Pattison's discussion of moment-to-moment interactions, or Dixon et al.'s discussion of agency).

This includes challenging if, how, and why one may be fully being welcomed with a rightful presence in a science learning community because of how power operates through the normative discourses and practices of these spaces. Powerful, justice-oriented, non-agnostic identity work happens when such normative discourses and practices are disrupted. This is because such disruptions allow youth to engage in becoming in ways that open up possibilities for re-creating what it means to be an expert in science. These possibilities center and amplify the cultural repertoires of practices that youth bring to learning because of who they are, where and with whom they grow up, and their imaginations for the present and future.

This political struggle is made manifest through interactions and emergent tensions between youth and the sociocultural/historicized narratives of science learning spaces/institutions that shape daily discourse, practices, and power relations of the people therein. This is where local contentious practice becomes a useful construct towards understanding the kinds of identity work that supports youths' rightful presence through political struggle.

The political struggle to be and become as rightfully present in science can be seen in the how local contentious practices makes visible how the space between person- and institution-in-history becomes and functions as a "space of refusal," and a "space of resistance" (Calabrese Barton et al., 2020). Such contention is disruptive/transformative for both the people as they author identities in relation to history-as-person as well as for the institutions to structurally re-orient towards radical change.

Political struggle is an important part of identity work, in part, because it helps us to see the need for disruptions. We suggested above that the normative discourses, practices, values and ideologies of a learning space need to be disrupted. We also suggest that the construct of identity itself requires disruption. Consider what Luis Urrieta (2019, p. 10) writes in relation to the colonizing nature of identity studies:

> I want to revisit and critique the concept of identity as a Western notion centered on self-awareness and individualism. The Western relational aspect of identity often centers on the contrast between Self/Other; the Self often implying the Western center, while the subaltern usually the marginal Other (Sarup, 1996)... The construction of identity as always premised on exclusion, what one is not, is therefore a Euro-Western set-up, an en-"trap"ment for the maintenance of power hierarchies and the preservation of whatever is constructed as normative (i.e., mestizo, ladino, white supremacy, etc.)

Urrieta's (2019) argument is critical to orienting the field towards new approaches to identity studies that work towards justice. Identities center difference. They are signatural. They not only claim who one is, but also who one is not. When youth engage in identity work to bring into school science discourse and practices that which has been silenced and erased towards a new social future in schooling and in science that is supportive of who they want to be, they are doing so from a space of the imaginary. Likewise, powered authority figures such as teachers and facilitators also need to operate from the space of the imaginary in figuring out how to recognize and support such rightful presence seeding identity work.

Acts of becoming or of identity work can be understood as involving not only how one sees oneself as an individual becoming, but also in how they coordinate with others in the present, but drawing from historicized pasts and hoped for futures, towards collective understanding, disruption, and action of and for new social-spatial futures. We can think of recognition of identities here as not so much of the individual in as much as it is of the new social order created by youth. For many youth, identity work is less an individual – "I" – process of becoming in relation to the Other, but rather a humanizing, critical and collective endeavor – "I/We" (Urrieta, 2019) to transform the practice of becoming for themselves and their communities. This turn asks us to consider youth as already-experts, with ample ability to leverage upon past and present narratives towards re-orienting what the future may look like and how they can shape it. This I/We framing shifts the gaze from the individual to the collective endeavor to remediate the systems of science and schooling that have worked to maintain self/other and in so doing erase youth as already-experts.

16.3 Discussion

The political struggle of rightful presence calls attention to the ways in which identity work – especially for young people who have been marginalized by dominant discourses, practices and relationalities – operate as a process of reclaiming erased histories and relationalities, and of transformative resistance towards how one is legitimized in orienting the field itself (Calabrese Barton et al., 2020). We view youth's political struggle to reclaim the processes of identity work in science as an effort to reorganize the spaces of learning as sites "of preserving and redefining collective history and community cultural wealth" (Rosario-Ramos et al., 2017, p. 223). This differs from the re-organization of the self. We refer to this kind of political struggle as one of reclaiming science because who one is and what it means to be and become in science is a contentious process of renegotiating the liminal space between people-in-history and institutional struggle; an active contestation of the terrain of becoming in science.

Reclaiming science acknowledges that while young people's lived lives and community wisdom are integral to the social, physical and representational dimensions of science engagement, they historically have not been centered in the design of science learning experiences and they have historically been erased from science. Overlaying identity work investigations with overtly justice-oriented frameworks can be a productive way forward to gaining a more nuanced understanding on how to design for and sustain consequential science learning experiences for historically underserved youth.

References

Bang, M., & Medin, D. (2010). Cultural processes in science education: Supporting the navigation of multiple epistemologies. *Science Education, 94*(6), 1008–1026. https://doi.org/10.1002/sce.20392

Calabrese Barton, A., & Tan, E. (2019). Designing for rightful presence in STEM: The role of making present practices. *Journal of the Learning Sciences, 28*(4–5), 616–658. https://doi.org/10.1080/10508406.2019.1591411

Calabrese Barton, A., & Tan, E. (2020). Beyond equity as inclusion: A framework of "rightful presence" for guiding justice-oriented studies in teaching and learning. *Educational Researcher, 49*(6), 433–440. https://doi.org/10.3102/0013189X20927363

Calabrese Barton, A., Kang, H., Tan, E., O'Neill, T., & Brecklin, C. (2012). Crafting a future in science: Tracing middle school girls' identity work over time and space. *American Education Research Journal, 50*(1), 37–75. https://doi.org/10.3102/0002831212458142

Calabrese Barton, A., Tan, E., & McDaniel, A. (2020). Becoming in STEM: Developing a culture of criticality in the space between person and institution. *Frontiers in Education, 5*, 37. https://doi.org/10.3389/feduc.2020.00037

Carlone, H. B., & Johnson, A. (2007). Understanding the science experiences of successful women of color: Science identity as an analytic lens. *Journal of Research in Science Teaching: The Official Journal of the National Association for Research in Science Teaching, 44*(8), 1187–1218. https://doi.org/10.1002/tea.20237

Elwood, S., & Mitchell, K. (2013). Another politics is possible: Neogeographies, visual spatial tactics, and political formation. *Cartographica: The International Journal for Geographic Information and Geovisualization, 48*(4), 275–292. https://doi.org/10.3138/carto.48.4.1729
Gutiérrez, K. D., & Penuel, W. R. (2014). Relevance to practice as a criterion for rigor. *Educational Researcher, 43*(1), 19–23. https://doi.org/10.3102/0013189X13520289
Hall, S. (2015). Cultural identity and diaspora. In *Colonial discourse and post-colonial theory* (pp. 392–403). Routledge.
Hazari, Z., Sonnert, G., Sadler, P., & Shanahan, M. (2010). Connecting high school physics experiences, outcome expectations, physics identity, and physics career choice: A gender study. *Journal of Research in Science Teaching, 47*(8), 978–1003. https://doi.org/10.1002/tea.20363
Holland, D., & Lave, J. (2009). Social practice theory and the historical production of persons. *Actio: An International Journal of Human Activity Theory*, (2), 1–15.
Lave, J., & Wenger, E. (1991). *Situated learning: Legitimate peripheral participation*. Cambridge University Press.
Morales-Doyle, D. (2017). Justice-centered science pedagogy: A catalyst for academic achievement and social transformation. *Science Education, 101*(6), 1034–1060. https://doi.org/10.1002/sce.21305
Nasir, N. (2011). *Racialized identities: Race and achievement among African American youth.* Stanford University Press.
Rosario-Ramos, E. M., Tucker-Raymond, E., & Rosario, M. (2017). Diasporican art as a space for identity building, cultural reclamation, and political reimagining. *Journal of Latinos and Education, 16*(3), 217–228. https://doi.org/10.1080/15348431.2016.1229615
Tan, E., & Calabrese Barton, A. (2020). Hacking a path in and through STEM: How youth navigate and transform the landscapes of STEM. *Teachers College Record*. https://doi.org/10.1177/016146812012200211
Tan, E., Kang, H., O'Neill, T., & Calabrese Barton, A. (2013). Desiring a career in STEM-related fields: How middle school girls articulate and negotiate between their narrated and embodied identities in considering a STEM trajectory. *Journal of Research in Science Teaching, 50*(10), 1143–1179. https://doi.org/10.1002/tea.21123
Tuck, E. (2009). Suspending damage: A letter to communities. *Harvard Educational Review, 79*(3), 409–428. https://doi.org/10.17763/haer.79.3.n0016675661t3n15
Urrieta, L. (2019). Indigenous reflections on identity, trauma, and healing: Navigating belonging and power. *Genealogy, 3*(2), 26. https://doi.org/10.3390/genealogy3020026
Urrieta, L., & Noblit, G. (2018). *Cultural constructions of identity*. Oxford University Press.
Wortham, S. (2006). *Learning identity: The joint emergence of social identification and academic learning*. Cambridge University Press.

GPSR Compliance
The European Union's (EU) General Product Safety Regulation (GPSR) is a set of rules that requires consumer products to be safe and our obligations to ensure this.

If you have any concerns about our products, you can contact us on

ProductSafety@springernature.com

In case Publisher is established outside the EU, the EU authorized representative is:

Springer Nature Customer Service Center GmbH
Europaplatz 3
69115 Heidelberg, Germany

www.ingramcontent.com/pod-product-compliance
Ingram Content Group UK Ltd.
Pitfield, Milton Keynes, MK11 3LW, UK
UKHW020233290926
15470UKWH00002B/510
9783031176418